Advances in Experimental Medicine and Biology

Proteomics, Metabolomics, Interactomics and Systems Biology

Sub-Series Editor

Daniel Martins-de-Souza
University of Campinas (UNICAMP)
Institute of Biology
Laboratory of Neuroproteomics
Campinas, Brazil

Advances in Experimental Medicine and Biology presents multidisciplinary and dynamic findings in the broad fields of experimental medicine and biology. The wide variety in topics it presents offers readers multiple perspectives on a variety of disciplines including neuroscience, microbiology, immunology, biochemistry, biomedical engineering and cancer research.

Advances in Experimental Medicine and Biology has been publishing exceptional works in the field for over 30 years and is indexed in Medline, Scopus, EMBASE, BIOSIS, Biological Abstracts, CSA, Biological Sciences and Living Resources (ASFA-1), and Biological Sciences. The series also provides scientists with up-to-date information on emerging topics and techniques.
2020 Impact Factor: 2.622 5 Year Impact Factor: 2.324

This series of volumes focuses on concepts, techniques and recent advances in the field of proteomics, interactomics, metabolomics and systems biology. Recent advances in various 'omics' technologies enable quantitative monitoring of myriad various biological molecules in a high-throughput manner, and allow determination of their variation between different biological states on a genomic scale. Now that the sequencing of various genomes, from prokaryotes to humans, has provided the list and linear sequence of proteins and RNA that build living organisms, defining the complete set of interactions that sustain life constitutes one of the key challenges of the postgenomic era. This series is intended to cover experimental approaches for defining protein-protein, protein-RNA, protein-DNA and protein-lipid interactions; as well as theoretical approaches dealing with data analysis, integration and modeling and ethical issues.

More information about this series at http://www.springer.com/series/15040

Ana Valéria Colnaghi Simionato
Editor

Separation Techniques Applied to Omics Sciences

From Principles to Relevant Applications

Springer

Editor
Ana Valéria Colnaghi Simionato
Analytical Chemistry Department
University of Campinas, Institute of Chemistry
Campinas, São Paulo, Brazil

Proteomics, Metabolomics, Interactomics and Systems Biology
ISSN 0065-2598 ISSN 2214-8019 (electronic)
Advances in Experimental Medicine and Biology
ISBN 978-3-030-77251-2 ISBN 978-3-030-77252-9 (eBook)
https://doi.org/10.1007/978-3-030-77252-9

This Springer imprint is published by the registered company Springer Nature Switzerland AG
The registered company address is: Gewerbestrasse 11, 6330 Cham, Switzerland

To my beloved children, Isabela and Rafael: my love for you is the size of the sky!

Preface

Omic sciences have been of ultimate importance to comprehend the complex biochemical reactions and related events that occur in a biological system. The classical central dogma of molecular biology, which states that genetic information flows unidirectionally from DNA to RNA and then to proteins, has been gradually replaced and complemented by the systems biology approach. This multidisciplinary approach tries to explain the biological system as a whole, where the entire organism is influenced by a variety of internal events as well as the environment, showing that each level of the biological information flux may influence the previous or the subsequent one.

Separation techniques constitute the first primordial dimension to obtain comprehensive data on biological samples analyses. The second dimension method has often been the hybridization of separation techniques with mass spectrometry and, more rarely, nuclear magnetic resonance.

This book presents liquid chromatography, gas chromatography, and capillary electrophoresis, the three main separation techniques lately available, applied to key omic sciences, such as proteomics, metabolomics, peptidomics, glycomics, and foodomics. Additionally, important directions on proteomics and metabolomics large set of data analyses are also approached. The fundamentals of each technique will not be covered herein. Instead, the recent advances in such techniques will be presented focusing on the application to omics analyses and unique aspects in each case. Therefore, this book intends to offer wide ranging options available to researchers on omics sciences, and how to integrate them in order to achieve the comprehension of a biological system as a whole.

Campinas, São Paulo, Brazil — Ana Valéria Colnaghi Simionato

Acknowledgments

Firstly, I would like to thank Daniel Martins de Souza for the invitation to edit this book and for trusting my ability to fulfill the expected quality of this series. The acknowledgments are extended to Springer's editorial team.

To all authors and co-authors who have accepted my invitation to contribute for the completion of this work, my sincere acknowledgment. I had devised initial topics in my mind and you have made them come true with excellence.

To my students who have directly or indirectly assisted me, either co-authoring some of the chapters with me, or performing their research projects with the expected quality so that our group has been acquiring more and more practical knowledge about metabolomics and separation techniques.

To all my teachers and professors (from kindergarten to graduation): it is not possible nor fair to cite the names of all of them (cause maybe I could forget some), but each one has profoundly contributed to my professional achievements and to my personal formation.

To my parents (Itamar Simionato and Maria Clélia Colnaghi Simionato), my daughter (Isabela Simionato Cantu), and my son (Rafael Simionato Cantu) for all support and love.

To the National Council for Scientific and Technological Development (CNPq), São Paulo Research Foundation (FAPESP), and Coordination for the Improvement of Higher Education Personnel (CAPES), who have supported research in Brazil, and, especially, my research group.

Contents

The New Omics Era into Systems Approaches: What Is the Importance of Separation Techniques?

Flávia da Silva Zandonadi, Fábio Santos Neves, Elisa Castañeda Santa Cruz, Alessandra Sussuilini, and Ana Valéria Colnaghi Simionato

Abstract

Omics sciences have been facing challenges in different fields, especially in life sciences. One of these challenges involves assessing biology into systems interpretation. With the advance of genomics, molecular biology has been projected into the realm of systems biology. In a different direction, systems approaches are making definitive strides toward scientific understanding and biotechnological applications. Separation techniques provided meaningful progress in the omics era, conducting the classical molecular biology to contemporary systems biology. In this introductory chapter, the relevance of these techniques to the development of different omics sciences, within the systems biology context, will be discussed.

Keywords

Systems biology · Genomics · Proteomics · Metabolomics · Separation techniques

F. da Silva Zandonadi · F. S. Neves · E. C. S. Cruz
Department of Analytical Chemistry, Institute of Chemistry, University of Campinas, Campinas, SP, Brazil

A. Sussuilini
Department of Analytical Chemistry, Institute of Chemistry, University of Campinas, Campinas, SP, Brazil

National Institute of Science and Technology for Bioanalytics, Institute of Chemistry, University of Campinas, Campinas, SP, Brazil

A. V. C. Simionato (✉)
Analytical Chemistry Department, University of Campinas, Institute of Chemistry, Campinas, São Paulo, Brazil
e-mail: avsimionato@unicamp.br

1 Systems Biology

How are biologists building strategies to understand life? This question has been methodically surveyed by exploring the characteristics of living organisms in different ways. Moreover, as a reflection of the necessity in deciphering the biological dynamics, the progress of cost-effective technologies able to comprehensively assess DNA, RNA, protein, and metabolites, molecules that orchestrate all the biological dynamics, has also been promoted. Before defining systems biology and omics sciences, it is important to describe the main approaches applied in life sciences since the beginning of these studies.

In 2004 Westerhoff and Palsson developed a series of arguments about two scientific schools considering their origin in the expansion of molecular biology to genome-wide analyses [1]. The idea of biology using integration tools is not new. The first regulatory circuit within the molecular biology context was described more than 40 years ago [1,

A. V. Colnaghi Simionato (ed.), *Separation Techniques Applied to Omics Sciences*, Advances in Experimental Medicine and Biology 1336, https://doi.org/10.1007/978-3-030-77252-9_1

2]. These studies were conducted focusing on the regulatory mechanisms, admittedly on a small scale. Molecular biologists began to apply systems approaches to unravel the molecular components and the logic that underlie cellular processes, often in parallel with the characterization of individual macromolecules. High-throughput technologies have made the scale of such inquiries much larger, enabling the view of the genome, for example, as the "system" in study [3–5].

As mentioned before, the dynamics of life is assessed by the studies of DNA, RNA, proteins, and metabolites. Behind these molecules, studies were addressed under investigation of multiple escalation levels, i.e., molecular, cellular, organism, and ecological organization. Survey by reductions, as defined by the classical scientific method, explores complexity in its individualized parts. The biological system has been dissected into their constituent parts and explained according to the chemical basis of numerous living processes [6, 7], thus producing multifaceted and disconnected knowledge. Beyond the philosophy of biology, reductionism method could be encompassed, according to Brigandt and Love [7]:

> … a set of ontological, epistemological, and methodological claims about the relations between different scientific domains. The basic question of reduction is whether the properties, concepts, explanations, or methods from one scientific domain (typically at higher levels of organization) can be deduced from or explained by the properties, concepts, explanations, or methods from another domain of science (typically at lower levels of organization).

From this definition, the multifaceted and disconnected knowledge of the biological dynamics could no longer be questioned under reductionist pragmatism. Automation, miniaturization, and multiplexing of various assays led to the generation of additional omics data types [8]. These enormous amounts of information that come from the omics sciences (such as genomics, proteomics, and metabolomics) no longer could be interpreted under compartmentalization contexts but as an integrated system.

For this reason, a more formal and mechanistic framework was required to analyze multiple high-throughput data types systematically [9, 10]. At this moment, the structure of scientific theories, the relations between scientific disciplines, the nature of explanation, the diversity of methodology, and aspects of biological complexity, especially the regulation process under environmental influences, changed to system status. Not only biology but other areas reached the center of the status of a new concept, the science of complex systems.

The knowledge from classical chemistry has provided human and technological resources, from analytical techniques to recent developments in high-throughput approaches and bioinformatics. The biological investigation from the bewildering diversity of interactions and regulatory networks has produced a formal and mechanistic framework necessary to analyze multiple high-throughput data types systematically [10], as represented in a summarized timeline inserting omics sciences into systems biology (Fig. 1).

Systems biology describes changes and connects variables over time, taking the chaotic, unpredictable, or counterintuitive contrasting with much simpler linear system properties from the reductionist methods. Defined as a new level of understanding and capturing of the dynamics of large sets of interacting components, the field combined the molecular and cell biology approaches with a stronger commitment toward quantitative experimentation under physiological conditions and with formal mathematical modeling (e.g., R-Theory) [11].

Unquestionably, the mathematical models are crucial for handling the associated complexity as formal representations to system-level understanding, but the main idea of the next topics is to describe the progress of main separation techniques, developments, and important milestones of the evolution of molecular biology into systems biology.

2 Omics Sciences and the Systems Biology Era

The identification, qualification, and application of diagnostic and prognostic biomarkers remain the holy grail of the current omics paradigm. Genomics, moving on to proteomics and metabolomics, premise, and promise of systems biology, has provided a powerful motivation for scientists

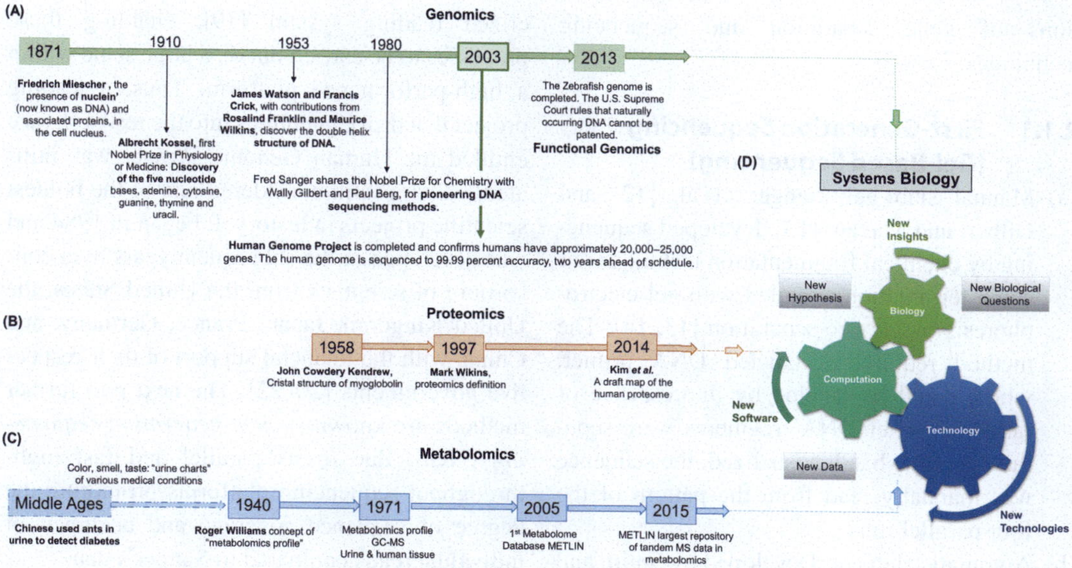

Fig. 1 High-throughput omics timeline to systems biology. From classical to modern science, how the high-throughput studies and multidisciplinary interpretations provide the resurgence of entering biology as a system. Classical genetics to genomics (**a**), protein biochemistry to proteomics (**b**), metabolomics and emergence of the next field (**c**), systems biology (**d**)

to combine the data generated from multiple omics approaches (e.g., genomics, transcriptomics, proteomics, and metabolomics) to create a holistic understanding of cells, organisms, and communities, relating to their growth, adaptation, development, and progression to disease. In this section, some tools and advanced techniques in genomics, proteomics, and metabolomics are summarized. These techniques are essential tools to the new challenges in life sciences, highlighting the systems biology field into the paradigm from omics sciences.

2.1 Genomics to Systems Biology, Post-Genomics Era

Around many definitions related to the genomics field, the most classic and simplistic one is the study of the complete genome of organisms, aiming at decoding and identifying relationships among the gene set, growth, and development of the organisms. This science is dedicated to determine the complete sequence of organisms' DNA, or mapping a smaller genetic scale.

DNA sequencing history began in 1977 when Frederick Sanger and colleagues described a methodology for determining the sequential order of nucleotides that make up the structure of DNA, based on the principle of controlled termination of dideoxynucleotide replication [12]. In its original version, this method was not a suitable tool for sequencing complex genomes, covering thousands, millions, or even billions of base pairs in complex organisms (e.g., mammals). Genome assembly was used to reconstruct the exact gene disposition and to locate other genome components in the chromosome, since there was no computer software available to analyze the generated sequences and sort them correctly.

One of the most significant challenges of these first steps into the genomics field has been the development of new DNA sequencing going through different achievements along all over the next 30 years. From these first 10 years, genome studies have made rapid progress in three generations of gene separation and sequencing techniques.

2.1.1 First-Generation Sequencing (Gel-Based Sequencing)

(a) Manual slab gel: Sanger et al. [12] and Gilbert and Maxam [13] developed sequencing by chemical fragmentation techniques or chain termination, coupled with gel electrophoresis-based size separation [13, 14]. The method required a labeled DNA primer, which could be labeled by fluorescence or radiation. After DNA fragments were separated and the bands visualized, the sequence was manually read from the pattern of the four parallel runs.
(b) Automatic slab gel: Developed by Smith and colleagues [15]. The key differences between this method and the former one were the tagging and the number of reactions. By tagging four different dyes of different fluorescent emission wavelengths instead of one, the four reactions were reduced to a single one [15], which demonstrated optimization in gel (such as diameter and length), electrophoresis conditions (such as current and temperature), optics and electronics used in data acquisition, and software used in data reduction (72 Kb/h/slab).
(c) Capillary gel electrophoresis: Cohen et al. [16] demonstrated the use of polyacrylamide gel-filled in capillaries used to reach single-nucleotide separation of DNA oligonucleotide markers by UV detection [16], which was improved by using ultrasensitive fluorescence by Swerdlow and Gesteland [17], verifying enhanced speed, resolution, and efficiency comparing with the former methods [17]. Furthermore, DNA sequencing using capillary array electrophoresis was developed by Huang et al. [18], which performs rapid, parallel separation followed by on-column detection using multicolor, confocal fluorescence scanner.

Genomics began only in the late 1980s, after Sanger's method was modified to allow automated sequencing and integration with a computerized reading system [19], enabling these processes to be carried out on a large scale within a high-performance platform. Thus, a leading project that drove genomics into the research race entitled the Human Genome Project was initiated, which is still considered one of the boldest scientific projects in history. It began in 1984 and was developed in the subsequent years by a consortium of scientists from the United States, the United Kingdom, Japan, France, Germany, and China, with the financial support of their respective governments [20–22]. The next two further methods are known as *next-generation sequencing* (NGS) due to its parallel and fast high-throughput sequencing platforms promoting the degree of sequence coverage and accuracy of individual reads compared to Sanger's one.

2.1.2 Second-Generation Sequencing

The sequence by synthesis (SBS) method requires the direct action of DNA polymerase to produce the visible result. Considering Sanger's method is also based on sequence by synthesis, Nyrén [23] developed a new technique by using the luminescent method for measuring pyrophosphate synthesis known as pyrosequencing [23]. A few years later, Ronaghi et al. [24] and Nyrén [23] performed real-time sequencing by synthesis within a proper choice of enzyme and substrate in a solid-phase format [24]. This approach was then bought by 454 Life Science (2005), nowadays Roche, which developed an emulsion method for DNA amplification and an instrument for sequencing by synthesis using pyrosequencing protocol for solid support and picoliter scale volumes [25].

After that, in 2005, Turcatti et al. developed the "Illumina" sequencing platform composed of four companies, Solexa and Illumina among them. This approach is based on fluorescent reversible terminators for sequencing [26, 27].

2.1.3 Third-Generation Sequencing

Braslavsky and colleagues [28] developed a single-molecule sequencer (SMS), later commercialized by Helicos BioSciences (2009). This technology is capable of sequencing single molecules without DNA amplification through polymerase chain reaction (PCR). Moreover, it enhanced the sequencing speed and reduced costs [28, 29]. Puglisi and collaborators developed single-molecule sequencing in real time (SMRT) in 2010. This process enables the observation of DNA synthesis as it occurs in real time, generating very long reads of sequences up to 10 kilobases long, which is useful for de novo genome assemblies [30].

Although the nanopore DNA sequencing methodology started within the second-generation sequencers in the 1990s [31], Oxford Nanopore announced a third-generation single-molecule platform in 2012 that represented a clear step onto sequencing of single DNA molecules using this technology, based on the principle of minute changes in electric current across the nanopore immersed in a conducting fluid with voltage applied when a moving nucleotide, or DNA strand, passes through it and the ion current probes the base identity [32]. All these approaches have brought the cost of human genome sequencing down from US$ 300 million in 2001 to US$1500 in 2015 [33, 34]. Within the improvements in DNA sequencing, physicians can identify a particular type of cancer, enabling them to make better choices for treatments.

In the last three decades, the field was markedly characterized by the proliferation and evolution of technologies, especially for those technologies able to provide new possibilities to decipher the genome from several species and further locate and identify regulatory patterns in gene code. The genomics in their structural fields starts to dive into the necessity in how to elucidate their functions aspects within biological systems and to begin to understand the mechanisms that control interactions [35]. At this time, the gene pool is immediately associated to the conception of a practically static set, while its products, represented by a messenger RNA (transcriptome) and protein (proteome), have a dynamic character, showing continuous changes in response to internal and external stimuli [36]. However, it is known that there is a complex process of regulation, and even at the advanced genome sequencing stage, new research platforms have begun to emerge to integrate individual functional genes and their products (RNAs and proteins) into a global context – a new field in biology named functional genomics [37].

Functional genomics attempts to describe functions and interactions regarding encoded genes and proteins by making use of genome-wide approaches, in contrast to the gene-by-gene approach of classical molecular biology techniques [38]. Biological high-throughput methods were probably the first start for the studies of genes and their regulatory molecules, by applying a data combination derived from the various processes related to DNA sequence, gene expression, and protein function, such as coding and noncoding transcription, protein translation, protein-DNA, protein-RNA, and protein-protein interactions [39].

The classical genomics toward functional genomics studies opened investigations based on the interaction between genes and their products. As mentioned before, high-throughput analytical techniques were the most important tools that allowed biology to have a new perspective as systems rather than study their elements one by one or few at a time, or even as structural elements. Proteins, among the encoded gene products, are vital to living organisms, as they comprise the machinery required for the operation of metabolic pathways.

Genomics and transcriptomics research have progressed due to advances in microarray technology, but protein studies (proteomics) coined to describe the set of proteins encoded by the genome. It was evident, among the omics, that proteomics emerged during the genomics progress as a complementary field regarding to the functional genomics approach. Even though mass spectrometry (MS) is the most common technique used for the detection of analytes in proteomics and metabolomics research, the microarray-based expression [40, 41] and small molecule-based array [42, 43] techniques have been widely used to integrate gene and protein information.

2.2 Proteomics and the Advance in Systems Biology Studies

Proteins, the molecular products of genes, are vital to living organisms, as they comprise the machinery required for the operation of metabolic pathways. Protein expression depends on cellular and environmental conditions. For nearly two decades, proteomics research has attempted to provide the identity and expression level of a large numbers of proteins in different physiological states in cells, body fluids, or tissues. The expectation is that this information will improve the understanding of biological functions and provide molecular signatures for particular health and disease states.

In contrast to mRNA expression analysis, proteomics indicates actual, rather than potential, functional states of a biological system. The bottleneck in proteomics is that there is no amplification step, like a PCR amplification for DNA. The low abundance and high dynamic range of proteins in biological samples, as well as data acquisition and analysis time, remain a challenge in this area. Therefore, proteomics approach drives the continuous development of analytical techniques and bioinformatics tools aiming to deepen the biological functions comprehension.

Proteomics was first defined in 1995, as the protein content complementary to a genome [44]. This concept and new field of science started after the human genome was almost fully sequenced and 24,000 encoded genes was reported [45]. This information brought new challenges into this science area, especially involving the measurement and identification of the amount of gene products, considering the complexity and different regulatory processes. Besides, the gene expression step can be modulated from transcription to the posttranslational modification (PTM) of proteins. Considering this intricate regulatory mechanism, the development of a technique that is able to identify the entire proteome in a single analysis is very challenging [46, 47].

From the definitions established in 1995 to our days, proteomics not only complements the genome but also provides a better biological, phenotypical, and functional understanding of the entire physiology. Considering the emerging omics fields, since proteomics encompasses the quantitative, functional, structural, and PTM characterization of proteins, new areas that make possible the determination of the protein relationships (interactome) and systems biology were developed [48].

In the beginning, a typical proteomics experiment resulted in a list of identified proteins, with no information regarding abundance, distribution, or stoichiometry. Abundance information is critical considering the expression regulation dynamics, reflecting the balance among the entire biochemistry in order to comprise the life dynamics. All the gaps and questions, especially for the stoichiometry balances [49], have been answered according to the advanced analytical capabilities. MS-based proteomics has been extensively used to identify the components of biological systems, and it is the method of choice to consistently quantify the effects of network perturbation in time and space [50–52]. Besides the protein level, life biochemistry operates in multidimensional space. In this way, two important questions have been opened in omics fields: (1) how is this balance achieved and (2) to what extent each of these processes contributes to the regulation of cellular protein abundances. Moreover, back to the regulation events, the biochemistry of living organisms builds up an inside-outside regulatory molecular function system that is the main reservoir for these questions for the research community.

Due to the high complexity of the proteome, there is no particular standard method for sample preparation [53–56]. For the sample preparation procedure based on proteomics, the first and conventional methods used for sample fractionation from complex mixtures are one- and two-dimensional electrophoresis (1-DE and 2-DE, respectively) [56]. Although they are widely used as fractionation techniques, at least in the early days, both suffer from disadvantages including limited dynamic range [57], poor solubilization leading to poor resolution at extremes of isoelectric point (pI) [58], and inability to identify proteins in low abundance [59].

1-DE is traditionally used as a protein fractionation method, mainly when molecular mass is used as separation factor. The technique uses sodium dodecyl sulfate (SDS), a detergent that solves the poor protein solubilization issue. This method is simple, fast, and reproducible and can separate proteins in a broad spectrum based on the molecular mass. 1-DE technique has been used as a step for protein fractionation [60], even though the resolving power is limited when used for separation of high complex protein mixtures. Historically, MS-based proteomics began with the use of 2-DE separation. 2-DE uses two physicochemical proteins properties to separate complex samples obtained from cells and tissues among other biological sources. Thus, this technique occurs in two steps: in the first step, the isoelectric focusing (IEF) separates proteins according to the respective pI; while in the second step, polyacrylamide gel electrophoresis with SDS (SDS-PAGE) separates proteins according to the respective molecular masses (MM). Thousands of proteins can be separated simultaneously. Moreover, information on pI, MM, and relative abundance can be obtained, as well as posttranslational modifications, since they generally cause altered electrophoretic mobility [61]. In general, the generated spots correspond to a single polypeptide chain present in the sample. The development of the gel may determine the number of polypeptide chains, and the amount of each one may be defined with dyes and subsequent densitometric analysis [62].

Advances in this technique to separate proteins were the development of narrow ranges of pH, besides enhanced software tools for correlation analysis of proteins with pH [63–66]. Afterwards, 2-DE was combined to fluorescence probe labeling techniques, giving rise to 2-DE-DIGE (differential in-gel electrophoresis) [67]. The improvement of separation techniques in 2-DE gels was based on the need to minimize the experimental effects, as well as facilitating the comparison step between the samples. 2-DE-DIGE is a technique based on label sample preparation. Samples are labeled separately with different fluorescence probes (Cy2, Cy3, or Cy5) and combined in the same vial so that the run is performed in the 2D gel, thus minimizing the experimental variation and facilitating the comparison between the same protein from different samples (spot matching) [68].

Since the beginning of proteomics, it has heavily relied on 2-DE for the separation and visualization of proteins. Correlated techniques applied to optimize this separation technique still show many inherent drawbacks. 2-DE is costly, insensitive to low copy proteins, scarcely reproducible, and cannot be used for the entire proteome at the dynamic range view [69].

Over the years, several gel-free proteomics techniques have been developed to either fill the gaps left by 2-DE or to entirely abolish the gel-based techniques. Performing proteomics without gel separation of 2-DE electrophoresis gave rise to the term shotgun protein analysis in 1998. Developed by Yates et al., shotgun proteomics consists of the combination of enzymatic protein digestion, followed by separation by liquid chromatography (LC) coupled to sequential (tandem) mass spectrometry (MS/MS) [70]. This technique provided a significant gain in efficiency and sensitivity in the analysis of complex protein mixtures since it automated the process of protein separation, minimizing sample loss, in addition to using nano-flow scale chromatography [71].

The idea of sample complexity, especially for proteins extracted from biological tissues and the wide dynamic concentration range, is still the main challenge for the technique improvements. In this direction, the selective fractionation of complex proteome is an efficient strategy to optimize the identification coverage in complex proteomes. Among the improvements, the multidimensional methods or systems promote the development and comprehension of the proteomics field. The multidimensional methods [72] means that the idea of combining different separation techniques is fundamental [73]. According to the proteome experts, such methods are the way to understand the inherent challenges in gaining insight beyond the "tip of the proteoberg" (The Multidimensional Future of Proteomics, 2016).

Furthermore, the multidimensional techniques based on chromatography allow the separation of these complex mixtures by using multiple columns with different stationary phases, coupled orthogonally, which means that fractions from the first column can be selectively transferred to other columns for additional separation.

Any liquid chromatography separation mode can be used at the protein level, including ion exchange [74, 75], reverse phase [76, 77], hydrophilic interaction [78], or size exclusion [79], prior to digestion. One of the best known multidimensional techniques, also developed by the Yates group, was Multidimensional Protein Identification Technology (MudPIT). Briefly, this technology uses two liquid chromatography separations modes: in the first dimension, proteins are separated in cation exchange columns, according to analytes charge density, while in the second dimension, the separation occurs in reverse phase columns, based on proteins hydrophobicity [79, 80]. This methodology has brought significant improvement in the dynamic range and coverage of the proteome studies. However, the increased process complexity, the low reproducibility, the longer analysis time, and the high cost of analysis are still the main limitations of its use [80, 81].

In the first decade of 2000, important labeling techniques were developed, minimizing the proteome complexity and the analysis time once the different samples were submitted to the same analysis, reducing the variability and the steps in the process [82]. Besides, liquid chromatography coupled with mass spectrometry (LC-MS) also optimizes the proteolytic product separation (e.g., truncated polypeptides), expanding the protein range identification. However, among the analytical techniques, LC has been following the progress and the necessity to improve proteomics separation resolving power, analysis coverage, sensitivity, and throughput. Especially in proteomics, LC is still the best option regarding fractionation of peptide mixtures to enable and maximize identification and quantification of the component peptides by MS, as identification technique.

Implementation of liquid phase separations before MS analysis reduces the number of analytes entering the mass spectrometer at any given time, which minimizes ionization suppression where a nominally detectable species is not detected due to detector dynamic range limitations, and under-sampling in ion selection for MS/MS analysis in shotgun measurements. Furthermore, analytes can be focused within narrow zones (or peaks) during the liquid phase separation steps, which concentrate them and benefits MS detection sensitivity. Within liquid phase separation techniques, LC, especially in its capillary format, has significantly advanced over the past decade to make it a prevalent technique in modern-day proteomics analyses as the physicochemical properties (e.g., mass, charge, and hydrophobicity) of peptides make them amenable to efficient LC separation.

2.3 Metabolomics

Recently, the biology studies used to be discussed focusing on the gene and products of the gene coding, mainly related to transcriptomics and proteomics, whose technologies and tools are very well established and widely applied. However, because of the necessity to fully understand the phenotype diseases caused by DNA mutations, attack of pathogens, and environmental conditions, metabolomics have been an expanding field of research to develop and integrate transcriptomics, proteomics, and metabolomics in a fully systems biology approach [83].

In the post-genomics era, metabolomics became crucial due to its strong relation to phenotype, besides integration with comprehensive transcriptomics and/or proteomics systems to discover specific biomarkers and validate biomarkers significance [84]. In this context, metabolomics has been used to validate and verify the regulation of genes and/or proteins in a biological system and their corresponding variations (e.g., upregulation, downregulation, concentration, or intensity levels) under specific experimental conditions (e.g., different times, gene mutations, biotic or abiotic stress, phenotype alterations).

Moreover, it has been a relevant field to investigate functional integration of gene expression to transcriptomics and proteomics, as well as the emerging fields of "phenomics" and "fluxomics," contributing to the development of biological system networks, the identification of unknown gene/protein functions, abnormal gene-metabolite relationships due to knowns/unknowns gene mutations, and the analysis of metabolic pathways to explore biochemical activities [85]. Metabolite levels can reflect the closer integration of gene expression and protein synthesis, considering the influence of the environmental conditions and/or other organism's interactions as well as to control gene expression through allosteric interactions of transcriptions factors related to specific metabolites [86]. Therefore, untargeted metabolomics tries to measure all metabolites, which can be assessed by a multiplatform study (in a cell, tissue, or organism) within a specific design, reflecting a snapshot of all the physiological events as a response of gene expression and environmental conditions [85]. While metabolomics contributes to the obtention of a snapshot of the biological system under investigation, integration between transcriptomics, proteomics, and metabolomics led to visualization of a well-detailed picture or network with the respective biological mechanisms and their association to diseases from gene mutation or phenotypic alterations, constituting the so-called systems biology approach (Fig. 2).

Metabolomics plays an important role in the analysis of gene function or loss of function. It contributes with information about biological systems, as products from biochemical processes in living systems, which are influenced by abiotic (environment and stage of development) and biotic factors (transcription, mRNA degradation, posttranslational modification, protein dynamics, metabolite concentrations, and fluxes) [87, 88]. As an emerging and developing field, various concepts and definitions for the term "metabolomics" are found in the scientific literature. Drexler et al. define metabolomics as the "qualitative and quantitative (relative or absolute) analysis of the entire endogenous metabolome (metabolites with masses less than 1500 Da)" [89].

Metabolome analysis has recently been used in systems biology studies to quite comprehensively investigate the metabolic changes originated from genetic, environmental, and organism differences factors by comparing the basal levels of metabolites with those produced after alteration [90]. Autism is a genetic disease into intellectual disability (ID) diseases most commonly caused by fragile X syndrome (FXS). FXS is caused by the mutation of the X-linked fragile X mental retardation 1 (*fmr1*) gene resulting in their hypermethylation. The metabolic signature and biomarker identification associated with FXS due to *fmr1* gene inactivation results in brain metabolism alterations related to neurotransmitter levels, osmoregulation, energy metabolism, and oxidative stress response. Cardiovascular and metabolic disorders are also strongly related to gene mutations. Systems biology has been used to understand the impact of these genetic disorders in human metabolism through identification of variance in the promoter of specific genes as *fads1*, *elovl2*, *acads*, *acadm*, *acadl*, *sptlc3*, *etfdh*, and *slc16a9* that are responsible to enhanced change in the conversion rate of several metabolites associated with dyslipidemia, obesity, and diabetes. The *elovl2* and *slc16a9* genes have been associated to lipid concentration, as well as arachidonic acid, and cholesterol/triglyceride levels have been related to variant of the *fads1* gene that encodes a fatty acid desaturase enzyme [91, 92]. Therefore, information of the organism's phenotype can be explained as a consequence of the genome mutation and/or environment on the metabolome.

The two most commonly used strategies for metabolomics analysis are "targeted" and "untargeted." Targeted analyses are directed to certain classes of compounds related to known and specific metabolic pathways for their quantification by using specialized extraction protocols, separation methods, and detection techniques. On the other hand, the untargeted analysis is directed to establish the fingerprint of metabolites present in a sample, constituting a qualitative analysis of crude metabolite mixtures [89].

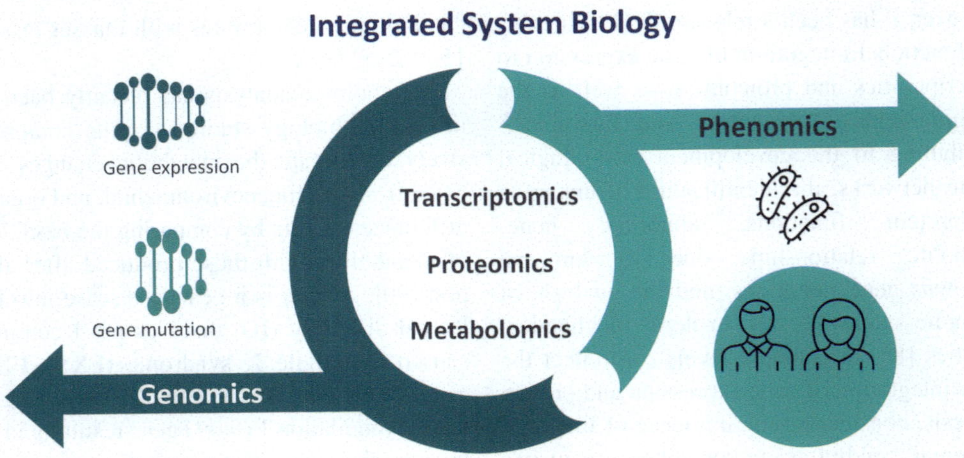

Fig. 2 Integrated systems biology from genomics to phenomics through transcriptoms, proteomics, and metabolomics sciences

A targeted approach is applied to detect few or several metabolites, if not all, according to analytical technique limitations, included in a targeted pathway. Primary metabolites including sugars, amino acids, and tricarboxylic acids that are involved in primary metabolic processes, such as respiration and photosynthesis, and secondary metabolites including alkaloids, phenolics, steroids, lignins, and tannins can be detected and quantified [93–95]. It requires optimization of the selected metabolite extraction step to reduce matrix effect and maximize the recovery and overall sensitivity for detection, generally using specific internal standards [96]. On the other hand, an untargeted approach focuses on the analysis of all metabolites (or most of them) of a biological system, not requiring a prior knowledge of the metabolites that will be altered under gene expression or mutation, causing phenotypical changes from transcriptomics, proteomics, or metabolomics alterations. Moreover, they can be detected using multivariate statistical analysis tools, such as principal component analysis (PCA), hierarchical cluster analysis (HCA), and discriminant analysis (DA) as partial least squares regression (PLS), *PLS* discriminant *analysis* (PLS-DA), Orthogonal-PLS-DA (O-PLS-DA) [97]. An untargeted analysis is usually applied prior to the targeted one to set a specific pathway to be studied, as well as the metabolites that vary under any genetic or phenotypic conditions [85, 93].

The main analytical techniques employed in recent metabolomics studies are nuclear magnetic resonance (NMR), one-dimensional and comprehensive two-dimensional gas chromatography coupled to mass spectrometry (GC-MS and GC × GC-MS, respectively), and liquid chromatography coupled to mass spectrometry (LC-MS), although other techniques such as capillary electrophoresis coupled to mass spectrometry (CE-MS) have also been employed. All these techniques and methods can be applied to the analyses of samples containing a wide variety of metabolites. However, each of them has different advantages and disadvantages considering detection and quantification limits, sensitivity, selectivity, interference, resolution, repeatability, reproducibility, and physicochemical characteristics of the compounds [89, 98]. In untargeted or targeted metabolomics, there is no single protocol capable of identifying and quantifying all possible metabolites in a single analysis, and due to a high degree of structural diversity, molecular mass, and main polarity differences between primary and secondary metabolites, there is no single protocol capable of identifying and quantifying all possible metabolites in a single analysis. Therefore, generally biphasic or triphasic extraction protocols with organic/aqueous solvents are performed to reduce sample complexity, generating high polar, low polar, and protein fractions. Afterwards, sample fractions are analyzed by different separation techniques using suitable stationary phases and instrument platforms to obtain the best metabolite profile [99, 100].

Analytical instrumentations are commonly associated with metabolomics to measure numerous metabolites (from hundreds to thousands of compounds) in order to evaluate metabolic changes in response to external stimuli, such as attack of pathogens or changes of environmental conditions, and elucidate metabolic pathways of the gene expression or mutations. However, simultaneous separation and detection of all metabolites in a biological sample with appropriate analytical sensitivity and resolution in a single analysis has not been achieved yet, due to high sample complexity (numerous metabolites with different chemical classes in a wide range of concentration levels), pointing to the demand of analytical techniques improvement [101].

Ion mobility mass spectrometry (IMS) has gained popularity over the last few years due to high selectivity and resolution power for several isomeric compounds variety. Differential mobility spectrometry (DMS) has been utilized in targeted metabolomics to separate small molecules, although it has a great potential to be used in untargeted metabolomics as well [102, 103]. DMS addresses high selectivity through unique mass-to-charge and migration time combinations, allied to high accuracy. The DMS-MS analysis is typically four times faster than a typical LC-MS one, showing potential to the screening of numerous metabolites in order to address system biological tasks [103].

IMS resolution, selectivity, and accuracy improvement, mainly related to recent mass spectrometry advances, allowed detection of hundreds to thousands features in a single analysis, requiring advanced data analysis tools. Big data has become a fundamental aspect of systems biology to elucidate the complex networks by which gene expression, gene mutation, pathogen-host interaction, or environmental-host interaction is developed. The big data tools, such as machine learning and deep learning algorithms, and neural networks programmable in Python, Java, MATLAB, and R languages, allows mapping and modeling pathways to identify underappreciated gene, RNA, proteins, or metabolite functions and connections [104]. Diseases are driven by genetic and epigenetic factors and environmental factors such as attack of pathogens, leading to disturbance in immunological balance [105].

Systems biology and the main omics, approached in this book, provide new perspectives in science, rescuing the idea of how to understand biology as a system. Under molecular and cellular biology, during the last centuries, the studies were conducted by simplified methodologies, as the reductionism. Nevertheless, from the first classical sequencing techniques, the progress of the high-throughput techniques has moved forward the biology fields into the systems approach.

3 Perspectives

It is clear that the open questions in science have conducted the scientists into incredible knowledge fields. Classical science, even with the simplified methods (reductionism), has driven fields as genomics, proteomics, and metabolomics into systems biology progress. Looking back to the literature, the evolution of analytical techniques plays a key role in biology, especially in molecular biology areas (DNA/RNA, proteins, and metabolites), where the separation and characterization methods use analytical chemistry tools for comprehensive analyses of biological systems. However, as originated from classical science, these techniques allowed the progress and evolution of the theoretical and technological advances in nano-biotechnology, robotics, genetics, mathematics, and computational biology, among others, determining factors that allowed and facilitated integrative approaches, which constitute the main purpose of systems biology.

References

1. Westerhoff HV, Palsson BO (2004) The evolution of molecular biology into systems biology. Nat Biotechnol 22(10):1249–1252
2. Lodish H, Berk A, Zipursky SL, Matsudaira P, Baltimore D, Darnell J (2000) Molecular cell biology, 4th edn. National Center for Biotechnology Information, Bookshelf, 9
3. D'Argenio V (2018) The high-throughput analyses era: are we ready for the data struggle? High-Throughput 7(1):8
4. Boisvert S, Laviolette F, Corbeil J (2010) Ray: simultaneous assembly of reads from a mix of high-throughput sequencing technologies. J Comput Biol 17(11):1519–1533
5. Reuter JA, Spacek DV, Snyder MP (2015) High-throughput sequencing technologies. Mol Cell 58(4):586–597
6. Van Regenmortel MHV (2019) Editorial: Biological Complexity Emerges from the Ashes of Genetic Reductionism. In: HIV/AIDS: Immunochemistry, Reductionism and Vaccine Design. Springer, Cham. https://doi.org/10.1007/978-3-030-32459-9_7
7. Brigandt I, Love AC (2008) Reductionism in biology. In E. N. Zalta (Ed.), The Stanford Encyclopedia of Philosophy (Fall 2008 Edition). <http://plato.stanford.edu/archives/fall2008/entries/reduction-biology>
8. Uetz P, Giot L, Cagney G, Mansfield TA, Judson RS, Knight JR, et al. (2000) A comprehensive analysis of protein–protein interactions in Saccharomyces cerevisiae. Nature 403(6770):623–627
9. Ge H, Walhout AJ, Vidal M (2003) Integrating 'omic' information: a bridge between genomics and systems biology. Trends Genet 19(10):551–560
10. Palsson B (2002) In silico biology through "omics". Nat Biotechnol 20(7):649–650
11. Meng TC, Somani S, Dhar P (2004) Modeling and simulation of biological systems with stochasticity. In Silico Biol 4(3):293–309
12. Sanger F, Nicklen S, Coulson AR (1977) DNA sequencing with chain-terminating inhibitors. Proc Natl Acad Sci 74(12):5463–5467
13. Gilbert W, Maxam A (1973) The nucleotide sequence of the lac operator. Proc Natl Acad Sci 70(12):3581–3584
14. Sanger F, Coulson AR (1975) A rapid method for determining sequences in DNA by primed synthesis with DNA polymerase. J Mol Biol 94(3):441–448
15. Smith LM, Sanders JZ, Kaiser RJ, Hughes P, Dodd C, Connell CR et al (1986) Fluorescence detection in automated DNA sequence analysis. Nature 321(6071):674–679
16. Cohen A, Najarian D, Paulus A, Guttman A, Smith JA, Karger B (1988) Rapid separation and purification of oligonucleotides by high-performance capillary gel electrophoresis. Proc Natl Acad Sci 85(24):9660–9663
17. Swerdlow H, Gesteland R (1990) Capillary gel electrophoresis for rapid, high resolution DNA sequencing. Nucleic Acids Res 18(6):1415–1419
18. Huang XC, Quesada MA, Mathies RA (1992) DNA sequencing using capillary array electrophoresis. Anal Chem 64(18):2149–2154
19. Venter JC, Adams MD, Myers EW, Li PW, Mural RJ, Sutton GG et al (2001) The sequence of the human genome. Science 291(5507):1304–1351
20. Frazer KA (2012) Decoding the human genome. Genome Res 22(9):1599–1601
21. Consortium IHGS (2001) Initial sequencing and analysis of the human genome. Nature 409(6822):860
22. She X, Jiang Z, Clark RA, Liu G, Cheng Z, Tuzun E et al (2004) Shotgun sequence assembly and recent segmental duplications within the human genome. Nature 431(7011):927–930
23. Nyrén P (1987) Enzymatic method for continuous monitoring of DNA polymerase activity. Anal Biochem 167(2):235–238
24. Ronaghi M, Karamohamed S, Pettersson B, Uhlén M, Nyrén P (1996) Real-time DNA sequencing using detection of pyrophosphate release. Anal Biochem 242(1):84–89
25. Margulies M, Egholm M, Altman WE, Attiya S, Bader JS, Bemben LA et al (2005) Genome sequencing in microfabricated high-density picolitre reactors. Nature 437(7057):376–380
26. Fedurco M, Romieu A, Williams S, Lawrence I, Turcatti G (2006) BTA, a novel reagent for DNA attachment on glass and efficient generation of solid-phase amplified DNA colonies. Nucleic Acids Res 34(3):e22
27. Turcatti G, Romieu A, Fedurco M, Tairi A-P (2008) A new class of cleavable fluorescent nucleotides: synthesis and optimization as reversible terminators for DNA sequencing by synthesis. Nucleic Acids Res 36(4):e25
28. Braslavsky I, Hebert B, Kartalov E, Quake SR (2003) Sequence information can be obtained from single DNA molecules. Proc Natl Acad Sci 100(7):3960–3964
29. Bowers J, Mitchell J, Beer E, Buzby PR, Causey M, Efcavitch JW et al (2009) Virtual terminator nucleotides for next-generation DNA sequencing. Nat Methods 6(8):593
30. Uemura S, Aitken CE, Korlach J, Flusberg BA, Turner SW, Puglisi JD (2010) Real-time tRNA transit on single translating ribosomes at codon resolution. Nature 464(7291):1012–1017
31. Deamer DW, Akeson M (2000) Nanopores and nucleic acids: prospects for ultrarapid sequencing. Trends Biotechnol 18(4):147–151
32. Schneider GF, Dekker C (2012) DNA sequencing with nanopores. Nat Biotechnol 30(4):326
33. Kaiser J (2008) A plan to capture human diversity in 1000 genomes. Science 319(5868):1336

34. Mardis ER (2017) DNA sequencing technologies: 2006–2016. Nat Protoc 12(2):213
35. Woychik RP, Klebig M, Justice M, Magnuson T, Avrer E (1998) Functional genomics in the post-genome era. Mutat Res Fundam Mol Mech Mutagen 400(1–2):3–14
36. Greenbaum NL, Mundoma C, Peterman DR (2001) Probing of metal-binding domains of RNA hairpin loops by laser-induced lanthanide (III) luminescence. Biochemistry 40(4):1124–1134
37. Attur M, Dave M, Tsunoyama K, Akamatsu M, Kobori M, Miki J et al (2002) "A system biology" approach to bioinformatics and functional genomics in complex human diseases: arthritis. Curr Issues Mol Biol 4:129–146
38. Bunnik EM, Le Roch KG (2013) An introduction to functional genomics and systems biology. Adv Wound Care 2(9):490–498
39. Jansen R, Greenbaum D, Gerstein M (2002) Relating whole-genome expression data with protein-protein interactions. Genome Res 12(1):37–46
40. Shoemaker DD, Linsley PS (2002) Recent developments in DNA microarrays. Curr Opin Microbiol 5(3):334–337
41. LaBaer J, Nicolini C (2019) Functional proteomics and nanotechnology-based microarrays. CRC Press
42. Kuruvilla FG, Shamji AF, Sternson SM, Hergenrother PJ, Schreiber SL (2002) Dissecting glucose signalling with diversity-oriented synthesis and small-molecule microarrays. Nature 416(6881):653–657
43. Peng B, Yu C, Du S, Liew SS, Mao X, Yuan P et al (2018) MSN-on-a-chip: cell-based screenings made possible on a small-molecule microarray of native natural products. Chembiochem 19(9):986–996
44. Wilkins MR, Sanchez J-C, Gooley AA, Appel RD, Humphery-Smith I, Hochstrasser DF et al (1996) Progress with proteome projects: why all proteins expressed by a genome should be identified and how to do it. Biotechnol Genet Eng Rev 13(1):19–50
45. Avise JC (2010) Footprints of nonsentient design inside the human genome. Proc Natl Acad Sci 107(Supplement 2):8969–8976
46. Bock C, Farlik M, Sheffield NC (2016) Multi-omics of single cells: strategies and applications. Trends Biotechnol 34(8):605–608
47. Hasin Y, Seldin M, Lusis A (2017) Multi-omics approaches to disease. Genome Biol 18(1):83
48. Bensimon A, Heck AJ, Aebersold R (2012) Mass spectrometry–based proteomics and network biology. Annu Rev Biochem 81:379–405
49. Göth M, Pagel K (2017) Ion mobility–mass spectrometry as a tool to investigate protein–ligand interactions. Anal Bioanal Chem 409(18):4305–4310
50. Lamond AI, Uhlen M, Horning S, Makarov A, Robinson CV, Serrano L et al (2012) Advancing cell biology through proteomics in space and time (PROSPECTS). Mol Cell Proteomics 11(3):O112.017731
51. Altelaar AM, Munoz J, Heck AJ (2013) Next-generation proteomics: towards an integrative view of proteome dynamics. Nat Rev Genet 14(1):35–48
52. Varusai TM, Nguyen LK (2018) Dynamic modelling of the mTOR signalling network reveals complex emergent behaviours conferred by DEPTOR. Sci Rep 8(1):643
53. Walker JM (2005) The proteomics protocols handbook. Humana press
54. Von Hagen J (2008) Proteomics sample preparation. John Wiley & Sons
55. Andrecht S, von Hagen J (2008) The need for standards in proteomics sample preparation. In: Proteomics sample preparation. John Wiley & Sons. 5–20
56. Westermeier R, Naven T, Höpker H-R (2008) Proteomics in practice: a guide to successful experimental design. John Wiley & Sons
57. Corthals GL, Wasinger VC, Hochstrasser DF, Sanchez JC (2000) The dynamic range of protein expression: a challenge for proteomic research. Electrophoresis 21(6):1104–1115
58. Willard KE, Giometti CS, Anderson NL, O'Connor TE, Anderson NG (1979) Analytical techniques for cell fractions: XXVI. A two-dimensional electrophoretic analysis of basic proteins using phosphatidyl choline/urea solubilization. Anal Biochem 100(2):289–298
59. Santoni V, Rabilloud T, Doumas P, Rouquié D, Mansion M, Kieffer S et al (1999) Towards the recovery of hydrophobic proteins on two-dimensional electrophoresis gels. Electrophoresis 20(4–5):705–711
60. Shevchenko A, Tomas H, Havli J, Olsen JV, Mann M (2006) In-gel digestion for mass spectrometric characterization of proteins and proteomes. Nat Protoc 1(6):2856
61. Görg A, Drews O, Lück C, Weiland F, Weiss W (2009) 2-DE with IPGs. Electrophoresis 30(S1):S122–SS32
62. O'Farrell PH (1975) High resolution two-dimensional electrophoresis of proteins. J Biol Chem 250(10):4007–4021
63. Görg A, Obermaier C, Boguth G, Weiss W (1999) Recent developments in two-dimensional gel electrophoresis with immobilized pH gradients: wide pH gradients up to pH 12, longer separation distances and simplified procedures. Electrophoresis 20(4–5):712–717
64. Görg A, Obermaier C, Boguth G, Harder A, Scheibe B, Wildgruber R et al (2000) The current state of two-dimensional electrophoresis with immobilized pH gradients. Electrophoresis 21(6):1037–1053
65. Görg A, Weiss W, Dunn MJ (2004) Current two-dimensional electrophoresis technology for proteomics. Proteomics 4(12):3665–3685
66. Øye OK, Jørgensen KM, Hjelle SM, Sulen A, Ulvang DM, Gjertsen BT (2013) Gel2DE-A software tool for correlation analysis of 2D gel electrophoresis data. BMC Bioinf 14(1):215

67. Ünlü M, Morgan ME, Minden JS (1997) Difference gel electrophoresis. A single gel method for detecting changes in protein extracts. Electrophoresis 18(11):2071–2077
68. Larbi NB, Jefferies C (2009) 2D-DIGE: comparative proteomics of cellular signalling pathways. In: Toll-like receptors. Springer, pp 105–132
69. Koomen JM, Haura EB, Bepler G, Sutphen R, Remily-Wood ER, Benson K et al (2008) Proteomic contributions to personalized cancer care. Mol Cell Proteomics 7(10):1780–1794
70. Yates JR III (1998) Mass spectrometry and the age of the proteome. J Mass Spectrom 33(1):1–19
71. Yates JR III (2013) The revolution and evolution of shotgun proteomics for large-scale proteome analysis. J Am Chem Soc 135(5):1629–1640
72. Wolters DA, Washburn MP, Yates JR (2001) An automated multidimensional protein identification technology for shotgun proteomics. Anal Chem 73(23):5683–5690
73. Larance M, Lamond AI (2015) Multidimensional proteomics for cell biology. Nat Rev Mol Cell Biol 16(5):269–280
74. Shi Y, Xiang R, Horváth C, Wilkins JA (2004) The role of liquid chromatography in proteomics. J Chromatogr A 1053(1–2):27–36
75. Essader AS, Cargile BJ, Bundy JL, Stephenson JL Jr (2005) A comparison of immobilized pH gradient isoelectric focusing and strong-cation-exchange chromatography as a first dimension in shotgun proteomics. Proteomics 5(1):24–34
76. Shen Y, Zhao R, Belov ME, Conrads TP, Anderson GA, Tang K et al (2001) Packed capillary reversed-phase liquid chromatography with high-performance electrospray ionization Fourier transform ion cyclotron resonance mass spectrometry for proteomics. Anal Chem 73(8):1766–1775
77. Xu P, Duong DM, Peng J (2009) Systematical optimization of reverse-phase chromatography for shotgun proteomics. J Proteome Res 8(8):3944–3950
78. Boersema PJ, Mohammed S, Heck AJ (2008) Hydrophilic interaction liquid chromatography (HILIC) in proteomics. Anal Bioanal Chem 391(1):151–159
79. Lecchi P, Gupte AR, Perez RE, Stockert LV, Abramson FP (2003) Size-exclusion chromatography in multidimensional separation schemes for proteome analysis. J Biochem Biophys Methods 56(1–3):141–152
80. Horvatovich P, Hoekman B, Govorukhina N, Bischoff R (2010) Multidimensional chromatography coupled to mass spectrometry in analysing complex proteomics samples. J Sep Sci 33(10):1421–1437
81. Yates JR, Ruse CI, Nakorchevsky A (2009) Proteomics by mass spectrometry: approaches, advances, and applications. Annu Rev Biomed Eng 11:49–79
82. Walther TC, Mann M (2010) Mass spectrometry–based proteomics in cell biology. J Cell Biol 190(4):491–500
83. Kuo T-C, Tian T-F, Tseng YJ (2013) 3Omics: a web-based systems biology tool for analysis, integration and visualization of human transcriptomic, proteomic and metabolomic data. BMC Syst Biol 7(1):64
84. Nam H, Chung BC, Kim Y, Lee K, Lee D (2009) Combining tissue transcriptomics and urine metabolomics for breast cancer biomarker identification. Bioinformatics 25(23):3151–3157
85. Kumar M, Kuzhiumparambil U, Pernice M, Jiang ZJ, Ralph PJ (2016) Metabolomics: an emerging frontier of systems biology in marine macrophytes. Algal Res 16:76–92
86. Lempp M, Farke N, Kuntz M, Freibert SA, Lill R, Link H (2019) Systematic identification of metabolites controlling gene expression in E. coli. Nat Commun 10(1):1–9
87. Fiehn O (2002) Metabolomics – the link between genotypes and phenotypes. Plant Mol Biol 48(1–2):155–171
88. Goodacre R (2005) Metabolomics – the way forward. Metabolomics 1(1):1–2
89. Drexler DM, Reily MD, Shipkova PA (2011) Advances in mass spectrometry applied to pharmaceutical metabolomics. Anal Bioanal Chem 399(8):2645–2653
90. Davidovic L, Navratil V, Bonaccorso CM, Catania MV, Bardoni B, Dumas M-E (2011) A metabolomic and systems biology perspective on the brain of the fragile X syndrome mouse model. Genome Res 21(12):2190–2202
91. Illig T, Gieger C, Zhai G, Römisch-Margl W, Wang-Sattler R, Prehn C et al (2010) A genome-wide perspective of genetic variation in human metabolism. Nat Genet 42(2):137
92. Adamski J (2012) Genome-wide association studies with metabolomics. Genome Med 4(4):34
93. Broadhurst D, Goodacre R, Reinke SN, Kuligowski J, Wilson ID, Lewis MR et al (2018) Guidelines and considerations for the use of system suitability and quality control samples in mass spectrometry assays applied in untargeted clinical metabolomic studies. Metabolomics 14(6):17
94. Koal T, Deigner H-P (2010) Challenges in mass spectrometry based targeted metabolomics. Curr Mol Med 10(2):216–226
95. Griffiths WJ, Koal T, Wang Y, Kohl M, Enot DP, Deigner HP (2010) Targeted metabolomics for biomarker discovery. Angew Chem Int Ed 49(32):5426–5445
96. Kueger S, Steinhauser D, Willmitzer L, Giavalisco P (2012) High-resolution plant metabolomics: from mass spectral features to metabolites and from whole-cell analysis to subcellular metabolite distributions. Plant J 70(1):39–50

97. Shalhoub J, Sikkel MB, Davies KJ, Vorkas PA, Want EJ, Davies AH (2014) Systems biology of human atherosclerosis. Vasc Endovascular Surg 48(1):5–17
98. Almstetter MF, Oefner PJ, Dettmer K (2012) Comprehensive two-dimensional gas chromatography in metabolomics. Anal Bioanal Chem 402(6):1993–2013
99. Cajka T, Smilowitz JT, Fiehn O (2017) Validating quantitative untargeted lipidomics across nine liquid chromatography-high-resolution mass spectrometry platforms. Anal Chem 89(22):12360–12368
100. Cajka T, Fiehn O (2016) Increasing lipidomic coverage by selecting optimal mobile-phase modifiers in LC-MS of blood plasma. Metabolomics 12(2):1–11
101. Gonzalez-Dominguez A, Duran-Guerrero E, Fernandez-Recamales A, Lechuga-Sancho AM, Sayago A, Schwarz M et al (2017) An overview on the importance of combining complementary analytical platforms in metabolomic research. Curr Top Med Chem 17(30):3289–3295
102. Nichols CM, Dodds JN, Rose BS, Picache JA, Morris CB, Codreanu SG et al (2018) Untargeted molecular discovery in primary metabolism: collision cross section as a molecular descriptor in ion mobility-mass spectrometry. Anal Chem 90(24):14484–14492
103. Wernisch S, Pennathur S (2019) Application of differential mobility-mass spectrometry for untargeted human plasma metabolomic analysis. Anal Bioanal Chem 411(24):6297–6308
104. Kuenzi BM, Ideker T (2020) A census of pathway maps in cancer systems biology. Nat Rev Cancer 20(4):233–246
105. Tang HH, Sly PD, Holt PG, Holt KE, Inouye M (2020) Systems biology and big data in asthma and allergy: recent discoveries and emerging challenges. Eur Respir J 55(1):1900844

Biological Applications for LC-MS-Based Proteomics

Bradley J. Smith and Daniel Martins-de-Souza

Abstract

Since its inception, liquid chromatography-mass spectrometry (LC-MS) has been continuously improved upon in many aspects, including instrument capabilities, sensitivity, and resolution. Moreover, the costs to purchase and operate mass spectrometers and liquid chromatography systems have decreased, thus increasing affordability and availability in sectors outside of academic and industrial research. Processing power has also grown immensely, cutting the time required to analyze samples, allowing more data to be feasibly processed, and allowing for standardized processing pipelines. As a result, proteomics via LC-MS has become popular in many areas of biological sciences, forging an important seat for itself in targeted and untargeted assays, pure and applied science, the laboratory, and the clinic. In this chapter, many of these applications of LC-MS-based proteomics and an outline of how they can be executed will be covered. Since the field of personalized medicine has matured alongside proteomics, it has also come to rely on various mass spectrometry methods and will be elaborated upon as well. As time goes on and mass spectrometry evolves, there is no doubt that its presence in these areas, and others, will only continue to grow.

Keywords

Proteome · Proteomics · LC-MS-based proteomics · Personalized medicine · Liquid chromatography-mass spectrometry

1 Introduction

To study the many aspects of cell biology, a dizzying amount of equipment and tool sets are available or even required for different applica-

B. J. Smith
Laboratory of Neuroproteomics, Department of Biochemistry and Tissue Biology, Institute of Biology, University of Campinas (UNICAMP), Campinas, Brazil

D. Martins-de-Souza (✉)
Laboratory of Neuroproteomics, Department of Biochemistry and Tissue Biology, Institute of Biology, University of Campinas (UNICAMP), Campinas, Brazil

Instituto Nacional de Biomarcadores em Neuropsiquiatria (INBION), Conselho Nacional de Desenvolvimento Científico e Tecnológico, São Paulo, Brazil

Experimental Medicine Research Cluster (EMRC), University of Campinas, Campinas, SP, Brazil

D'Or Institute for Research and Education (IDOR), São Paulo, Brazil
e-mail: dmsouza@unicamp.br

A. V. Colnaghi Simionato (ed.), *Separation Techniques Applied to Omics Sciences*, Advances in Experimental Medicine and Biology 1336, https://doi.org/10.1007/978-3-030-77252-9_2

tions. Liquid chromatography is one such tool that has many individual uses; however, for many assays, including proteomics, it is better combined with another methodology, such as tandem mass spectrometry (MS/MS). One reason for this is due to the sheer complexity of protein-containing samples [1] and the wide dynamic range of proteins in living organisms [2]. Attempting to separate proteins by chromatographic methods alone would not be sufficient for the identification and quantitation of the thousands of different proteins and their isoforms and modified states that are present in a cell or tissue.

This problem is recognized – and greatly reduced – by coupling LC with mass spectrometry (MS) or tandem mass spectrometry (MS/MS), which can perform global identifications as well as other assays like the extremely sensitive quantitation of predetermined proteins. For these applications, separation by LC also plays a vital role in improving mass spectrometry methods by reducing the complexity of the sample at any given time point, additionally improving reproducibility from online coupling [3]. Over the past few decades, advances in LC-MS technology by way of sensitivity, resolution, reproducibility, and automation, as well as reductions in the costs to purchase and operate the systems [4], have caused their role in a clinical setting to steadily increase [5–7].

In this chapter, different methods of carrying out several LC-MS- and LC-MS/MS-based proteomic assays will be briefly discussed before their various overall applications are covered, which can range from laboratorial research to personalized medicine, as well as some specific examples of their uses. Since the field is expanding quickly, some applications that are still under research or going through testing phases will also be mentioned, many of which intend to bring more applications to the clinic in personalized medicine.

2 Proteomics via LC-MS

Mass spectrometry coupled with liquid chromatography (LC-MS) has become a widely used platform for many types of research projects. With the advent of online coupling between chromatographic separation and identification by mass spectrometry, many samples can now be set up in a queue for separation and any identification with reduced experimenter-based errors. Quantitation steps and fine tuning the various possible configurations also allow for its use in a widening range of experimental designs.

One of the first mass spectrometer designs required "eyeball" identification of molecules via obtained spectra, limiting both sample complexity and number [8]. However, over 25 years ago, advancements in computer systems and informatics allowed for the genesis of computer-based identification [9] and the eventual introduction of more varied and complex samples. When proteomics was first paired with mass spectrometry, one groundbreaking method to obtain large-scale proteomic data was to run the sample in a two-dimensional gel, whereupon individual spots with lower complexity could be excised, digested, and injected into the MS system [10].

Even after such a separation or in other types of purified samples, a single injection could still contain many proteins and even more peptides after digestion. Liquid chromatography was found to be extremely useful in such cases since a sample could be fractionated before being analyzed, reducing its complexity. In an attempt to reduce experimenter error, increase reproducibility, and decrease the time necessary to complete an analysis, a liquid chromatography system was coupled online to a mass spectrometer, generating a continuous flow of data across the sample's elution time [11]. Further improving upon the LC-MS design, multidimensional liquid chromatography systems have been since incorporated and optimized, utilizing multiple columns to quickly trap and separate peptides with reproducibility and high-resolution peak separation [12–14].

When brought together, liquid chromatography and mass spectrometry now play a key role at the center of many types of omic studies [15], including proteomics, due to its high resolution and sensitivity and the reproducibility of modern equipment. A proteomic profile changes both with cell type and with a wide range of stimuli and regulatory processes, and understanding a proteomic snapshot of a cell, as well as how the proteome changes in response to a stimulus or condition, provides unique insight into the inner workings of cells. Since there are 24 standard amino acids with dozens of possible modifications, mass spectrometers must go beyond an initial mass/charge (*m/z*) reading of a peptide, which would be insufficient to identify a peptide, and subsequently a protein, especially in a complex mixture.

In assays that require such a level of detail, the peptides that are ionized at the mass spectrometer ionization source are focused and optionally filtered before peptide fragmentation, with collision-induced dissociation (CID) being the most widely used method for proteomics [16]. The newly formed fragments (or more specifically the transitions of the precursor ions) can then be optionally filtered again before their *m/z* values are registered. Reading multiple combinations of fragments of a single peptide allows a mass spectrometer to determine, at least partially, the sequence of amino acids present, which becomes crucial when performing assays to identify which proteins are present in a sample [17].

Proteomic assays can be divided into two fundamental groups: targeted and untargeted studies. In a targeted assay, a specific list of known precursors and transitions is focused upon to detect the presence of those peptides and quantify them. When performing targeted assays, dynamic range can exceed four orders of magnitude, and sensitivity can be extremely high; sub-fmol/mg-of-sample sensitivity has already been obtained [18]. Quantitation is also exceptionally accurate in targeted studies [19], though the actual accuracy also depends on experimenter technique and the capabilities of the spectrometer. The limit of detection has also rapidly decreased, moving past the attomole level [20] into the zeptomole range [21]. Such sensitivity has also received attention for clinical applications that use extremely small sample volumes. In contrast with targeted studies, untargeted studies trade some of the aforementioned sensitivity for the ability to identify the peptides in the sample. By doing so, thousands of proteins [22, 23] and proteoforms [24–26] can be identified and quantified in a single injected sample.

Though different experiments can call for modifications to these suggestions, Fig. 1 depicts a flow chart to assist in determining which methods may be compatible with a project that plans to involve proteomics. Each of these categories will be discussed further in the following sections. It is always important to keep in mind what equipment is available for an experiment as well. Due to inherent equipment capabilities and limitations, not every LC-MS setup is able to perform both targeted and untargeted experiments [16], and not every spectrometer can perform both data-dependent and data-independent acquisition, for example.

3 Targeted Proteomics

Targeted proteomics is especially useful for the detection and quantitation of proteins/polypeptides with low abundance, the visualization of rare isoforms or posttranslational modifications (PTMs), and the validation of other identification or quantitation methods. A targeted assay has two main steps: a selection step and an acquisition step. In the selection step, theoretical mass data or data acquired from discovery studies is used to create a method for the spectrometer that determines what precursor ions will be filtered at the MS1 level and optionally fragmented at the MS2 level. In MS2 methods, the resulting transitions (peptide fragments) can also be preselected so they can reach the mass analyzer for identification and quantitation, referred to as single reaction monitoring (SRM) or multiple reaction monitoring (MRM). If all transitions are allowed to reach the mass analyzer for a given precursor, this method is called parallel reaction monitoring (PRM) and requires a high-resolution mass

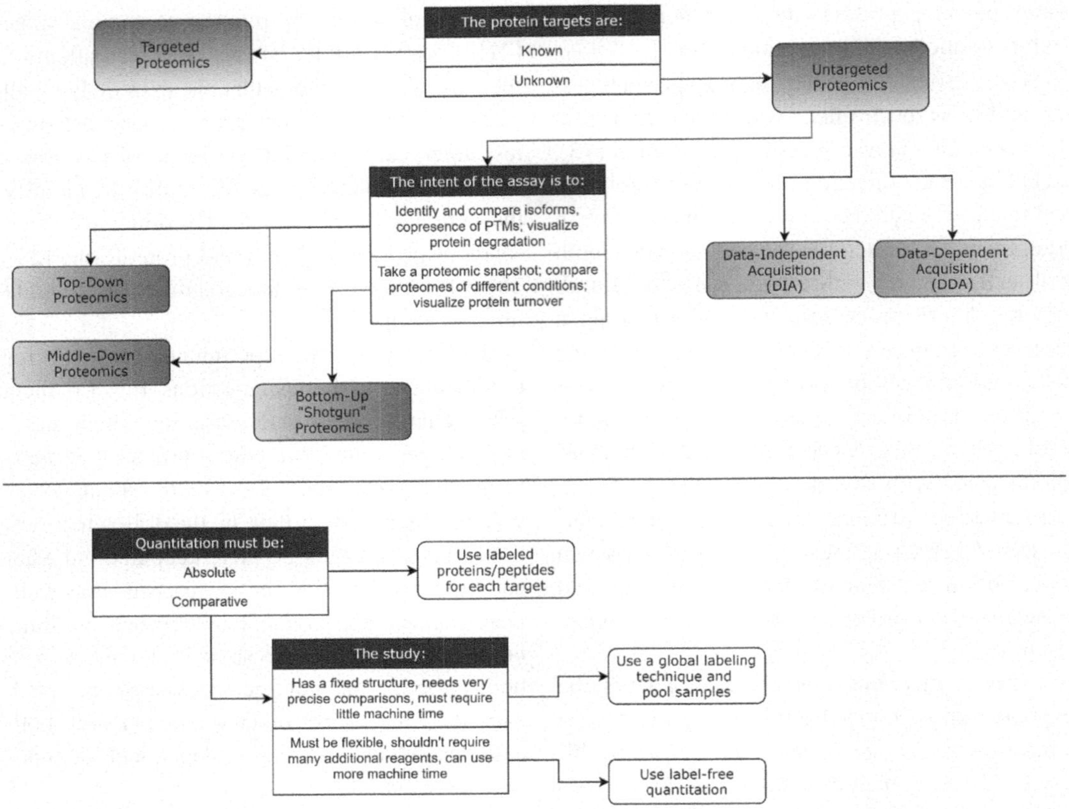

Fig. 1 A brief guide for novice users to aid in the choice of MS methods for proteomic analyses (top). A method for distinguishing between three basic divisions of quantitation is also included (bottom)

spectrometer but offers comparatively better results [27]. Once a method is created, the acquisition step can then be carried out to detect the transitions chosen in the selection step. A method created in the selection step has an added benefit of being translatable to other machines and setups with little to no additional optimization, making it perfect for clinical and commercial assays and applications.

The mass analyzers that perform these analyses must be able to accurately measure peptide ions and their transitions to ensure proper identification when comparing to the preselected *m/z* values [28]. Triple quadrupole (QQQ) mass spectrometers – a decades-old technology that has been repeatedly repurposed and improved upon for newer uses – are especially useful for this application [29] due to dual mass filters, before and after the collision cell. This placement allows only certain precursor ions to enter the collision cell and only a specific subset of fragmentation products to reach the mass analyzer, significantly reducing noise and undesired signals. The combination of quadrupole technology with ion traps (QqLIT), such as the QTrap system [30], can increase resolution and scan rate at the cost of sensitivity and *m/z* range [16]. The number of transitions that are selected and used for quantitation can be increased by applying a methodology called parallel reaction monitoring (PRM), useful when performing analyses on complex mixtures, though it requires a high-resolution mass spectrometer [31].

4 Untargeted Proteomics

Untargeted proteomics, in contrast with targeted proteomics, is well suited for discovery assays or when studying global changes and dysregulations in biological pathways. Untargeted proteomics itself is further divided into two main groups: top-down [32–34] and bottom-up [35, 36] (also referred to as shotgun proteomics). In top-down proteomics, intact proteins and polypeptides are submitted to MS/MS analysis, providing data such as protein degradation, isoforms, and copresence of posttranslational modifications (PTMs). Top-down assays, however, are still only capable of identifying a few hundred proteins in a whole-proteome sample and around a thousand proetoforms [37]. In contrast, bottom-up proteomics relies on a digestion step after protein extraction. The resulting, smaller molecules are able to be separated by high- or ultra-performance liquid chromatography (HPLC or UPLC), and high sensitivity is possible for even whole-proteome samples. A relatively recent combination of these two methods, called middle-down proteomics, has been suggested to cover smaller polypeptide sequences, following protocols similar to bottom-up proteomics while keeping the ability to identify co-occurring posttranslational modifications [38, 39]. For more detailed information about these three classifications and their differences, see Lermyte et al. [40].

The way that untargeted assays are carried out can also be divided into two categories, depending on how the mass spectrometer collects spectra, namely, data-dependent acquisition (DDA) and data-independent acquisition (DIA). In DDA, the software controlling the spectrometer makes an on-the-fly selection to determine which precursor ions are to be fragmented. This is the method of choice for lower-resolution instruments such as time-of-flight (TOF) and triple quadrupole (QQQ) mass spectrometers due to instrument limitations. However, this does not inherently exclude high-resolution spectrometers from being useful in DDA [41]. In contrast, DIA requires extremely high resolving power since all the precursor ions reaching the fragmentation cell are fragmented and sent to the mass detector, resulting in a large amount of nearly simultaneous data. This can be done either by filtering for windows of *m/z* values in a stepwise manner (sequential window acquisition of all theoretical mass spectra (SWATH-MS)) [42] or by fragmenting the entire set of precursor ions (MS^E) while rapidly alternating between high and low collision energies [43].

Due to the increased number of signals that are simultaneously acquired in DIA, data analysis is by far more complicated than in DDA [44] and the convoluted spectra that are obtained make for a complex mathematical problem that must be carefully dealt with [45]. As such, the vast amount of data that can be obtained in a single injection and the ability to reanalyze DIA data with different parameters make it indispensable for large-scale analyses of complex samples, especially in the research sector.

5 Quantitation

When a proteomic assay requires quantitation, there are two overarching categories to accomplish this: label-free and labeled techniques. Both are compatible with targeted and untargeted assays, and each has its benefits and drawbacks for different experimental designs. At its core, in label-free quantitation, the mass spectrometer counts the number of times different precursors, or transition ions are measured in the LC-MS system. That data can then be compared between runs to estimate the relative amount of a protein in one sample compared to another. One drawback of label-free techniques is that this comparison can only be made with the same protein between samples unless absolute quantitation is applied since different peptide sequences, PTMs, and other factors affect ionization and fragmentation efficiency [46]. Preparation for label-free quantitation is procedurally simple, though after data acquisition, software processing is more resource-intensive due to the convoluted spectra. Some software additionally aligns chromatographic data from different runs to improve the quality of comparisons between samples; however, this lengthens processing time with growing sample number. Lastly, physical conditions, the LC system, the mass spectrometer, and the contents of the sample itself can vary, making raw data poorly translatable to other studies. Nonetheless, the cost and laboratory setup required for label-free analyses are relatively low after the initial infrastructure investment, and the number of possible samples or conditions is limited only by the time available to perform the experiment.

In contrast, labeled quantitation requires the use of stable isotopes or isobaric mass tags. These can label or be incorporated into proteins or peptides at various stages of sample preparation, ranging from during cell or animal growth to just before the samples are injected. Using labels is comparatively costly, though it is an investment that can decrease variations due to physical/environmental conditions and experimenter error, therefore increasing the data quality of the comparison. One major limitation of labeled quantitation is that sample multiplexing – when multiple, labeled samples are mixed and analyzed in a mass spectrometer as a single injection – limits the number of conditions that can be simultaneously run, a number inherent to the method that is chosen. For the same reasons as label-free quantitation, labeled quantitation protocols provide only relative quantitation data unless an absolute quantitation method is used.

Absolute quantitation is compatible with both label-free and labeled quantitation. Having an absolute value for a protein is an important aspect to many assays, especially in a clinical or industrial setting. To perform absolute quantitation, some additional steps are necessary: (1) the protein(s) of interest must be selected in advance and optionally, depending on the labeling method, peptides that are most commonly digested, ionized, and fragmented must be identified and selected; (2) labeled versions of those proteins or peptides must be obtained in a pure form; and lastly, (3) a known quantity of the labeled proteins/peptides must be "spiked" into each sample to be analyzed. Relative quantitation can then be performed between the known value of the spiked peptides/proteins and the biological sample. Since the quantity of the spiked protein is precisely known, the quantity of the protein in the original sample can be calculated with a high degree of accuracy, limited by the sensitivity and resolution of the mass spectrometer.

6 Applications for Proteomics

As the cost to perform proteomic analyses has fallen, the feasibility of using mass spectrometry has increased, opening up numerous opportunities for unparalleled applications in various sectors. With rising availability, more efforts have been made to apply proteomics and LC-MS to experimental designs and projects that have not only spread through the research field but have also entered the clinic, ranging from commercial, standardized testing to applications for personalized medicine (Fig. 2).

6.1 Pure and Applied Research

A traditional application of proteomics in research projects is to visualize how a cell or organism responds to a specific stimulus or condition. mRNA levels are not always a good representation of the proteomic state of a cell [47–49] and can therefore fail to be an accurate representation of the cell's phenotype, especially when posttranslational modifications are taken into consideration as well. Therefore, LC-MS offers insight into the changes that occur in cells over time, in response to pharmacological agents, when exposed to risk factors for diseases and disorders, when oxidative stress occurs, or in response to changes in diet, among many others. Some techniques can also answer other specific questions, though there are too many to cover all of them in this chapter. One example is the protein labeling technique called pSILAC (pulsed stable isotope labeling by amino acids in cell culture), able to provide data about protein turnover rates by measuring how quickly labeled amino acids are incorporated into newly translated proteins [50].

LC-MS is also indispensable in studies involving posttranslational modifications (PTMs). Even 15 years ago, there were already over 200 known covalent modifications of proteins [51], each conferring specific structural or functional changes to a corresponding protein. A single shotgun LC-MS/MS run can be analyzed and reanalyzed using different parameters

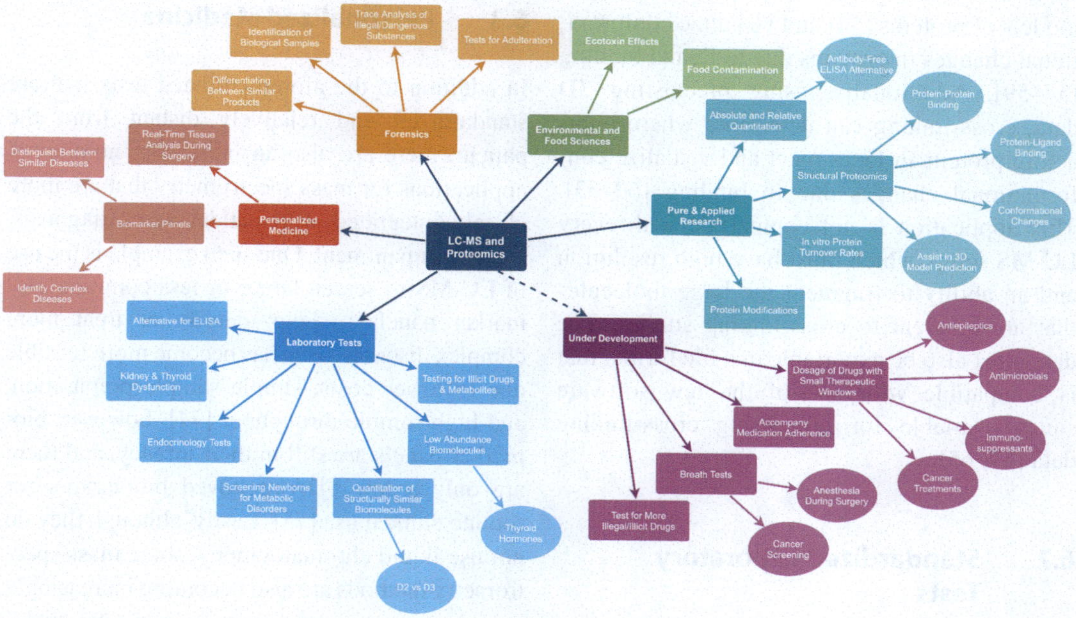

Fig. 2 A map of the generalized applications of LC-MS and proteomics and the categories to which each belongs. Application categories are distinguished by color. Circles represent subgroups of the applications in rectangles

to reveal any number of these PTMs. In many software pipelines, this can become problematic, since this exponentially increases the number of possibilities to identify each ion that reached the detector; however, this issue and problems with high false discovery rates have been recognized by PeaksPTM [52], which attempts to identify all possible modifications on proteins. Top-down proteomics can also provide unique PTM data, such as co-present modifications [53], and have gained popularity due to its ability to visualize the various modified states of contractile proteins, especially important when studying the underlying causes of heart disease [54].

Another attractive application for proteomics is to replace ELISA tests or to complement them. Immuno-based protein quantitation can be highly sensitive, but there are some caveats, especially in certain situations, such as when dealing with protein isoforms [55]. In most clinical settings, ELISA tests are still very practical since a mass spectrometer and specialist are not required to perform the assay; however, in a research setting, developing a completely new ELISA test, creating an antibody, and validating its affinity and specificity can be quite time-consuming. Even upon establishing a test for a specific protein or isoform, some questions cannot be answered with immune-based assays, like when investigating PTMs or when dealing with novel protein mutants. Though the initial setup costs can be difficult to overcome, LC-MS has immense potential for this type of assay and has already shown up to zeptomole [10–21] sensitivity.

Structural proteomics (mapping data about protein structures and protein-protein interactions) is also possible with LC-MS/MS. While mass spectrometry is not comparable to X-ray crystallography or cryo-EM at mapping protein structures, it can become useful for proteins or protein complexes that are not crystallizable or are otherwise incompatible with these methods. Cross-linkers are chain molecules that are designed to covalently bind to certain amino acid residues on the surfaces of proteins. Since the molecule length is a known value, two peptide sequences bound to the ends of a cross-linker will have a calculatable distance. For example, it is possible to use cross-linking technology to confirm or complement simulated 3D models of proteins [56] and visualize conformational changes in proteins due to ligand binding [57–59]. Additionally, using preexisting 3D data, cross-linking can determine where interacting protein surfaces meet and visualize conformational changes due to binding [60–63]. This application is not compatible with every LC-MS setup, which must have high resolution and an ability to fragment the large molecules that are inherent to cross-linking studies. The data must also be exportable in a file format that is compatible with one of the few software suites available for processing cross-linking data [64, 65].

6.2 Standardized Laboratory Tests

As the equipment to perform LC-MS-based tests has become more accessible, more ways to use the technology have been researched. Due to the high sensitivity of the method and its ability to distinguish between extremely similar molecules, it is well suited for high-throughput tests for illicit drugs and their metabolites [66]. Additionally, specialized tests have been standardized for specific biomarkers to screen newborns for inborn errors of metabolism, perform endocrinology tests, and identify diseases that would be otherwise difficult to detect or confirm, such as pheochromocytoma, a type of hormone-secreting tumor in the adrenal glands [6, 67]. LC-MS is also employed in some medical laboratories to measure thyroglobulin and iothalamate for thyroid and kidney dysfunction in patients, respectively [33, 68]. Structurally similar biomolecules like vitamin D2 and D3 [69, 70] and low-abundance biomolecules like thyroid hormones [71–73] are also good targets for quantitative LC-MS methods, since issues with specificity and cross-reactivity, which are inherently possible in immunoassays, are reduced and sensitivity is increased. In some laboratories with the proper funding and personnel, ELISA tests can be complemented with or replaced by LC-MS as well, as discussed in the previous section.

6.3 Personalized Medicine

In addition to the aforementioned tests that are standardized and relatively distant from the patient, there are also an increasing number of applications for mass spectrometry that are more closely integrated with patient care, diagnosis, and accompaniment. One such example is the use of LC-MS to screen larger or less common biomarker panels to help identify or treat more complex diseases. This has become more feasible due to lower costs, simple sample preparation, and high sample throughput [74]; however, biomarker panels are still in their infancy, and there are only a few FDA-approved biomarkers for routine clinical use [75]. Lastly, although they do not use liquid chromatography, some mass spectrometry methods are also becoming increasingly useful in the operating room for real-time analysis of tissue during cancer excision surgeries by using matrix-assisted laser desorption/ionization (MALDI) imaging [76–78], picosecond infrared laser mass spectrometry (PIRL-MS) [79], or a unique nondestructive technique called iKnife (using rapid evaporative ionization MS) [80, 81] that has successfully been integrated into the da Vinci Surgical System [82].

6.4 Other Sectors

Due to the high resolution and sensitivity of LC-MS methods, other sectors have also noticed potential applications for specific biological and chemical assays. In ecotoxicology, mass spectrometry can be utilized to document the interactions between the environment, various organisms and microorganisms, and chemical toxins [83] or antibiotics [84]. In some research projects, "sentinel species" are studied in detail to extrapolate toxicity across many species [85]. LC-MS can also be used for determining the composition of complex sample matrices, like the assays carried out in the petroleum industry for decades [86], with updates, making them still used to this day [87]. There are also many applications in the food industry to detect various forms of biological and nonbiological contamination [88, 89]. Forensic science also relies on LC-MS to identify biological sample sources [90], perform trace analyses of illegal and dangerous substances [91], and analyze various other chemometrics in diverse cases [92]. In a fusion between history and art, LC-MS has also been even used to determine the origins of different paints [93].

6.5 Applications Under Development

Many additional applications for LC-MS are at various stages of research and implementation. One very attractive option is to use mass spectrometry to accompany a patient's treatment in a more dynamic manner, such as when regulating the dosage of medications with small therapeutic windows like immunosuppressants [94, 95], antimicrobials [96], antiepileptics [97], and cancer treatments [98]. Many of these, however, have not yet left proof-of-concept stages and lack a clinical level of standardization. Along these lines, MS has also been proposed to assist in verifying treatment adherence, which would even cover unintentional nonadherence due to adulterated medication [99]. Newborn screening is also being constantly expanded to detect more disorders and diseases with fewer false positives and false negatives [100]. Additionally, thousands of potential biomarkers are undergoing testing, such as trimethylamine N-oxide, a well-studied biomarker for cardiovascular disease [101–105] and over 1000 biomolecules for cancer alone [106–108]. Lastly, the detection of an even wider range of illicit and pharmaceutical drugs has been undergoing standardization [109]. Though not coupled with liquid chromatography, mass spectrometry has also been investigated to detect the various compounds that are present in exhaled breath, which holds potential for various forms of cancer screening [110–112] and can be used to accompany anesthesia dosing during surgery [113–115]. It is not far-fetched to imagine that in the near future, a wide range of biomarkers could be routinely analyzed to predict or diagnose a wide range of disorders and diseases at an affordable cost to the patient.

7 Conclusion

Overall, mass spectrometry and liquid chromatography have proven to be a formidable pair to study a wide variety of biological and nonbiological questions. They have spanned across omic sciences, gaining an unparalleled role in proteomics, offering both targeted and untargeted approaches. The ever-increasing sensitivity and resolution of mass spectrometers strongly suggest that this technique will be present and evolving for many years to come. Many other applications have been discovered for LC-MS as well in environmental and business sectors. Assays to identify and quantify synthetic molecules, toxins, metabolites, proteins, and organisms themselves are now possible to accompany environmental states, perform quality control in various industries, and carry out certain tests in forensic science.

In a clinical setting, applications of LC-MS have also long begun to directly play a role in patient care, offering fast and highly sensitive tests for protein and metabolite biomarkers for various diseases and disorders. Bringing the bench closer to the bedside, personalized medicine has also found a partner in LC-MS, and development has begun to work with drugs with narrow therapeutic windows, screen for complex diseases and disorders, and assist during tumor excision surgery. As the cost of this methodology continues to become more affordable and as its availability, sensitivity, reproducibility, and resolution increase, liquid chromatography-mass spectrometry will undoubtedly play an even more significant role in laboratorial, industrial, environmental, clinical, and personalized medicine settings.

Acknowledgments The authors would like to thank the Coordination for the Improvement of Higher Education Personnel (Coordenação de Aperfeiçoamento de Pessoal de Nível Superior, CAPES; grant number 88887.495565/2020-00) and the São Paulo Research Foundation (Fundação de Amparo à Pesquisa do Estado de São Paulo, FAPESP; grant numbers 2014/10068-4, 2016/07948-8, 2017/25588-1, 2019/00098-7 and 2019/05747-3) for financial support.

Conflict of Interest The authors declare that there are no conflicts of interest.

References

1. Ponomarenko EA et al (2016) The size of the human proteome: the width and depth. Int J Anal Chem 2016:7436849
2. Zubarev RA (2013) The challenge of the proteome dynamic range and its implications for in-depth proteomics. Proteomics 13:723–726
3. Ishihama Y (2005) Proteomic LC–MS systems using nanoscale liquid chromatography with tandem mass spectrometry. J Chromatogr A 1067:73–83
4. Bu X, Regalado EL, Hamilton SE, Welch CJ (2016) The emergence of low-cost compact mass spectrometry detectors for chromatographic analysis. TrAC Trends Anal Chem 82:22–34
5. Dooley KC (2003) Tandem mass spectrometry in the clinical chemistry laboratory. Clin Biochem 36:471–481
6. Grebe SK, Singh RJ (2011) LC-MS/MS in the clinical laboratory – where to from here? Clin Biochem Rev 32:5–31
7. Seger C, Salzmann L (2020) After another decade: LC–MS/MS became routine in clinical diagnostics. Clin Biochem 82:2–11. https://doi.org/10.1016/j.clinbiochem.2020.03.004
8. Gohlke RS, McLafferty FW (1993) Early gas chromatography/mass spectrometry. J Am Soc Mass Spectrom 4:367–371
9. Eng JK, McCormack AL, Yates JR (1994) An approach to correlate tandem mass spectral data of peptides with amino acid sequences in a protein database. J Am Soc Mass Spectrom 5:976–989
10. Rabilloud T, Lelong C (2011) Two-dimensional gel electrophoresis in proteomics: a tutorial. J Proteome 74:1829–1841
11. Arpino PJ (1982) On-line liquid chromatography/mass spectrometry? An odd couple! TrAC Trends Anal Chem 1:154–158
12. Stoll DR et al (2007) Fast, comprehensive two-dimensional liquid chromatography. J Chromatogr A 1168:3–43
13. Stoll DR (2017) Chapter 7 – Introduction to two-dimensional liquid chromatography – theory and practice. In: Holčapek M, Byrdwell WC (eds) Handbook of advanced chromatography/mass spectrometry techniques. AOCS Press, Urbana, IL, pp 227–286. https://doi.org/10.1016/B978-0-12-811732-3.00007-8
14. Sarrut M, D'Attoma A, Heinisch S (2015) Optimization of conditions in on-line comprehensive two-dimensional reversed phase liquid chromatography. Experimental comparison with one-dimensional reversed phase liquid chromatography for the separation of peptides. J Chromatogr A 1421:48–59
15. Di Girolamo F, Lante I, Muraca M, Putignani L (2013) The role of mass spectrometry in the "omics" era. Curr Org Chem 17:2891–2905
16. Han X, Aslanian A, Yates JR (2008) Mass spectrometry for proteomics. Curr Opin Chem Biol 12:483–490
17. Cottrell JS (2011) Protein identification using MS/MS data. J Proteome 74:1842–1851
18. Schoenherr RM et al (2012) Multiplexed quantification of estrogen receptor and HER2/Neu in tissue and cell lysates by peptide immunoaffinity enrichment mass spectrometry. Proteomics 12:1253–1260
19. Liebler DC, Zimmerman LJ (2013) Targeted quantitation of proteins by mass spectrometry. Biochemistry 52:3797–3806
20. Ruppen-Cañás I et al (2012) An improved quantitative mass spectrometry analysis of tumor specific mutant proteins at high sensitivity. Proteomics 12:1319–1327
21. Swaminathan J et al (2018) Highly parallel single-molecule identification of proteins in zeptomole-scale mixtures. Nat Biotechnol 36:1076–1082. https://doi.org/10.1038/nbt.4278
22. Griffin NM et al (2010) Label-free, normalized quantification of complex mass spectrometry data for proteomic analysis. Nat Biotechnol 28:83–89
23. Chen EI, McClatchy D, Park SK, Yates JR (2008) Comparisons of mass spectrometry compatible surfactants for global analysis of the mammalian brain proteome. Anal Chem 80:8694–8701
24. Catherman AD et al (2013) Large-scale top-down proteomics of the human proteome: membrane proteins, mitochondria, and senescence. Mol Cell Proteomics 12:3465–3473
25. Fornelli L et al (2017) Advancing top-down analysis of the human proteome using a benchtop quadrupole-orbitrap mass spectrometer. J Proteome Res 16:609–618
26. Tran JC et al (2011) Mapping intact protein isoforms in discovery mode using top-down proteomics. Nature 480:254–258
27. Borràs E, Sabidó E (2017) What is targeted proteomics? A concise revision of targeted acquisition and targeted data analysis in mass spectrometry. Proteomics 17:1700180
28. Schaeffer-Reiss C (2008) A brief summary of the different types of mass spectrometers used in proteomics. In: Thompson JD, Ueffing M, Schaeffer-Reiss C (eds) Functional proteomics: methods and protocols. Humana Press, Clifton, NJ, pp 3–16. https://doi.org/10.1007/978-1-59745-398-1_1
29. Hopfgartner G et al (2004) Triple quadrupole linear ion trap mass spectrometer for the analysis of small molecules and macromolecules. J Mass Spectrom 39:845–855
30. Stahl-Zeng J et al (2007) High sensitivity detection of plasma proteins by multiple reaction monitoring of N-glycosites. Mol Cell Proteomics 6:1809–1817
31. Peterson AC, Russell JD, Bailey DJ, Westphall MS, Coon JJ (2012) Parallel reaction monitoring for high resolution and high mass accuracy quantitative, targeted proteomics. Mol Cell Proteomics 11:1475–1488

32. Padula MP et al (2017) A comprehensive guide for performing sample preparation and top-down protein analysis. Proteomes 5:11
33. Fornelli L et al (2018) Top-down proteomics: where we are, where we are going? J Proteome 175:3–4
34. Cupp-Sutton KA, Wu S (2020) High-throughput quantitative top-down proteomics. Mol Omics 16:91–99
35. Gillet LC, Leitner A, Aebersold R (2016) Mass spectrometry applied to bottom-up proteomics: entering the high-throughput era for hypothesis testing. Annu Rev Anal Chem 9:449–472
36. Smith BJ, Martins-de-Souza D, Fioramonte M (2019) A guide to mass spectrometry-based quantitative proteomics. In: Pre-clinical models. Springer, Humana Press, New York, NY, pp 3–39
37. Cleland TP et al (2017) High-throughput analysis of intact human proteins using UVPD and HCD on an orbitrap mass spectrometer. J Proteome Res 16:2072–2079
38. Cristobal A et al (2017) Toward an optimized workflow for middle-down proteomics. Anal Chem 89:3318–3325
39. Pandeswari PB, Sabareesh V (2018) Middle-down approach: a choice to sequence and characterize proteins/proteomes by mass spectrometry. RSC Adv 9:313–344
40. Lermyte F, Tsybin YO, O'Connor PB, Loo JA (2019) Top or middle? Up or down? Toward a standard lexicon for protein top-down and allied mass spectrometry approaches. J Am Soc Mass Spectrom 30:1149–1157
41. Kalli A, Smith GT, Sweredoski MJ, Hess S (2013) Evaluation and optimization of mass spectrometric settings during data-dependent acquisition mode: focus on LTQ-orbitrap mass analyzers. J Proteome Res 12:3071–3086
42. Gillet LC et al (2012) Targeted data extraction of the MS/MS spectra generated by data-independent acquisition: a new concept for consistent and accurate proteome analysis. Mol Cell Proteomics 11:O111.016717
43. Plumb RS et al (2006) UPLC/MSE; a new approach for generating molecular fragment information for biomarker structure elucidation. Rapid Commun Mass Spectrom 20:1989–1994
44. Ludwig C et al (2018) Data-independent acquisition-based SWATH-MS for quantitative proteomics: a tutorial. Mol Syst Biol 14:e8126
45. Marcotte EM (2007) How do shotgun proteomics algorithms identify proteins? Nat Biotechnol 25:755–757
46. Lundgren DH, Hwang S-I, Wu L, Han DK (2010) Role of spectral counting in quantitative proteomics. Expert Rev Proteomics 7:39–53
47. Koussounadis A, Langdon SP, Um IH, Harrison DJ, Smith VA (2015) Relationship between differentially expressed mRNA and mRNA-protein correlations in a xenograft model system. Sci Rep 5:10775
48. Vogel C, Marcotte EM (2012) Insights into the regulation of protein abundance from proteomic and transcriptomic analyses. Nat Rev Genet 13:227–232
49. de Sousa Abreu R, Penalva LO, Marcotte EM, Vogel C (2009) Global signatures of protein and mRNA expression levels. Mol BioSyst 5:1512–1526
50. Schwanhäusser B, Gossen M, Dittmar G, Selbach M (2009) Global analysis of cellular protein translation by pulsed SILAC. Proteomics 9:205–209
51. Walsh CT, Garneau-Tsodikova S, Gatto GJ (2005) Protein posttranslational modifications: the chemistry of proteome diversifications. Angew Chem Int Ed 44:7342–7372
52. Han X, He L, Xin L, Shan B, Ma B (2011) PeaksPTM: mass spectrometry-based identification of peptides with unspecified modifications. J Proteome Res 10:2930–2936
53. Siuti N, Kelleher NL (2007) Decoding protein modifications using top-down mass spectrometry. Nat Methods 4:817–821
54. Cai W, Tucholski TM, Gregorich ZR, Ge Y (2016) Top-down proteomics: technology advancements and applications to heart diseases. Expert Rev Proteomics 13:717–730
55. Denburg MR et al (2016) Comparison of two ELISA methods and mass spectrometry for measurement of vitamin D-binding protein: implications for the assessment of bioavailable vitamin D concentrations across genotypes. J Bone Miner Res 31:1128–1136
56. Schneider M, Belsom A, Rappsilber J (2018) Protein tertiary structure by crosslinking/mass spectrometry. Trends Biochem Sci 43:157–169
57. Becker JM, Naider F (2015) Cross-linking strategies to study peptide ligand-receptor interactions. Methods Enzymol 556:527–547
58. Kim KM, Yi EC, Kim Y (2012) Mapping protein receptor-ligand interactions via in vivo chemical crosslinking, affinity purification, and differential mass spectrometry. Methods 56:161–165
59. Rannversson H et al (2016) Genetically encoded photocrosslinkers locate the high-affinity binding site of antidepressant drugs in the human serotonin transporter. Nat Commun 7(11261)
60. Lössl P et al (2014) Analysis of nidogen-1/laminin γ1 interaction by cross-linking, mass spectrometry, and computational modeling reveals multiple binding modes. PLoS One 9:e112886
61. Lima DB et al (2018) Characterization of homodimer interfaces with cross-linking mass spectrometry and isotopically labeled proteins. Nat Protoc 13:431–458
62. O'Reilly FJ, Rappsilber J (2018) Cross-linking mass spectrometry: methods and applications in structural, molecular and systems biology. Nat Struct Mol Biol 25:1000–1008
63. Chavez JD, Bruce JE (2019) Chemical cross-linking with mass spectrometry: a tool for systems structural biology. Curr Opin Chem Biol 48:8–18
64. Liu F, Rijkers DTS, Post H, Heck AJR (2015) Proteome-wide profiling of protein assemblies by

cross-linking mass spectrometry. Nat Methods 12:1179–1184
65. Iacobucci C et al (2018) A cross-linking/mass spectrometry workflow based on MS-cleavable cross-linkers and the MeroX software for studying protein structures and protein-protein interactions. Nat Protoc 13:2864–2889
66. Wissenbach DK et al (2011) Drugs of abuse screening in urine as part of a metabolite-based LC-MSn screening concept. Anal Bioanal Chem 400:3481–3489
67. Abou-Diwan C (2011) LC-MS/MS in the clinical lab: strengths, applications, challenges. MLO Med Lab Obs 43:20
68. Seegmiller JC et al (2010) Iothalamate quantification by tandem mass spectrometry to measure glomerular filtration rate. Clin Chem 56:568–574
69. Gilliland DL, Black CK, Denison JE, Seipelt CT (2012) Simultaneous determination of vitamins D2 and D3 by LC-MS/MS in infant formula and adult nutritionals: first action 2011.13. J AOAC Int 95:583–587
70. Eyles D et al (2009) A sensitive LC/MS/MS assay of 25OH vitamin D3 and 25OH vitamin D2 in dried blood spots. Clin Chim Acta 403:145–151
71. Spencer CA (2000) Assay of thyroid hormones and related substances. In: Feingold KR et al (eds) Endotext, South Dartmouth, MA. MDText.com, Inc.
72. Bowerbank SL, Carlin MG, Dean JR (2019) A direct comparison of liquid chromatography-mass spectrometry with clinical routine testing immunoassay methods for the detection and quantification of thyroid hormones in blood serum. Anal Bioanal Chem 411:2839–2853
73. Sakai H et al (2015) Correlation between serum levels of 3,3′,5′-Triiodothyronine and thyroid hormones measured by liquid chromatography-tandem mass spectrometry and immunoassay. PLoS One 10:e0138864
74. Anjo SI, Santa C, Manadas B (2017) SWATH-MS as a tool for biomarker discovery: from basic research to clinical applications. Proteomics 17:1600278
75. Center for Drug Evaluation and Research (2020) List of qualified biomarkers. FDA. https://www.fda.gov/drugs/cder-biomarker-qualification-program/list-qualified-biomarkers
76. Brégeon F et al (2014) MALDI-ToF mass spectrometry for the rapid diagnosis of cancerous lung nodules. PLoS One 9:e97511
77. Basu SS et al (2019) Rapid MALDI mass spectrometry imaging for surgical pathology. npj Precis Oncol 3:1–5
78. Calligaris D et al (2015) MALDI mass spectrometry imaging analysis of pituitary adenomas for near-real-time tumor delineation. PNAS 112:9978–9983
79. Woolman M et al (2020) In situ tissue pathology from spatially encoded mass spectrometry classifiers visualized in real time through augmented reality. Chem Sci 11:8723. https://doi.org/10.1039/D0SC02241A
80. Zhang J et al (2017) Nondestructive tissue analysis for ex vivo and in vivo cancer diagnosis using a handheld mass spectrometry system. Sci Transl Med 9(406). https://doi.org/10.1126/scitranslmed.aan3968
81. Sans M et al (2019) Performance of the MasSpec Pen for rapid diagnosis of ovarian cancer. Clin Chem 65:674–683
82. Keating MF et al (2020) Integrating the MasSpec Pen to the da Vinci surgical system for in vivo tissue analysis during a robotic assisted porcine surgery. Anal Chem 92:11535–11542
83. Groh KJ, Suter MJ-F (2014) Mass spectrometry in environmental toxicology. Chimia 68:140–145
84. Batt AL, Aga DS (2005) Simultaneous analysis of multiple classes of antibiotics by ion trap LC/MS/MS for assessing surface water and groundwater contamination. Anal Chem 77:2940–2947
85. Gouveia D et al (2019) Ecotoxicoproteomics: a decade of progress in our understanding of anthropogenic impact on the environment. J Proteome 198:66–77
86. Dark WA, McFadden WH (1978) The role of HPLC and LC-MS in the separation and characterization of coal liquefaction products. J Chromatogr Sci 16:289–293
87. Han Y, Zhang Y, Xu C, Hsu CS (2018) Molecular characterization of sulfur-containing compounds in petroleum. Fuel 221:144–158
88. Gallart-Ayala H, Núñez O, Lucci P (2013) Recent advances in LC-MS analysis of food-packaging contaminants. TrAC Trends Anal Chem 42:99–124
89. Picó Y, Barceló D (2008) The expanding role of LC-MS in analyzing metabolites and degradation products of food contaminants. TrAC Trends Anal Chem 27:821–835
90. Van Steendam K, De Ceuleneer M, Dhaenens M, Van Hoofstat D, Deforce D (2013) Mass spectrometry-based proteomics as a tool to identify biological matrices in forensic science. Int J Legal Med 127:287–298
91. Wood M et al (2006) Recent applications of liquid chromatography–mass spectrometry in forensic science. J Chromatogr A 1130:3–15
92. Kumar R, Sharma V (2018) Chemometrics in forensic science. TrAC Trends Anal Chem 105:191–201
93. La Nasa J et al (2015) Novel application of liquid chromatography/mass spectrometry for the characterization of drying oils in art: elucidation on the composition of original paint materials used by Edvard Munch (1863–1944). Anal Chim Acta 896:177–189
94. McShane AJ, Bunch DR, Wang S (2016) Therapeutic drug monitoring of immunosuppressants by liquid chromatography-mass spectrometry. Clin Chim Acta 454:1–5
95. Molnár PM et al (2011) Procedure for determination of immunosuppressive drugs in whole blood with liquid chromatography-isotope dilution mass spectrometry. Clin Lab 57:983–992

96. Neugebauer S, Wichmann C, Bremer-Streck S, Hagel S, Kiehntopf M (2019) Simultaneous quantification of nine antimicrobials by LC-MS/MS for therapeutic drug monitoring in critically ill patients. Ther Drug Monit 41:29–37
97. Yin L et al (2017) A parallel-column LC–MS/MS method for high-throughput analysis of eight antiepileptic drugs in clinical therapeutic drug monitoring. Chromatographia 80:137–143
98. Becher F et al (2017) A simple and rapid LC-MS/MS method for therapeutic drug monitoring of cetuximab: a GPCO-UNICANCER proof of concept study in head-and-neck cancer patients. Sci Rep 7:1–11
99. Tanna S, Ogwu J, Lawson G (2020) Hyphenated mass spectrometry techniques for assessing medication adherence: advantages, challenges, clinical applications and future perspectives. Clin Chem Lab Med 58:643–663
100. Ombrone D, Giocaliere E, Forni G, Malvagia S, la Marca G (2016) Expanded newborn screening by mass spectrometry: new tests, future perspectives. Mass Spectrom Rev 35:71–84
101. Suzuki T, Heaney LM, Bhandari SS, Jones DJL, Ng LL (2016) Trimethylamine N-oxide and prognosis in acute heart failure. Heart 102:841–848
102. Suzuki T, Heaney LM, Jones DJL, Ng LL (2017) Trimethylamine N-oxide and risk stratification after acute myocardial infarction. Clin Chem 63:420–428
103. Barrea L et al (2018) Trimethylamine-N-oxide (TMAO) as novel potential biomarker of early predictors of metabolic syndrome. Nutrients 10:1971
104. Janeiro MH, Ramírez MJ, Milagro FI, Martínez JA, Solas M (2018) Implication of trimethylamine N-oxide (TMAO) in disease: potential biomarker or new therapeutic target. Nutrients 10:1398
105. Fu BC et al (2020) Associations of plasma trimethylamine N-oxide, choline, carnitine, and betaine with inflammatory and cardiometabolic risk biomarkers and the fecal microbiome in the Multiethnic Cohort Adiposity Phenotype Study. Am J Clin Nutr 111:1226–1234
106. Polanski M, Anderson NL (2006) A list of candidate cancer biomarkers for targeted proteomics. Biomark Insights. https://doi.org/10.1177/117727190600100001
107. Belczacka I et al (2019) Proteomics biomarkers for solid tumors: current status and future prospects. Mass Spectrom Rev 38:49–78
108. Sajic T et al (2019) A new class of protein biomarkers based on subcellular distribution: application to a mouse liver cancer model. Sci Rep 9:6913
109. Trefz P et al (2017) Drug detection in breath: non-invasive assessment of illicit or pharmaceutical drugs. J Breath Res 11:024001
110. Chandran D et al (2019) The use of selected ion flow tube-mass spectrometry technology to identify breath volatile organic compounds for the detection of head and neck squamous cell carcinoma: a pilot study. Medicina (Kaunas) 55(6):306
111. Chin S-T, Romano A, Doran SLF, Hanna GB (2018) Cross-platform mass spectrometry annotation in breathomics of oesophageal-gastric cancer. Sci Rep 8:5139
112. Singh KD et al (2019) Standardization procedures for real-time breath analysis by secondary electrospray ionization high-resolution mass spectrometry. Anal Bioanal Chem 411:4883–4898
113. Berchtold C, Bosilkovska M, Daali Y, Walder B, Zenobi R (2014) Real-time monitoring of exhaled drugs by mass spectrometry. Mass Spectrom Rev 33:394–413
114. Beck O, Olin A-C, Mirgorodskaya E (2016) Potential of mass spectrometry in developing clinical laboratory biomarkers of nonvolatiles in exhaled breath. Clin Chem 62:84–91
115. Dong H et al (2017) Simultaneous on-line monitoring of propofol and sevoflurane in balanced anesthesia by direct resistive heating gas chromatography. J Chromatogr A 1506:93–100

The Role of Chromatographic and Electromigration Techniques in Foodomics

Javier González-Sálamo, Diana Angélica Varela-Martínez, Miguel Ángel González-Curbelo, and Javier Hernández-Borges

Abstract

Foodomics is the discipline aimed at studying the prevention of diseases by food, identifying chemical, biological and biochemical food contaminants, determining changes in genetically modified foods, identifying biomarkers able to confirm the authenticity and quality of foods or studying the safety, quality and traceability of foods, among other issues. It is mainly based on the use of genomic, transcriptomic, proteomic and metabolomic tools, among others, in order to understand the effect of food on animals and humans at the level of genes, messenger ribonucleic acid, proteins and metabolites. Since the first definition of Foodomics, a reasonable number of works have shown the extremely high possibilities of this discipline, which is highly based on the use of advanced analytical hyphenated techniques – especially for proteomics and metabolomics. This book chapter aims at providing a general description of the role of chromatographic and electromigration techniques that are currently being applied to achieve the main objectives of Foodomics, particularly in the proteomic and metabolomic fields, since most published works have been focused on these approaches, and to highlight relevant applications.

Keywords

Foodomics · Chromatography · Electrophoresis · Mass spectrometry · Hyphenated techniques

J. González-Sálamo
Departamento de Química, Unidad Departamental de Química Analítica, Facultad de Ciencias, Universidad de La Laguna (ULL), San Cristóbal de La Laguna, Spain
e-mail: jgsalamo@ull.edu.es

D. A. Varela-Martínez
Departamento de Química, Unidad Departamental de Química Analítica, Facultad de Ciencias, Universidad de La Laguna (ULL), San Cristóbal de La Laguna, Spain

Departamento de Ciencias Básicas, Facultad de Ingeniería, Universidad EAN, Bogotá D.C., Colombia
e-mail: davarela@universidadean.edu.co

M. Á. González-Curbelo
Departamento de Ciencias Básicas, Facultad de Ingeniería, Universidad EAN, Bogotá D.C., Colombia
e-mail: magonzalez@universidadean.edu.co

J. Hernández-Borges (✉)
Departamento de Química, Unidad Departamental de Química Analítica, Facultad de Ciencias, Universidad de La Laguna (ULL), San Cristóbal de La Laguna, Spain

Instituto Universitario de Enfermedades Tropicales y Salud Pública, Universidad de La Laguna (ULL), San Cristóbal de La Laguna, Spain
e-mail: jhborges@ull.edu.es

A. V. Colnaghi Simionato (ed.), *Separation Techniques Applied to Omics Sciences*, Advances in Experimental Medicine and Biology 1336, https://doi.org/10.1007/978-3-030-77252-9_3

Abbreviations

2D	Two-dimensional
ACN	Acetonitrile
C_{18}	Octadecylsilane
CE	Capillary electrophoresis
CSIC	National Research Council of Spain
DAD	Diode array detector
ddPCR	Droplet digital polymerase chain reaction
DNA	Deoxyribonucleic acid
EOF	Electroosmotic flow
ESI	Electrospray ionization
FT	Fourier transform
GC	Gas chromatography
HILIC	Hydrophilic interaction liquid chromatography
HPLC	High-performance liquid chromatography
HRM	High-resolution melting
HS-SPME	Headspace solid-phase microextraction
ICR	Ion cyclotron resonance
IT	Ion trap
LAMP	Loop-mediated isothermal amplification
LC	Liquid chromatography
LCxLC	Two-dimensional liquid chromatography
m/z	Mass/charge ratio
MALDI	Matrix-assisted laser desorption ionization
miRNA	Micro-ribonucleic acid
mRNA	Messenger ribonucleic acid
MS	Mass spectrometry
MS/MS	Tandem mass spectrometry
NMR	Nuclear magnetic resonance
PAGE	Polyacrylamide gel electrophoresis
PCR	Polymerase chain reaction
PTM	Posttranslational modification
Q	Quadrupole
QqQ	Triple quadrupole
RNA	Ribonucleic acid
rRNA	Ribosomal ribonucleic acid
SDS	Sodium dodecyl sulphate
TOF	Time-of-flight
tRNA	Transfer ribonucleic acid
UHPLC	Ultra high-performance liquid chromatography

1 Introduction

Food science, which studies food from a global outlook, has become a discipline of outmost importance nowadays, since food is absolutely essential for human survival. Food science approaches the study of food from different perspectives: those related with food production, food quality and safety, nutritional content, health issues, etc. This results in a relatively complex research environment, which requires the application of different strategies.

In 2009, Dr. Alejandro Cifuentes, from the National Research Council of Spain (CSIC), defined a new term for those approaches that were being developed for certain years and that are also becoming more and more important nowadays: Foodomics, as it was first defined, which is "the discipline that studies the food and nutrition domains through the application and integration of advanced omic technologies to improve consumers well-being, health, and knowledge" [1–4]. Therefore, food and omic techniques are joined at a major scale. Such new discipline is clearly the result of the evolution and development of advanced analytical techniques with the growth of a comprehensive strategy, in which different disciplines are being integrated: from food chemistry to bioinformatics, passing by Biochemistry, Phytochemistry, Molecular Biology and even Analytical Chemistry, among others. Classical and monodisciplinary studies are being substituted by highly complex approaches, which include modern and advanced strategies.

Foodomics is mainly based on the use of genomic, transcriptomic, proteomic and metabolomic tools in order to understand the effect of food on animals and humans at the level of genes, messenger ribonucleic acid (mRNA), proteins and metabolites (see Fig. 1). Genomics was the first to be introduced and applied in food analysis, since it can also be considered a particular "biological system" [5]. It was followed by the application of proteomics in the mid-1990s and from that time on by the rest of the omic techniques [5].

The objectives of Foodomics include the application of such omic techniques for the

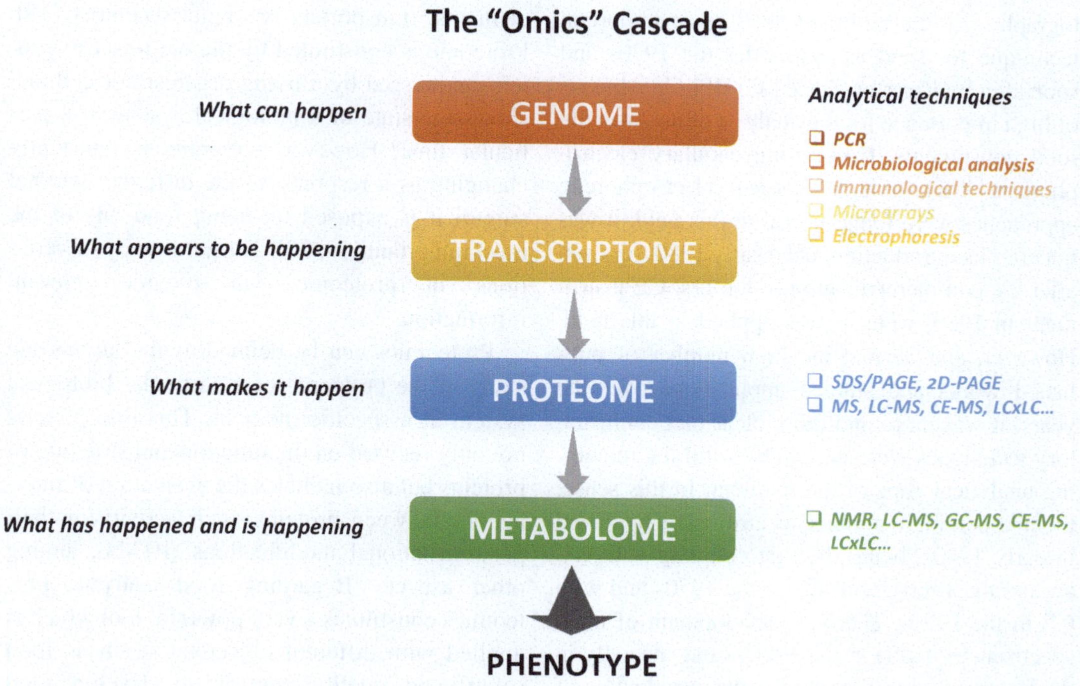

Fig. 1 Scheme of Foodomics platform, including analytical methodologies. (Redrawn from [44])

following general purposes [6, 7]: to study the prevention of diseases by food; to identify chemical, biological and biochemical food contaminants; to determine changes in genetically modified foods; to identify biomarkers able to confirm the authenticity and quality of foods; and, in general, to study the safety, quality and traceability of foods, among others.

It should be highlighted that any food is indeed an extremely complex system that can also have a very complex effect on humans health. Such effects are also highly difficult to discover and require the global approach previously mentioned. In fact, as a result of such issue, and within the Foodomics context, it has even been defined the term Foodome as "the collection of all compounds present in any investigated food sample and/or in any biological system interacting with the investigated food at a given time". Unravelling the Foodome, identifying it, its functions and effects, is part of the objectives of Foodomics.

Since the first definition of Foodomics, a reasonable number of works have shown the extremely high possibilities of this discipline, which is highly based on the use of advanced analytical hyphenated techniques – especially for proteomics and metabolomics. Some of these results have already been presented in different review articles [5, 8–11] – including special issues [12], book chapters and also a complete book [13], as well as in specific international conferences (i.e. http://www.foodomics.eu) devoted exclusively to the Foodomics field.

During the first half of the twentieth century, foods were analysed following classical approaches, which were called at that time "wet chemistry" laboratory methods [14]. After that period, they were gradually replaced by more modern instrumental techniques like potentiometry, spectrophotometry, chromatography, electrophoresis, etc., mainly as a result of the demands of the food and agriculture sectors [6]. Gas chromatography (GC), though born after liquid chromatography (LC), established itself as an important technique for food analysis after the 1970s and soon after high-performance LC (HPLC), which is of high importance for the analysis of non-volatile food constituents. Concerning capillary electrophoresis (CE), though classical electrophoretic approaches were found useful in this field, it was not after its introduction in the early 1980s, or even after the commercialization of the first CE instrument in 1989, when it was applied in this field. However, and despite the high number of published works and studied applications over the years, it was more and more clear that confirmatory techniques were necessary to fill the remaining analytical gaps of the moment. In this sense, the first commercial mass spectrometers appeared in early 1940s, being the first coupling with GC around the 1960s, with LC in the 1970s and with CE in the 1990s. The high development of mass spectrometry (MS) in the 1980s (and specially in the last two decades) as well as the introduction of comprehensive separation techniques [15–17] has come up with highly robust, sensitive and accurate hyphenated techniques. Such hyphenation plays nowadays a crucial role in many disciplines and, in particular, in the Foodomics field, especially in proteomic and metabolomic applications.

The aim of this book chapter is to provide a general description of the role of chromatographic and electromigration techniques that are currently being applied to achieve the main objectives of Foodomics, particularly in the proteomic and metabolomic fields, since most published works have been focused on these approaches, and to highlight relevant applications. For clarification purposes, such works have been divided taken into account these omic techniques, though it should be highlighted that in some cases, more techniques have been simultaneously combined.

2 Proteomic Approaches in Foodomics

Proteins, which are maybe the most peculiar food components, are involved in many crucial biological processes, acting as catalysers, structural elements, transporters or signal receptors [18]. Proteome is constituted by the entire set of proteins expressed by a living organism or cell and shows the state of a biological system at a particular time. However, proteome is constantly changing as a response to the different external stimuli it is exposed to, being food one of the most important [19, 20]. Thus, the study of variations in proteome can provide relevant information.

Proteomics can be defined as the large-scale study of the proteome of a particular biological system at a specific moment. This discipline is not only focused on the function and structure of proteins but also includes the evaluation of interactions between proteins, their activity or their posttranslational modifications (PTMs), among other aspects. Regarding food analysis, proteomics constitutes a very powerful tool which is applied with different objectives, such as food safety and quality control, to develop food authenticity or even to assess the effects of food ingredients on humans health [19, 21]. However, it is important to emphasize that the application of proteomic tools constitutes a very challenging task, because the proteome of each living organism may differ from that of any other, or even of itself depending on the media conditions [3]. Thus, the current need of quantifying proteins and the wide concentration ranges in which proteins are present in organisms can make difficult the quantification of low-abundance proteins, being necessary to establish analytical strategies that allow their proper separation, identification and quantification.

In this sense, and apart from the multiple extraction methodologies (solvent- and sorbent-based) applied for the extraction of proteins according to the nature of the analysed matrix [8, 22, 23], sample preparation methodologies used in proteomics are focused on a reduction of the proteome complexity (less soluble and low-abundance proteins are difficult to analyse [24]). For this purpose, three main approaches may be used: depletion of the high concentrations proteins (affinity depletion, immunoprecipitation, etc.); enrichment of the low-abundance proteins (affinity capture, affinity chromatography, etc.);

and fractionation, purification and/or separation steps using gel-based techniques (sodium dodecyl sulphate polyacrylamide gel electrophoresis-SDS-PAGE-, two-dimensional-PAGE -2D-PAGE-, etc.) or gel-free techniques (LC, strong cation exchange, CE, etc.) [8, 19, 20]. In many cases, a clean-up step is included after sample preparation in order to remove detergents, salts, contaminants, etc., which can interfere in proteins/peptides determination. Finally, it is important to mention that, although necessary, sample preparation procedures should be minimized in order to avoid protein modifications or losses [24]. Table 1 compiles some examples of works dealing with food analysis from a proteomic perspective [25–32].

The development of new technologies such as isotope labelling, the enhancement of peptide/protein separation, MS and computational data analysis has provided an important improvement in proteomics [33]. Among them, MS has become the most powerful tool in proteomics because it

Table 1 Some examples of works dealing with food analysis from a proteomic perspective

Analytes	Matrix	Analytical technique	Column/capillary	Sample preparation and comments	References
Proteins	Maize	CE-ESI-IT-MS CE-ESI-TOF-MS	Ethylpyrrolidine methacrylate/N,N-dimethylacrylamide copolymer-coated fused silica capillaries (90 cm × 50 μm i.d.)	A top-down approach was followed	[25]
Peptides	Soybean seeds	CE-ESI-TOF-MS	Uncoated fused silica capillaries (50 μm i.d., 90 cm total length)	A bottom-up approach was followed	[26]
Peptides	Soybean seeds	MALDI-QTOF-MS	2D-PAGE	A bottom-up approach was followed No additional chromatographic fractionation was necessary	[27]
Peptides	α-casein and dephos-phorylated α-casein	LC×LC-ESI-IT-TOF-MS	C_{18} (1: 150 × 2.1 mm, 2.7 μm; 2: 30 × 4.6 mm, 2.7 μm)	A shotgun approach was followed α-casein or dephosphorylated α-casein	[28]
Peptides	Human milk	Nano-LC-ESI-IT-FT-ICR-MS/MS	C_{12} capillary column (150 mm × 75 μm)	A shotgun approach was followed	[29]
Peptides	Rosemary	Nano-LC-nESI-orbitrap-MS/MS	C_{18}-packed uncoated fused silica emitters (150 mm × 75 μm i.d., 375 μm o.d., tip opening 5 ± 1 μm, 3 μm)	Study of the antiproliferative activity of polyphenol-enriched rosemary extract on colon adenocarcinoma HT-29 cells A shotgun approach was followed	[30]
Peptides	Milk	Nano-LC-nanoESI-IT-MS/MS	C_{18} capillary column (15 cm × 75 μm i.d., 3 μm)	A shotgun approach was followed	[31]
Peptides	Rosemary	Nano-LC-nanoESI-Orbitrap-MS/MS	C_{18}-packed uncoated fused silica emitters (150 mm × 75 μm i.d., 375 μm o.d., tip opening 5 ± 1 μm, 3 μm)	A shotgun approach was followed. Dimethyl labelling was performed	[32]

allows the identification of the proteins without previous protein information [3]. In this sense, developments on instrumentation have brought an important growth of MS-based proteomic methods, among which the introduction of soft ionization sources has played a crucial role, since they allow to transfer proteins and peptides to the gas phase with minimal degradation [34]. Thus, matrix-assisted laser desorption ionization (MALDI) and electrospray ionization (ESI) are the most used sources in food science proteomic applications. Apart from ion sources, mass analysers are one of the most important parts of a MS instrument since it allows separating and selecting ions based on their mass/charge ratios (*m/z*). In this sense, ion trap (IT), quadrupole (Q) and time-of-flight (TOF) analysers are some of the mass analysers which provide suitable sensitivity, mass accuracy and resolution, although their combination in the same mass spectrometer is the most usual, being triple quadrupole (QqQ), QTOF or TOF-TOF some of the most used for the analysis of proteins/peptides in Foodomic applications [3, 13].

Nowadays, the strategies followed in proteomic studies are usually classified in two main approaches, *top-down* and *bottom-up*, in which different separation modes (ion exchange, reversed phase, size exclusion, etc.) play a very important role. These approaches are usually developed in a similar way independently of the application field, being the sample treatment and protein extraction steps modified according to the nature of the sample matrix. A schematic representation of different approaches followed in proteomic analyses is shown in Fig. 2. *Bottom-up* proteomic approach is the most widely used in food analysis. In this approach, proteins extracted from the sample are enzymatically digested (using trypsin in most cases) to obtain their respective peptides, which are subsequently analysed by MS. Briefly, proteins are separated by two-dimensional gel electrophoresis (2D-GE) or SDS-PAGE (gel-based approaches), and then they are in-gel digested before MS analysis [20]. As an example of the separation efficiency of these techniques, Brandão et al. [27] carried out a proteomic study of transgenic and non-transgenic soybean seeds. For this purpose, proteins were extracted using a buffer solution at pH 8.8 and then separated by 2D-PAGE. After an image analysis of 2D gels, proteins were in-gel digested, and the resulting samples were prepared for direct analysis by MALDI-QTOF-MS, which allowed the successful identification of eight out of ten protein spots that showed altered expression. However, in most cases, an additional fractionation step using LC or CE is necessary, since the peptide mixture obtained from a complex protein mixture proteolysis continues to be highly complex [8]. In this sense, reversed-phase columns are the most selected ones for LC-MS applications since they provide high-resolution separations of the obtained peptides and allow the use of compatible solvents with the typically used ionization sources [20, 24, 35], although ion exchange mode has also been used [36]. Besides the described *bottom-up* approach, "shotgun" is a high-throughput alternative approach based on the same principles, in which the protein mixture is directly enzymatically digested into peptides in solution and separated by mono- or multidimensional LC or CE coupled to tandem MS (MS/MS) [20, 28, 37], avoiding the previous in-gel proteins separation step. As an example, Picariello and coworkers [29] carried out a gel-free proteomic analysis of human milk following a "shotgun" approach. Proteins were initially separated from the matrix, enabling distinction between milk fat globule membrane proteins and whey proteins which, after a reduction of the disulphide bridges and an alkylation process, were digested in solution using trypsin. The tryptic digests were separated by nano-LC using a dodecylsilane packed capillary coupled to an IT Fourier transform ion cyclotron resonance mass spectrometer (IT-FT-ICR-MS) by an ESI source. This "shotgun" approach allowed the correct identification of proteins associated with 301 different gene products, providing one of the largest protein inventories of human milk. Although this approach has become popular in the proteomic field, the higher complexity of the resulting peptide mixture makes necessary a highly efficient separation and a sensitive detection. Besides, information is lost during such protein digestion procedure (leading

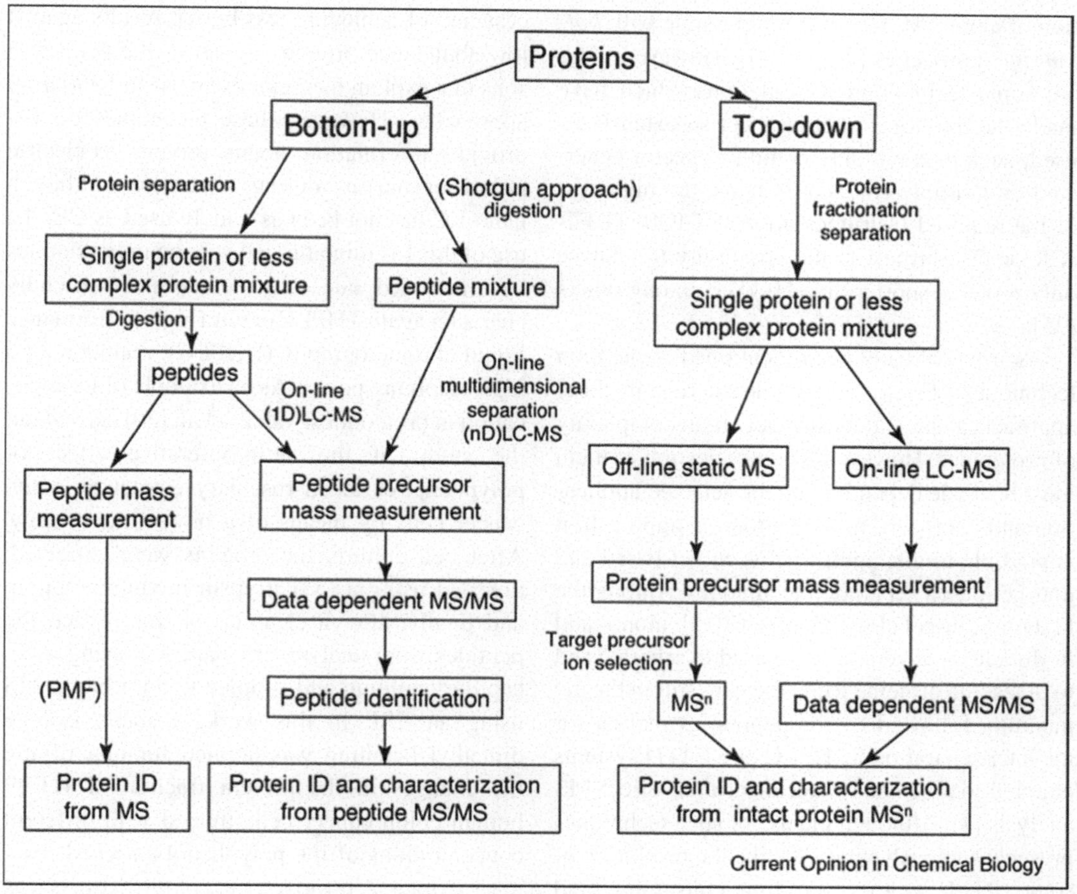

Fig. 2 Strategies for MS-based protein identification and characterization. Proteins extracted from biological samples can be analysed by bottom-up or top-down methods. In the bottom-up approach, proteins in complex mixtures can be separated before enzymatic (or chemical) digestion followed by direct peptide mass fingerprinting-based acquisition or further peptide separation on-line coupled to tandem MS. Alternatively, the protein mixture can be directly digested into a collection of peptides ("shotgun" approach), which are then separated by multidimensional chromatography on-line coupled to tandem mass spectrometric analysis. In the top-down approach, proteins in complex mixtures are fractionated and separated into pure single proteins or less complex protein mixtures, followed by off-line static infusion of sample into the mass spectrometer for intact protein mass measurement and intact protein fragmentation. An on-line LC-MS strategy can also be used for large-scale protein interrogation. (Reprinted from Han et al. [37] with permission from Elsevier)

to incorrect identifications), some peptides cannot be observed or correctly identified, and determination of low-abundance peptides is jeopardized [37].

However, the *bottom-up* approach has some limitations since only a portion of the total peptide of a protein is identified, losing information and making difficult the elucidation of the protein structure due to the multiple protein isoforms, originated from alternative splicing products and PTMs [35]. As a consequence, the *top-down* approach emerged as an alternative proteomic strategy to the abovementioned, based on the separation of intact proteins generally by 2D-GE or LC and their direct MS detection [37]. In this approach, no previous digestion is carried out and intact proteins are ionized in the MS source, and then the ions are fragmented inside the mass spectrometer [35, 37]. As the complete sequence of the proteins is obtained, it would be possible to achieve the protein identification in one step, as well as the location and characterization of any PTM, which is not possible with *bottom-up* approaches [24, 35, 37]. However, there are some technological limitations which have made the *top-down* approach not so extensively used, such as the highly complex spectra generated by multiple charged proteins, the high cost of the required instrumentation (FT-ICR, IT-FT-ICR or IT-orbitrap) or the possibility to analyse only proteins shorter than 50 kDa, among others [35].

As it has already been mentioned, separation techniques play a very important role in these approaches due to the extremely high complexity of proteomes. However, a good selection of them must be made depending on the selected application and approach. In the *bottom-up* approach, it is possible to distinguish between gel-based and non-gel-based separation techniques. Among the first ones, in-gel electrophoresis (both mono- and bi-dimensional versions) are used to separate and to digest the proteins while they are still in the gel medium, followed by the peptide extraction for the later separation by HPLC or CE [11] systems coupled to a MS/MS detector or their direct MS analysis [3]. Although the use of such techniques presents some advantages – like the resolution of some PTMs – they are time-consuming and labour-intensive and present problems to separate proteins with high hydrophobicity, extreme isoelectric points or high molecular weight and low recovery of some proteins [3, 35]. Alternatively, non-gel techniques (such as LC, nano-LC or CE) are used and can be applied in their multidimensional versions aiming a better separation of the peptides obtained from the in-solution digestion. As an example, Donato et al. [28] developed an automated reversed-phase (octadecylsilane -C_{18}-) two-dimensional LC (LCxLC) coupled to MS detection method for the separation of peptides resulting from α-casein and dephosphorylated α-casein digestion. For this purpose, significantly different pH values in the two dimensions (basic conditions for the first and acid for the second) were enough to obtain high peak capacity, although identical stationary phases were used. Figure 3 shows the LC×LC plots obtained for α-casein, and dephosphorylated α-casein digests separation. The possibility of automating the process and of achieving less biased results against low-abundance proteins is one of the main reasons that explain their extensive use in *bottom-up* approaches. However, these techniques do not provide information about protein isoelectric points or masses during separation. Although nano-LC has not been as widely used as CE, the use of this LC miniaturized version in Foodomics has increased due to its environment-friendly character against HPLC or ultra high-performance liquid chromatography (UHPLC), maintaining a high-resolving power. As an example of its application in proteomics, Valdés et al. [30] carried out the study of the antiproliferative effect of polyphenol-enriched rosemary extract on colon cancer cells by means of a proteomic strategy. After cell culture, the proteins were extracted, digested using a LysC/trypsin mixture solution, and finally, after a clean-up of the extract, the peptides were analysed by nano-LC using a C_{18} capillary column and coupled to an orbitrap MS using an ESI. In this work, a stable isotope dimethyl labelling was applied for quantifying the changes in the protein fraction of HT-29 human colon cancer cells treated with different concentrations of the polyphenol-enriched rosemary extract. A positive effect of the extracts was observed in the attenuation of colon cancer proliferation.

Most *top-down* approaches carry out an infusion of simple protein mixtures or even isolated proteins directly into the MS system due to the difficulty of coupling chromatographic systems to FT-ICR spectrometers. Gel-based methodologies can be used with the problematic derived from the extraction of proteins from the gel and the problems that detergents can produce on MS. Thus, reversed phase, size exclusion or ion exchange HPLC as well as CE has been used in these approaches with promising results [13, 20, 35]. In this sense, it is important to mention that the analysis of intact proteins using CE may lead to problems related with the adsorption of the proteins in the negatively charged inner wall of the capillary, which may cause changes on the electroosmotic flow (EOF) resulting in peaks broadening, tailing or poor migration [11]. In order to solve this problem, different coatings

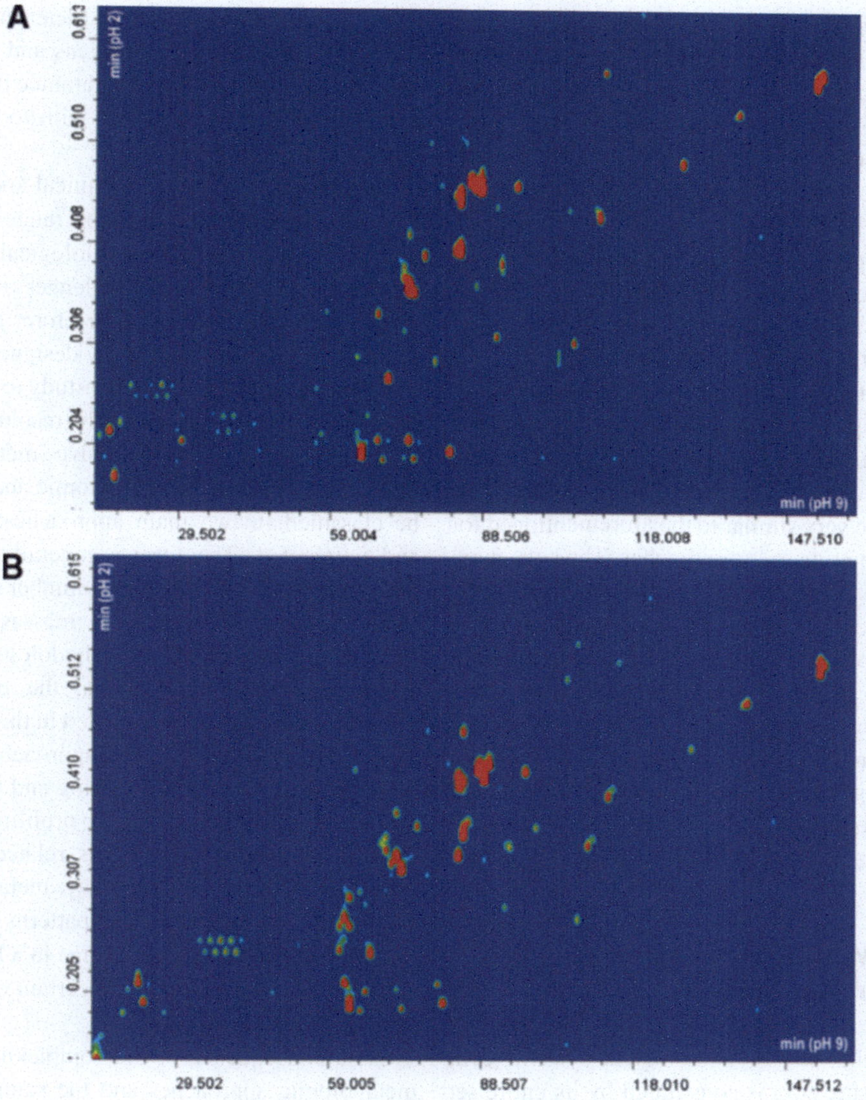

Fig. 3 LC×LC plot of α-casein (**a**) and dephosphorylated α-casein (**b**) tryptic digest separation. Colours are used to represent the value of absorbance at a given point. The relation between colours and absorbance is ruled by a colour map that can be configured by the user. (Reprinted from [28] with permission from ACS)

have been described [38]. As an example, Erny and co-workers [25] carried out a comparative study on a CE-ESI-IT-MS and a CE-ESI-TOF-MS systems in terms of sensitivity for the analysis of intact proteins. In this study, CE-ESI-TOF-MS revealed a higher number of proteins, while CE-ESI-IT-MS showed cleaner MS spectra. In both cases, ethylpyrrolidine methacrylate/N,N-dimethylacrylamide copolymer was used as physically adsorbed cationic coating for CE fused silica capillaries, requiring coating process after every run. After a proper optimization of CE-MS analysis conditions, the method applicability was explored by studying the zein protein composition of three natural maize lines and their corresponding transgenic lines, for which no significant differences were observed.

Finally, it is necessary to refer to peptidomics, another important omic discipline, which has occasionally been included as a subclass of proteomics, but it is focused on the analysis of another range of the proteome. Peptidomics addresses the qualitative, quantitative and functional description of all peptides and small proteins (0.5–15 kDa) with biological activity (biomarker peptides or active peptides) [39, 40]. The application of this discipline in the food science field has gained interest to study the effects that such peptides present in foods can have on humans health [40]. While these active peptides have been traditionally studied by immunoassay methods, the difficulties of these methods for the analysis of new peptides have become chromatographic separation followed by MS detection the most common alternative. The techniques used in peptidomics and the way in which they are applied are very similar to the aforementioned for bottom-up proteomic approaches. Thus, the main difference in the typical analytical workflow is the need of separating the free peptide fraction from the proteins present in the sample after the extraction step by means of filtration, selective precipitation and/or centrifugation. This procedure is generally followed by a pre-concentration step using conventional sorbent-based procedures in order to increase the concentration of bioactive peptides [40].

3 Metabolomic Approaches in Foodomics

The metabolome of a biological system (cell, organ, tissue, etc.) is constituted by its entire set of metabolites (endogenous or exogenous), which generally comprises a wide variety of small molecules (<1000–1500 Da) with very different physicochemical properties such as lipids, amino acids, vitamins or carbohydrates, among much others [3, 41, 42]. Since metabolites are considered as the final products of the genome, the study of the metabolome of a biological system can give an idea of how it operates and the effects that environmental conditions can have on it [3, 42]. Regarding food science, metabolomic studies aim at providing complementary information to genomic, transcriptomic and proteomic studies in order to shed light on the effects of the diet on the prevention of some diseases (food bioactivity), to ensure food safety (determination of chemical contaminants, pathogens and toxins or food allergens), as well as to guarantee food quality, evaluating its authenticity with tools which allow tracing it properly [40].

The broad spectrum of chemical species and the wide dynamic concentration range at which metabolites are present on biological systems constitute one of the main challenges of metabolomic studies [3, 42, 43]. Therefore, analytical procedures have to be carefully designed in relation to the characteristics of the study to be developed and to achieve a high resolution and sensitivity for a comprehensive metabolomic analysis. In this sense, metabolomic analysis can be classified in two main approaches: targeted and untargeted. The former generally involves the quantification of a reduced number of metabolites detecting their specific masses by MS/MS. Although this kind of methodologies can be useful in certain applications, the untargeted approaches are the most extended in the food science field. Among them, two approaches can be distinguished: metabolic profiling and metabolic fingerprinting. While metabolic profiling focuses on the analysis of metabolites related to each other or to a metabolic pathway, metabolic fingerprinting aims at identifying patterns or metabolite fingerprints that may change in a biological system under the influence of certain conditions [40, 42, 43].

As a consequence of the complexity of both metabolomic approaches and the samples analysed in these studies, an adequate analytical procedure has to be designed. As expected, the sample preparation step plays a very important role in metabolomics, allowing the robust and reproducible extraction of the metabolites [3]. This step will depend not only on the sample type but also on the nature of the metabolites, so the selected metabolomic approach must be considered. Thus, when targeted analysis is performed, selective extraction methods should be applied, while in untargeted approaches the sample preparation should maximize extraction of as much metabolites as possible. In any case, and as it has been previously indicated in this chapter, sample treatment is recommended to be reduced to the minimum in order to minimize the metabolites losses [3, 40]. In this sense, different solvent-based and sorbent-based extraction methodologies have also been applied for the extraction of metabolites depending on the nature of the sample matrix, the metabolites physicochemical properties and the followed approach [43, 44].

The aforementioned complexity of the metabolome of a living organism, which contains a large number of compounds with very different properties, is also an important challenge when the analytical platform to carry out the identification and/or quantification of the metabolites is selected. In this sense, analytical techniques currently used for metabolic profiling and fingerprinting (the ones most used in Foodomic applications) can be grouped in two different categories: nuclear magnetic resonance (NMR)-based and MS-based techniques. Although NMR has been highly used in metabolomic applications due to suitable reproducibility and robustness, the broad spectrum of compounds that can be analysed, the fact that it provides structural information with relatively simple sample preparation methods, it has a low sensitivity if it is compared to MS, which is most extensively used [42, 43], generally employing high-resolution mass analysers such as TOF, QTOF, TOF-TOF or Orbitrap, among others [3]. In any case, it is important to clarify that although both techniques are used alone in many applications, they provide complementary information, as presented in some works. An example of NMR and MS application in Foodomics is the work of Tomita and co-workers [45], in which a combination of NMR and GC-MS was used to reveal the compositional characteristics of *sunki*. In this case, water-soluble compounds were directly analysed by NMR, while volatile compounds were separated by GC-EI-Q-MS using a polar capillary column (60 m × 0.25 mm i.d. × 0.25 μm film thickness) after headspace solid-phase microextraction (HS-SPME) of the analytes. This study revealed changes in the chemical composition of sunki related with pH as well as the processing factory it came from and production year.

Despite the high potential of these detection techniques, the wide variety of analytes makes difficult their simultaneous analysis, so they are usually coupled to separation techniques in order to obtain higher resolution and sensitivity for low-abundance metabolites [3, 44]. Table 2 shows some examples of works dealing with food analysis from a metabolomic perspective [45–49]. On the one hand, the already known versatility of LC has made this technique one of the most extensively used for metabolomic studies in food science and nutrition, since it has demonstrated good performance on profiling of large, thermolabile and non-volatile/polar metabolites, such as secondary metabolites and complex lipids. Besides, the introduction of UHPLC has allowed an increment of the peak resolution and sensitivity, so important in metabolomics due to the large amount of analysed compounds [40, 42]. One of the key points that have allowed LC to become one of the main separation techniques in metabolomics is the wide variety of available stationary phases. In this kind of applications, reversed phase (non-polar or medium-polar metabolites) and hydrophilic interaction liquid chromatography (HILIC) (polar or ionic species) stationary phases have been used [40]. On the other hand, GC has shown high separation efficiency and reproducibility for the analysis of primary metabolites after chemical derivatization when it is coupled to MS detectors due to the use of EI sources [50]. Although less used, CE has proved to be an interesting alternative, especially for highly polar and charged thermolabile compounds, providing its already known high resolution and efficiency with a greener character in comparison to LC [51]. These separation techniques have been applied for the chemical fingerprinting of foods with protected designation of origin [46], metabolic profiling for detecting changes on chemical composition during food processing [47] or metabolic profiling for transgenic food detection [48], among others. It is important to mention that in certain applications, a higher separation power may be required in order to resolve the great amount of metabolites presents in the sample, so some of these techniques have also been applied in their multidimensional versions [40, 41]. As an example, Montero et al. [49] developed a metabolite profil-

Table 2 Some examples of works dealing with food analysis from a metabolomic perspective

Analytes	Matrix	Analytical technique	Column	Sample preparation and comments	References
54 water-soluble and 62 volatile compounds	Sunki	NMR GC-EI-Q-MS	Polar capillary column (60 m × 0.25 mm i.d., 0.25 μm)	Metabolite profiling	[45]
Lipids, amino acids and oligopeptides	Grana Padano cheese	UHPLC-ESI-QTOF-MS	C_{18} (50 × 2.1 mm, 1.8 μm)	Metabolite fingerprinting	[46]
62 non-volatile and 47 volatile compounds	Black tea	GC-EI-Q-MS	(5%-phenyl)-methylpolysiloxane coated capillary column (30 m × 0.25 mm i.d., 0.25 μm)	Metabolite profiling Non-volatile compounds were derivatized using N-methyl-N-(trimethylsilyl) trifluoroacetamide	[47]
27 metabolites	Maize	CE-ESI-TOF-MS	Uncoated fused silica capillary (80 cm × 50 μm i.d., 375 μm)	Metabolite profiling	[48]
89 metabolites	Liquorice	LC×LC-DAD-ESI-IT-MS/MS	First dimension: HILIC (150 × 1 mm, 3.5 μm) Second dimension: C_{18} (50 × 4.6 mm, 2.7 μm)	Metabolite profiling From 19 to 50 compounds were identified as patterns with geographical location and authentication purposes	[49]

ing of several liquorice samples from different locations by means of a LC×LC system coupled to a diode array detector (DAD) and to a IT-MS detector with ESI source. In this case, a HILIC column (150 × 1 mm, 3.5 μm) was employed in the first dimension using a mobile phase composed by ACN and 10 mmol/L ammonium acetate solution at pH 5.0, while a reversed-phase C_{18} column (50 × 4.6 mm, 2.7 μm) was selected for the second dimension with a mobile phase composed by water containing 0.1% (v/v) of formic acid and ACN. Two different gradients were employed in the second dimension in order to obtain the best separation performance, considering the eluting compounds from the first one. Figure 4 shows a two-dimensional profile obtained in this work. This study allowed the separation of up to 89 compounds for some of the analysed samples, allowing to assign the eluted compounds to a chemical family depending on their position on the 2D plots, which also allowed to create patterns of each sample. Thus, from 19 to 50 unique compounds were found in all samples and may be used to confirm the geographical origin and authenticity of unknown liquorice samples.

Finally, it is necessary to highlight that the analysis of the lipids present in a living organism constitutes a very challenging issue due to the wide diversity of lipids, which include glycerophospholipids, fatty acyls, sphingolipids, glycerolipids, sterol lipids, prenol lipids, polyketides and saccharolipids [52]. Such diversity of lipid classes has originated a new metabolomic subdiscipline focused on the analysis of the entire set of lipids of a cell, organ or biological system, known as lipidomics [53]. Lipidomic studies start with the extraction of lipids, where the use of mixtures of solvents has proven to be the most effective alternative. After the extraction, a chromatographic separation of the analytes (LC or GC) is carried out using MS detection as it has been previously detailed in this section for metabolomic approaches [40]. More details about lipidomics in food science can be found in the review articles of Chen et al. [54] or Hyötyläinen et al. [53].

Fig. 4 Two-dimensional HILIC×C_{18} liquorice metabolites profiles (280 nm) obtained for liquorice samples collected from China (**a**), Iran (**b**), Crotone (Italy, **c**), Azerbaijan (**d**) and Villapiana (Italy, **e**). (Reprinted from [49] with permission of Elsevier)

4 Genomic and Transcriptomic Approaches in Foodomics

As previously mentioned, genomics was the first of the omic tools to be introduced and applied in food analysis, which refers to the sequence, assembly and analysis of the structure and function of the complete set of nuclear deoxyribonucleic acid (DNA) that contains the specific information encoded in genes required to code proteins for the organism to build itself [55]. Genomic analysis in food science can be successfully used for the determination of food pathogens [56] and for the detection of transgenic foods [57], as well as for the determination of the authenticity of foods to avoid adulterations [58]. The latter is done based on the fact that the DNA provides highly specific biological information at every taxonomic level, remaining in a food matrix independently from an environmental influence, so it can be recovered and analysed [59]. For this purpose, classical polymerase chain reaction (PCR)-based methodologies have been used as routine techniques to easily amplify the amount of nucleic acid molecules [60]. However, food may contain compounds that can inhibit the PCR analysis, requiring some culture enrichment prior to PCR to avoid false negative results. Additionally, it is necessary to detect with a high sensitivity the presence of the target sequence after such amplification. For this purpose, conventional CE [61, 62] or lab-on-a-chip CE [63] has been combined with PCR-based methods. These techniques achieve even better results compared to the use of other analytical methods such as GC-MS [61] or HPLC [64]. As an example, Uncu et al. [61] compared PCR-CE (DNA analysis) with GC-MS (fatty acid profile analysis) for the analysis of olive oil to authenticate its botanical origin. While PCR-CE demonstrated equally efficient as GC-MS analysis in detecting adulteration with soybean, palm, rapeseed, sunflower, sesame, cottonseed and peanut oils, it was superior in revealing the adulterant species and detecting small quantities of corn and safflower oils in olive oil. Moreover, the PCR-CE analysis correctly identified the hazelnut oil adulteration, whereas it was not feasible to detect it through GC-MS analysis. Despite the above, the number of such technique's applications related to genomics is still relatively scarce. More recently, new genomic-based tools have also been used to study food authenticity, including high-resolution melting (HRM), droplet digital PCR (ddPCR), loop-mediated isothermal amplification (LAMP) and DNA barcoding [65]. However, genomics alone cannot furnish all the required information [66], giving rise to the postgenomic era.

As the next level, transcriptomics has been one of the most developed fields emerging from the genomic era. Transcriptomics studies and compares the complete set of ribonucleic acid (RNA) transcripts that are being expressed at a given moment in a particular cell, including messenger RNA (mRNA), micro-RNA (miRNA), transfer RNA (tRNA) and ribosomal RNA (rRNA) [67]. Transcriptomic tools have also been used for a wide variety of applications in the field of food science, including food production, quality assurance and health effects. As examples, it has been possible to decipher the genes responsible for an increase in the production under adverse environmental conditions [68] and the synthesis of nutrients and bioactive compounds [69]. Transcriptomic tools have also been used to identify toxin production mechanisms [70] and the genes responsible for the synthesis of biofilms [71]. Besides, they have been used to investigate the effects of bioactive food compounds [20] and for the systematic characterization of genetically modified foods [72]. However, even more than for genomics, the use of chromatographic and electromigration techniques has not been considered to quantify and profile the expression of the thousands of transcribed sequences. Contrary, in all these applications, the use of microarray technology, both on solid flat substrates (or microchips) and on spherical substrates (or particle microarrays) [73], and the use of massive sequencing technologies (also called RNA-Seq) [74] have been the followed procedures to date. As an important example, Valdés et al. [75] used a global Foodomic strategy based on genome-wide transcriptomics using microarrays and metabolomics using CE-TOF-MS and UHPLC-TOF-MS to study the potential therapeutic benefit of dietary polyphenols from rosemary on human leukaemia lines. More specifically, rosemary extracts rich in polyphenols such as rosmarinic acid, carnosol and carnosic acid were obtained from dried rosemary leaves using either supercritical fluid extraction or pressurized liquid extraction. Then, samples of two human erythroleukaemia lines, one showing a drug-sensitive phenotype (K562) and another exhibiting a drug-resistant phenotype (K562/R), were incubated with such polyphenols extracts for gene expression microarray analyses and were compared with their respective untreated controls. Similarly, comparative metabolomics with the cited instrumental analysis were also performed on treated and control leukaemia cells. Moreover, the Ingenuity Pathway Analysis® (IPA®) software was used for functional enrichment analysis as a previous step for a reliable data interpretation due to the massive data obtained using these transcriptomic and metabolomic platforms. The transcriptome microarray analysis showed that rosemary polyphenols altered the expression of approximately 1% of the genes covered in both leukaemia cell lines and metabolomic analysis suggested that such polyphenols affect differently the intracellular levels of some metabolites in two leukaemia cell sublines, showing that an integrative approach between different Foodomic techniques allows to achieve more comprehensive results. In fact, an expanded strategy using high-density gene expression microarray for transcriptomics, 2D-GE combined with MALDI-TOF-MS for proteomics and CE-TOF-MS and UPLC-Q/TOF-MS for metabolomics was applied to investigate the health benefits of rosemary polyphenols against colon cancer [76].

5 Conclusions and Trends

The application of omic techniques has demonstrated to be highly useful to understand food science as well as nutrition issues by combining multidisciplinary and complex approaches. However, and despite the good number of works and applications already published, we are only viewing the tip of the iceberg, since there are still a high number of challenges that have to be faced, which is a result of the high complexity of foods and their health effects. As an example, we have extremely poor information of the long-term effects of foods and ingredients in humans health or of the roles of nutritional compounds at the gene level. Besides, there is an important need of discovering biomarkers to detect hazardous products or of improving the detection of food safety problems before they increase and affect more people, among other issues.

Very frequently, a single "omic" approach is followed by laboratories working in Foodomics. Although such approach provides important data and is appropriate to reach conclusions, a more global approach in which several omic techniques are applied at the same time might be necessary. This clearly increases the complexity of the work and requires a multidisciplinary research.

The developments achieved in Foodomics are in parallel with the development of advanced analytical methodologies, since the use and development of high-resolving separation techniques as well as highly accurate mass spectrometers is essential to solve the complexity of foods and their effects. In particular, the applications of LC and CE (especially in the proteomic and metabolomic fields) have highly increased as a result of the application of reduced analytical time strategies (UHPLC, on-chip CE, etc.), the increased use of other separation mechanisms (i.e. HILIC) and the routine use of MS detectors. The increased application of heart-cutting and comprehensive techniques in recent years, which clearly suggest their need for the analysis of these complex samples (as a result of the improved peak capacity, selectivity and sensitivity), should also be highlighted. In the forthcoming future, more applications of multidimensional techniques will surely appear.

In the last years, we have also witnessed a tremendous progress in the development of mass spectrometry, especially in high-resolution MS methods with improved features and capabilities. As a result, their use has also increased in this field, especially in combination with chromatographic and electromigration techniques, and will surely continue to grow, since accurate and tandem mass spectrometers are highly necessary for a correct and unequivocal identification.

Despite the use of advanced separation and detection techniques, it should not be forgotten that a suitable sample preparation step is still highly necessary and should be carefully considered prior to the use of any separation and detection technique.

One of the main challenges of Foodomics is dealing with the enormous amount of data that the combination of chromatographic and electromigration techniques coupled to MS provides, which is produced at a high speed (this is no longer the bottleneck of the methods), many of which is noninformative and should be suitably eliminated. As a result, there is an important demand of integrating appropriate statistical tools able to discriminate information and to extract clear conclusions, since there is not a bioinformatic tool able to handle all the obtained data from the previously commented omic techniques yet.

In general, it should be indicated that Foodomics is an extremely interesting but complex and challenging field in which the combined used of omic techniques, i.e. genomics, transcriptomics, proteomics and metabolomics, among others, offers exciting opportunities and new answers in food science and nutrition. We are currently witnessing the very beginning of this discipline.

Acknowledgements J.G.S. would like to thank "Cabildo de Tenerife" for the Agustín de Betancourt contract at the Universidad de La Laguna.

References

1. Cifuentes A (2009) Food analysis and Foodomics. J Chromatogr A 1216(43):7109. Available from: https://www.sciencedirect.com/science/article/pii/S0021967309013545?via%3Dihub
2. Herrero M, García-Cañas V, Simo C, Cifuentes A (2010) Recent advances in the application of capillary electromigration methods for food analysis and Foodomics. Electrophoresis 31(1):205–228. Available from: https://doi.org/10.1002/elps.200900365
3. Herrero M, Simõ C, García-Cañas V, Ibáñez E, Cifuentes A (2012) Foodomics: MS-based strategies in modern food science and nutrition. Mass Spectrom Rev 31:49–69
4. García-Cañas V, Simó C, Herrero M, Ibáñez E, Cifuentes A (2012) Present and future challenges in food analysis: Foodomics. Anal Chem 84(23):10150–10159. Available from: https://doi.org/10.1021/ac301680q
5. Ferranti P (2018) The future of analytical chemistry in Foodomics. Curr Opin Food Sci 22:102–108. Available from: https://www.sciencedirect.com/science/article/pii/S221479931730098X
6. Gallo M, Ferranti P (2016) The evolution of analytical chemistry methods in Foodomics. J Chromatogr A 1428:3–15. Available from: https://www.sciencedirect.com/science/article/pii/S0021967315012728
7. Rychlik M, Kanawati B, Schmitt-Kopplin P (2017) Foodomics as a promising tool to investigate the mycobolome. TrAC Trends Anal Chem 96:22–30. Available from: https://www.sciencedirect.com/science/article/pii/S016599361730095X
8. Martinović T, Šrajer Gajdošik M, Josić D (2018) Sample preparation in foodomic analyses. Electrophoresis 39:1527–1542
9. Pimentel G, Burton KJ, Vergères G, Dupont D (2018) The role of Foodomics to understand the digestion/bioactivity relationship of food. Curr Opin Food Sci 22:67–73. Available from: https://www.sciencedirect.com/science/article/pii/S2214799317300826
10. Ibáñez C, Simó C, García-Cañas V, Acunha T, Cifuentes A (2015) The role of direct high-resolution mass spectrometry in Foodomics. Anal Bioanal Chem 407(21):6275–6287. Available from: https://doi.org/10.1007/s00216-015-8812-1
11. Ibáñez C, Simó C, García-Cañas V, Cifuentes A, Castro-Puyana M (2013) Metabolomics, peptidomics and proteomics applications of capillary electrophoresis-mass spectrometry in Foodomics: a review. Anal Chim Acta 802:1–13. Available from: https://www.sciencedirect.com/science/article/pii/S0003267013009884?via%3Dihub
12. Cifuentes A (2017) Special issue on Foodomics and modern food analysis. TrAC Trends Anal Chem 96:1–212
13. Cifuentes A (2013) Foodomics: advanced mass spectrometry in modern food science and nutrition. Wiley-VCH Verlag, Hoboken
14. McGorrin RJ (2009) One hundred years of progress in food analysis. J Agric Food Chem 57(18):8076–8088. Available from: https://doi.org/10.1021/jf900189s
15. Cacciola F, Dugo P, Mondello L (2017) Multidimensional liquid chromatography in food analysis. TrAC Trends Anal Chem 96:116–123. Available from: https://www.sciencedirect.com/science/article/pii/S016599361730078X
16. Nolvachai Y, Kulsing C, Marriott PJ (2017) Multidimensional gas chromatography in food analysis. TrAC Trends Anal Chem 96:124–137. Available from: https://www.sciencedirect.com/science/article/pii/S0165993617300973
17. Herrero M, Ibáñez E, Cifuentes A, Bernal J (2009) Multidimensional chromatography in food analysis. J Chromatogr A 1216(43):7110–7129. Available from: https://www.sciencedirect.com/science/article/pii/S0021967309011947?via%3Dihub
18. Nelson D, Cox M (2013) Lehninger principles of biochemistry, Fifth edn.
19. Gallardo JM, Carrera M, Ortea I (2013) Proteomics in food science. In: Foodomics. Wiley Online Books, New Jersey, pp 125–165. Available from: https://doi.org/10.1002/9781118537282.ch5
20. Valdés A, León C (2017) Foodomics evaluation of bioactive compounds in foods. TrAC Trends Anal Chem 96:2–13. Available from: https://www.sciencedirect.com/science/article/pii/S0165993617301541?via%3Dihub
21. Picariello G, Mamone G, Addeo F, Ferranti P (2012) Novel mass spectrometry-based applications of the "omic": sciences in food technology and biotechnology. Food Technol Biotechnol 50(3):286–305
22. Luthria DL, Maria John KM, Marupaka R, Natarajan S (2018) Recent update on methodologies for extraction and analysis of soybean seed proteins. J Sci Food Agric 98:5572–5580
23. Gilbert-López B, Mendiola JA, Ibáñez E (2017) Green Foodomics. Towards a cleaner scientific discipline. TrAC Trends Anal Chem 96:31–41. Available from: https://www.sciencedirect.com/science/article/pii/S0165993617301474?via%3Dihub#bib15
24. Mena MC, Albar JP (2013) Next generation instruments and methods for proteomics. In: Foodomics: advanced mass spectrometry in modern food science and nutrition. Wiley-VCH Verlag, Hoboken, pp 15–67
25. Emy GL, León C, Marina ML, Cifuentes A (2008) Time of flight versus ion trap MS coupled to CE to analyse intact proteins. J Sep Sci 31:1810–1818

26. Simó C, Domínguez-Vega E, Marina ML, García MC, Dinelli G, Cifuentes A (2010) CE-TOF MS analysis of complex protein hydrolyzates from genetically modified soybeans – a tool for Foodomics. Electrophoresis 31:1175–1183
27. Brandão AR, Barbosa HS, Arruda MAZ (2010) Image analysis of two-dimensional gel electrophoresis for comparative proteomics of transgenic and non-transgenic soybean seeds. J Proteomics 73(8):1433–1440. Available from: https://www.sciencedirect.com/science/article/pii/S187439191000028X?via%3Dihub#aep-section-id24
28. Donato P, Cacciola F, Sommella E, Fanali C, Dugo L, Dachà M et al (2011) Online comprehensive RPLC × RPLC with mass spectrometry detection for the analysis of proteome samples. Anal Chem 83(7):2485–2491. Available from: https://doi.org/10.1021/ac102656b
29. Picariello G, Ferranti P, Mamone G, Klouckova I, Mechref Y, Novotny MV et al (2012) Gel-free shotgun proteomic analysis of human milk. J Chromatogr A 1227:219–233. Available from: https://www.sciencedirect.com/science/article/pii/S0021967312000775?via%3Dihub#fig 0005
30. Valdés A, Artemenko KA, Bergquist J, García-Cañas V, Cifuentes A (2016) Comprehensive proteomic study of the antiproliferative activity of a polyphenol-enriched rosemary extract on colon cancer cells using nanoliquid chromatography-Orbitrap MS/MS. J Proteome Res 15(6):1971–1985
31. Nardiello D, Natale A, Palermo C, Quinto M, Centonze D (2017) Combined use of peptide ion and normalized delta scores to evaluate milk authenticity by ion-trap based proteomics coupled with error tolerant searching. Talanta 164:684–692. Available from: https://www.sciencedirect.com/science/article/pii/S0039914016308578?via%3Dihub#s0010
32. Valdés A, García-Cañas V, Pérez-Sánchez A, Barrajón-Catalán E, Ruiz-Torres V, Artemenko KA et al (2017) Shotgun proteomic analysis to study the decrease of xenograft tumor growth after rosemary extract treatment. J Chromatogr A 1499:90–100. Available from: https://www.sciencedirect.com/science/article/pii/S002196731730482X?via%3Dihub#bib0085
33. Zhang Y, Fonslow BR, Shan B, Baek M-C, Yates JR (2013) Protein analysis by shotgun/bottom-up proteomics. Chem Rev 113:2343–2394
34. Yates JR, Ruse CI, Nakorchevsky A (2009) Proteomics by mass spectrometry: approaches, advances, and applications. Annu Rev Biomed Eng 11:49–79
35. Wehr T (2006) Top-down versus bottom-up approaches in proteomics. LCGC N Am 24(9):1004–1010
36. Han G, Ye M, Zhou H, Jiang X, Feng S, Jiang X et al (2008) Large-scale phosphoproteome analysis of human liver tissue by enrichment and fractionation of phosphopeptides with strong anion exchange chromatography. Proteomics 8(7):1346–1361. Available from: https://doi.org/10.1002/pmic.200700884
37. Han X, Aslanian A, Yates JR (2008) Mass spectrometry for proteomics. Curr Opin Chem Biol 12(5):483–490. Available from: https://www.sciencedirect.com/science/article/pii/S1367593108001178#fig 1
38. Huhn C, Ramautar R, Wuhrer M, Somsen GW (2010) Relevance and use of capillary coatings in capillary electrophoresis-mass spectrometry. Anal Bioanal Chem 396(1):297–314
39. Crameri R (2005) The potential of proteomics and peptidomics for allergy and asthma research. Allergy 60(10):1227–1237. Available from: https://doi.org/10.1111/j.1398-9995.2005.00873.x
40. Gilbert-López B, Valdés A, Acunha T, García-Cañas V, Simó C, Cifuentes A (2017) Foodomics: LC and LC-MS-based omics strategies in food science and nutrition. Liq Chromatogr:267–299. Available from: https://www.sciencedirect.com/science/article/pii/B9780128053928000104
41. Belinato JR, Dias FFG, Caliman JD, Augusto F, Hantao LW (2018) Opportunities for green microextractions in comprehensive two-dimensional gas chromatography/mass spectrometry-based metabolomics – a review. Anal Chim Acta 1040:1–18. Available from: https://www.sciencedirect.com/science/article/pii/S0003267018310006?via%3Dihub
42. León C, Cifuentes A, Valdés A (2018) Foodomics applications. Compr Anal Chem 82:643–685. Available from: https://www.sciencedirect.com/science/article/pii/S0166526X18300643
43. Ibáñez C, Simó C (2013) MS-based metabolomics in nutrition and health research. In: Foodomics. Wiley Online Books, pp 245–270. Available from: https://doi.org/10.1002/9781118537282.ch9
44. Dettmer K, Aronov PA, Hammock BD (2007) Mass spectrometry-based metabolomics. Mass Spectrom Rev 26(1):51–78. Available from: https://doi.org/10.1002/mas.20108
45. Tomita S, Nakamura T, Okada S (2018) NMR- and GC/MS-based metabolomic characterization of sunki, an unsalted fermented pickle of turnip leaves. Food Chem 258:25–34. Available from: https://www.sciencedirect.com/science/article/pii/S0308814618304680#f0005
46. Rocchetti G, Lucini L, Gallo A, Masoero F, Trevisan M, Giuberti G (2018) Untargeted metabolomics reveals differences in chemical fingerprints between PDO and non-PDO Grana Padano cheeses. Food Res Int 113:407–413. Available from: https://www.sciencedirect.com/science/article/pii/S0963996918305672
47. Wu H, Huang W, Chen Z, Chen Z, Shi J, Kong Q et al (2019) GC–MS-based metabolomic study reveals dynamic changes of chemical compositions during black tea processing. Food Res Int 120:330–338. Available from: https://www.sciencedirect.com/science/article/pii/S0963996919301243
48. Levandi T, Leon C, Kaljurand M, Garcia-Cañas V, Cifuentes A (2008) Capillary electrophoresis time-of-flight mass spectrometry for comparative metabolomics of transgenic versus conventional maize. Anal

Chem 80(16):6329–6335. Available from: https://doi.org/10.1021/ac8006329
49. Montero L, Ibáñez E, Russo M, di Sanzo R, Rastrelli L, Piccinelli AL et al (2016) Metabolite profiling of licorice (Glycyrrhiza glabra) from different locations using comprehensive two-dimensional liquid chromatography coupled to diode array and tandem mass spectrometry detection. Anal Chim Acta 913:145–159. Available from: https://www.sciencedirect.com/science/article/pii/S0003267016301283
50. Valdés A, Simó C, Ibáñez C, García-Cañas V (2013) Foodomics strategies for the analysis of transgenic foods. TrAC Trends Anal Chem 52:2–15. Available from: https://www.sciencedirect.com/science/article/pii/S0165993613001945#s0025
51. Álvarez G, Montero L, Llorens L, Castro-Puyana M, Cifuentes A (2018) Recent advances in the application of capillary electromigration methods for food analysis and Foodomics. Electrophoresis 39(1):136–159. Available from: https://doi.org/10.1002/elps.201700321
52. Wang J, Wang C, Han X (2019) Tutorial on lipidomics. Anal Chim Acta 1061:28–41. Available from: https://www.sciencedirect.com/science/article/pii/S000326701930128X
53. Hyötyläinen T, Bondia-Pons I, Orešič M (2013) Lipidomics in nutrition and food research. Mol Nutr Food Res 57(8):1306–1318. Available from: https://doi.org/10.1002/mnfr.201200759
54. Chen H, Wei F, Dong X, Xiang J, Quek S, Wang X (2017) Lipidomics in food science. Curr Opin Food Sci 16:80–87. Available from: https://www.sciencedirect.com/science/article/pii/S2214799317300590?via%3Dihub#sec0015
55. Walczak J, Pomastowski P, Buszewski B (2016) Food quality control by hyphenated separation techniques. Heal Probl Civiliz 1(9):33–38
56. Andjelković U, Šrajer Gajdošik M, Gašo-Sokač D, Martinović T, Josić D (2017) Foodomics and food safety: where we are. Food Technol Biotechnol 55(3):290–307. Available from: https://www.ncbi.nlm.nih.gov/pubmed/29089845
57. Moura-Melo S, Miranda-Castro R, de-los-Santos-Álvarez N, Miranda-Ordieres AJ, Dos Santos Junior JR, da Silva Fonseca RA et al (2015) Targeting helicase-dependent amplification products with an electrochemical genosensor for reliable and sensitive screening of genetically modified organisms. Anal Chem 87(16):8547–8554. Available from: https://doi.org/10.1021/acs.analchem.5b02271
58. Creydt M, Fischer M (2018) Omics approaches for food authentication. Electrophoresis 39(13):1569–1581. Available from: https://doi.org/10.1002/elps.201800004
59. Agrimonti C, Vietina M, Pafundo S, Marmiroli N (2011) The use of food genomics to ensure the traceability of olive oil. Trends Food Sci Technol 22(5):237–244. Available from: https://www.sciencedirect.com/science/article/pii/S092422441100029X
60. Kumar S, Kahlon T, Chaudhary S (2011) A rapid screening for adulterants in olive oil using DNA barcodes. Food Chem 127(3):1335–1341. Available from: https://www.sciencedirect.com/science/article/pii/S0308814611001981
61. Uncu AT, Uncu AO, Frary A, Doganlar S (2017) Barcode DNA length polymorphisms vs fatty acid profiling for adulteration detection in olive oil. Food Chem 221:1026–1033. Available from: https://www.sciencedirect.com/science/article/pii/S0308814616318970?via%3Dihub
62. Spaniolas S, Bazakos C, Spano T, Zoghby C, Kalaitzis P (2010) The potential of plastid trnL (UAA) intron polymorphisms for the identification of the botanical origin of plant oils. Food Chem 122(3):850–856. Available from: https://www.sciencedirect.com/science/article/pii/S0308814610002232
63. Spaniolas S, Bazakos C, Awad M, Kalaitzis P (2008) Exploitation of the chloroplast trnL (UAA) intron polymorphisms for the authentication of plant oils by means of a lab-on-a-chip capillary electrophoresis system. J Agric Food Chem 56(16):6886–6891. Available from: https://doi.org/10.1021/jf8008926
64. Mayer HK (2005) Milk species identification in cheese varieties using electrophoretic, chromatographic and PCR techniques. Int Dairy J 15(6–9):595–604. Available from: https://www.sciencedirect.com/science/article/pii/S0958694604003176
65. Böhme K, Calo-Mata P, Barros-Velázquez J, Ortea I (2019) Recent applications of omics-based technologies to main topics in food authentication. TrAC Trends Anal Chem 110:221–232. Available from: https://www.sciencedirect.com/science/article/pii/S0165993618303625
66. Das M, Haberer G, Panda A, Das Laha S, Ghosh TC, Schäffner AR (2016) Expression pattern similarities support the prediction of orthologs retaining common functions after gene duplication events. Plant Physiol 171(4):2343–2357. Available from: https://www.ncbi.nlm.nih.gov/pubmed/27303025
67. Riedmaier I, Becker C, Pfaffl MW, Meyer HHD (2009) The use of omic technologies for biomarker development to trace functions of anabolic agents. J Chromatogr A 1216(46):8192–8199. Available from: https://www.sciencedirect.com/science/article/pii/S0021967309001952
68. Lodha TD, Basak J (2012) Plant–pathogen interactions: what microarray tells about it? Mol Biotechnol 50(1):87–97. Available from: https://doi.org/10.1007/s12033-011-9418-2
69. Annadurai RS, Neethiraj R, Jayakumar V, Damodaran AC, Rao SN, Katta MAVSK et al (2013) De novo transcriptome assembly (NGS) of Curcuma longa L. rhizome reveals novel transcripts related to anticancer and antimalarial terpenoids. PLoS One 8(2):e56217. Available from: https://dx.plos.org/10.1371/journal.pone.0056217
70. Brown DW, Butchko RAE, Busman M, Proctor RH (2012) Identification of gene clusters associated with fusaric acid, fusarin, and perithecial pigment produc-

tion in Fusarium verticillioides. Fungal Genet Biol 49(7):521–532. Available from: https://www.sciencedirect.com/science/article/pii/S1087184512000989

71. Tirumalai PS (2011) Expression of chitinase and chitin binding proteins (CBP's) by Listeria monocytogenes J0161 in biofilm and co-culture broths. Afr J Microbiol Res 5(29):5188–5193
72. Corujo M, Pla M, van Dijk J, Voorhuijzen M, Staats M, Slot M et al (2019) Use of omics analytical methods in the study of genetically modified maize varieties tested in 90 days feeding trials. Food Chem 292:359–371. Available from: https://www.sciencedirect.com/science/article/pii/S0308814618309300
73. Trachtenberg AJ, Robert J-H, Abdalla AE, Fraser A, He SY, Lacy JN et al (2012) A primer on the current state of microarray technologies BT – next generation microarray bioinformatics: methods and protocols. In: Wang J, Tan AC, Tian T (eds) . Humana Press, Totowa, pp 3–17. Available from: https://doi.org/10.1007/978-1-61779-400-1_1
74. Wang Z, Gerstein M, Snyder M (2009) RNA-Seq: a revolutionary tool for transcriptomics. Nat Rev Genet 10:57–63. Available from: https://doi.org/10.1038/nrg2484
75. Valdés A, Simó C, Ibáñez C, Rocamora-Reverte L, Ferragut JA, García-Cañas V et al (2012) Effect of dietary polyphenols on K562 leukemia cells: a Foodomics approach. Electrophoresis 33(15):2314–2327. Available from: https://doi.org/10.1002/elps.201200133
76. Ibáñez C, Valdés A, García-Cañas V, Simó C, Celebier M, Rocamora-Reverte L et al (2012) Global Foodomics strategy to investigate the health benefits of dietary constituents. J Chromatogr A 1248:139–153. Available from: https://www.sciencedirect.com/science/article/pii/S0021967312008746

CE-MS for Proteomics and Intact Protein Analysis

Valeriia O. Kuzyk, Govert W. Somsen, and Rob Haselberg

Abstract

This chapter aims to explore various parameters involved in achieving high-end capillary electrophoresis hyphenated to mass spectrometry (CE-MS) analysis of proteins, peptides, and their posttranslational modifications. The structure of the topics discussed in this book chapter is conveniently mapped on the scheme of the CE-MS system itself, starting from sample preconcentration and injection techniques and finishing with mass analyzer considerations. After going through the technical considerations, a variety of relevant applications for this analytical approach are presented, including posttranslational modifications analysis, clinical biomarker discovery, and its growing use in the biotechnological industry.

Keywords

CE-MS, Proteomics · Intact protein analysis · Sample concentration techniques · CE separation optimization · CE-MS interfacing · MS fragmentation · PTM analysis · Biopharmaceuticals analysis · Biomarker discovery

V. O. Kuzyk · G. W. Somsen · R. Haselberg (✉)
Division of Bioanalytical Chemistry, AIMMS: Amsterdam Institute of Molecular and Life Sciences, Vrije Universiteit Amsterdam, Amsterdam, The Netherlands
e-mail: r.haselberg@vu.nl

Abbreviations

ADC	Antibody-drug conjugate
BGE	Background electrolyte
CE	Capillary electrophoresis
CE-MS	Capillary electrophoresis-mass spectrometry
CID	Collision-induced dissociation
CRP	C-reactive protein
CZE	Capillary zone electrophoresis
EKI	Electrokinetic injection
EKS	Electrokinetic supercharging
EOF	Electroosmotic flow
EPO	Erythropoietin
ESI	Electrospray ionization
ETD	Electron transfer dissociation
FASS	Field-amplified sample stacking
FESI	Field-enhanced sample injection
FTICR	Fourier transform ion cyclotron resonance
HCD	Higher-energy collision dissociation
HDI	Hydrodynamic injection
IPA	Intact protein analysis
IT	Ion trap mass analyzer
ITP	Isotachophoresis
LB	Leading buffer
LC	Liquid chromatography

A. V. Colnaghi Simionato (ed.), *Separation Techniques Applied to Omics Sciences*, Advances in Experimental Medicine and Biology 1336, https://doi.org/10.1007/978-3-030-77252-9_4

LC-MS	Liquid chromatography-mass spectrometry
LE	Leading electrolyte
LOD	Limit of detection
LOQ	Limit of quantitation
LTQ-FTICR	Linear trap quadrupole-Fourier transform ion cyclotron resonance
mAbs	Monoclonal antibodies
MALDI	Matrix-assisted laser desorption/ionization
MCE	Microchip capillary electrophoresis
MS	Mass spectrometry
MSI	Multisegment injection
PTM	Posttranslational modification
QTOF	Quadrupole time of flight
RPLC	Reversed-phase liquid chromatography
SPE	Solid-phase extraction
TE	Terminal electrolyte
t-ITP	Transient isotachophoresis
TOF	Time-of-flight mass analyzer

1 Introduction

Proteomics as a scientific area encompasses large- and medium-scale studies of proteins structure and function within a living system. The proteome itself is a complex system with high dynamic range of its components abundancy, representation, and varieties. Whether researchers try to grasp the full diversity of this system (untargeted approach) or aim for a search and evaluation of certain components (targeted approach), a specific toolbox is an absolute must. The analytical techniques for proteome research are quite distinct from the traditional protein chemistry methods as they have to be sensitive enough to detect low-abundance proteins and to be sufficiently high-throughput for handling large amounts of protein species [1].

A high-end analysis in proteomics field – no matter whether peptides or whole proteins are being measured – is based on two immanent factors. To begin with, a precise mass determination with a sensitive detector is needed. Mass spectrometry (MS) is commonly acknowledged to be the supreme candidate for this task, especially if its configurations are able to fragment the analytes and gain more information about their primary structure.

However, for the complex samples that proteomics typically deals with, the value of accuracy and high resolution of mass spectrometers gets nearly lost without proper separation of the analytes prior to detection. A separation system that allows analysis of one compound at a time would be ideal. However, generally this is not realistic, so highest peak capacity allied to the best separation resolution should be pursued. Currently, high-performance liquid chromatography (or more precisely, reversed-phase liquid chromatography) is a widely recognized workhorse technology for proteomics applications in both laboratory and clinical settings mainly due to its efficiency, robustness, and reproducibility [2–4].

Despite numerous advantages that LC-MS hyphenation has to offer for proteomics, in some analytical applications, it may be beneficial to shift the balance towards capillary electrophoresis (CE) as separation step. Current developments in MS-based proteomics are primarily aimed at minimizing the analysis time while keeping the quality of separation and high sensitivity. For that, CE-MS is very suitable due to its short analysis times [5] without compromising the sensitivity [6]. Unlike most LC columns, CE capillaries do not need regeneration prior to new injection; therefore analytical throughput can be increased by multiple segmented injection [7]. Low sample consumption is an integral property of CE system, which is beneficial when dealing with amount- and volume-limited samples. Moreover, if the sample is rather diluted, numerous in-line and on-line preconcentration techniques will help to leverage the sensitivity of the analysis. It is also worth mentioning that, in contrast to RPLC, CE allows simultaneous analysis of very short or long peptides [8]. Another advantage is the low flow rate that helps to overcome ionization suppression phenomena [9] and allows the use of highly aqueous solutions. As hydrophilic molecules are commonly separated in aqueous buffers, that becomes a valuable characteristic of the method [10].

When combined with the high sensitivity of MS, CE indeed becomes a lucrative analytical solution. However, CE and RPLC in proteomics are not competitive but rather complementary techniques. This complementarity of the techniques has already been demonstrated for both peptides and proteins [11, 12], and the combination is even being referred to as "Swiss knife" for proteomics investigations [13]. Additionally, with the growing popularity of two-dimensional (2D) separation systems, it is not surprising to find hyphenated LC-CE-MS approaches being used for protein analysis [14].

This chapter aims to explore various parameters involved in achieving high-end CE-MS analysis of proteins, peptides, and their posttranslational modifications. The structure of the topics discussed in this book chapter can be conveniently mapped on the scheme of the system itself (Fig. 1), starting from sample preconcentration and injection techniques and finishing with mass analyzer considerations. After going through the technical considerations, a variety of relevant applications for this analytical approach are presented.

2 Technological Considerations

A CE-MS setup for the analysis of peptides and proteins does not conceptually differ from any other type of CE-MS application; however, a vast majority of the research for these analytes is done using capillary zone electrophoresis (CZE) as a separation mode. Electrospray ionization (ESI) is the most used option for the hyphenation with MS. Therefore, in this chapter we will be mainly focusing on this setup. Whereas CZE-ESI-MS is most frequently used, in every step of the analytical workflow (Fig. 1), specific considerations can make or break its success in peptide and protein analysis. Below, these issues are sequentially discussed.

2.1 Sample Introduction and Preconcentration

Hydrodynamic injection (HDI) is the most straightforward and the most used way of introducing any sample in the capillary. Here, the sample plug is pressure-pushed from the injection end of the capillary or is moved due to induced vacuum at the exit end of the capillary or is introduced with so-called siphoning by raising the inlet vial to a certain level above the outlet vial. The analyte portion will be injected as a fraction of the sample, and the amount of analyte will be proportional to the injection volume, independently on the charge of the analyte and the sample matrix. Alternatively, electrokinetic injection (EKI) is performed by applying a certain voltage over the capillary, introducing analytes based on their mobility and (potentially) the

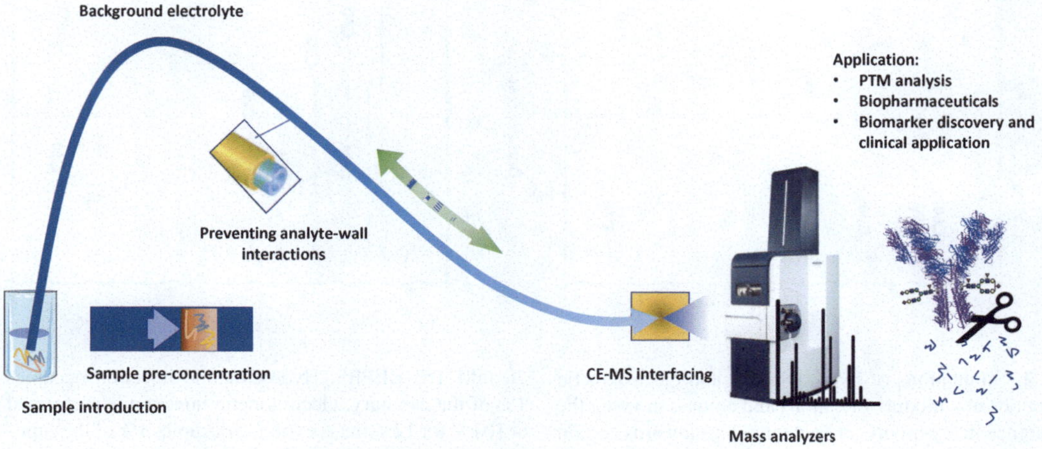

Fig. 1 Schematic representation of a CE-ESI-MS setup with the essential processes/components mapped

generated electroosmotic flow (EOF). The electrokinetic injection procedure is intrinsically more selective than hydrodynamic injection; however, it requires careful fit-for-purpose evaluation. It will not perform successfully on weakly charged analytes or in case of a highly charged matrix components. A more detailed discussion on the advantages and limitations for the two injection types for various applications from a mathematical modelling perspective can be found in the review of Breadmore [15]. He demonstrates the capacity of EKI to increase the signal intensity (Fig. 2) [15] and argues that this injection technique is superior over conventional HDI.

2.1.1 Electrophoretic Sample Concentration

One of the abovementioned advantages of CE-MS is the very small volume of sample needed for the analysis. However, that may become a serious drawback if the sample is at a very low concentration and/or is available in limited quantity. To overcome this issue, several electrophoretic preconcentration approaches have been developed. All of them make use of a difference between the composition of the sample solution and the bracketing solutions in the capillary, primarily background electrolyte (BGE), and additives. The concentration of the analyte takes place on the boundary of solution with different chemical properties. When the stacking event finishes, the analytes enter the BGE and undergo electrophoretic separation. To integrate these principles in CE-MS of proteins and peptides, one should consider two major restrictions. First, the stacking CE solutions must be MS-compatible. Second, the solution chemistry must be tuned to favor the analyte solubility and proper conformation. This is often crucial for intact protein analysis. However; it must not be neglected for the (glyco)peptide analysis either, since carbohydrates may shift the chemical properties of the peptide. Wet chemistry suitability for CE-MS is further covered in Sect. 2.3.

Isotachophoresis (ITP) is one of the most used techniques for increasing the sample loadability. It operates with discontinuous buffer systems of high ionic mobility leading electrolyte (LE) and low ionic mobility terminating electrolyte (TE). These two solutions bracket the sample plug that has an intermediate ionic mobility. LE and TE co-ions should have higher and lower effective mobilities than those of analytes, respectively. Subsequent application of an electric potential results in analytes being focused at the LE/TE interface, and the concentration of all the ions in the sample plug will adhere to the Kohlrausch adjustment of concentration. A form of isotacho-

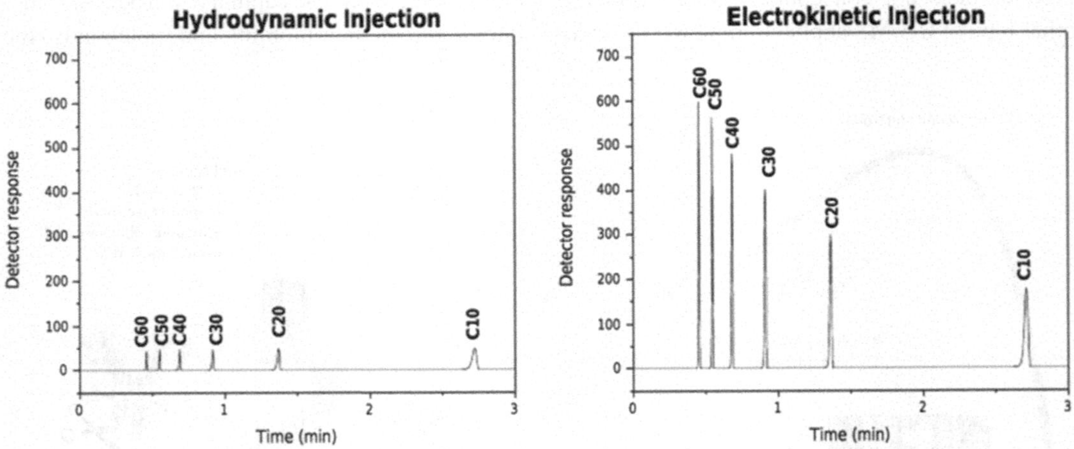

Fig. 2 Simulations of hydrodynamic and electrokinetic injections of a mixture of cations and anions, showing the difference in sensitivity and potential bias towards certain analytes. Separation conditions with no EOF, BGE 20 mM Tris-HEPES. Hydrodynamic injection occupies 1% of the capillary; electrokinetic injection is performed at 10 kV for 12 s (matrix volume occupies 1% of the capillary). Sample contains 1 μM of each analyte [15]

phoresis, that is frequently used for stacking in CE-MS, is transient isotachophoresis (t-ITP), where the concentration step is directly followed by an electrophoretic separation of ionizable analytes [16]. In that case a sample is dissolved in either TE or LE (the choice depends on the ion mobilities of the analytes) and is bracketed by the counterpart electrolyte. When the ITP focusing step is finished, ions of the TE are already mixed with the BGE; hence the electrophoretic separation is initiated. Typically this method has no adverse effect on separation efficiency and even offers an improvement on it [17]. Integration of t-ITP might increase the limit of detection (LOD) up to two orders of magnitude [18] and allows to inject up to 25% of the total capillary volume instead of standard 1–2% [19]. However, its application is limited by the choice of suitable electrolytes and buffers. Acetic acid is a well-known TE that is most often used for CE-MS of peptides and proteins. A great advantage of t-ITP is overcoming CE-MS incompatibility with non-volatile buffers: the ITP event can be used to separate the analytes in the nonvolatile conditions, and the following CZE separation will immanently transfer them to the volatile BGE, which does not interfere in the ESI ionization process.

The *dynamic pH junction* preconcentration mechanism is based on a large pH difference between the sample plug and the adjacent BGE. The most widely used buffer system for pH junction in CE-MS is basic leading buffer (LB) with the sample plug sandwiched between acidic BGE (e.g., acetic acid). Peptides and proteins often bear a negative charge in a basic environment; therefore, they migrate backwards to the anode until encountering the acidic BGE and hence acquire a positive charge when they pass the pH boundary. This makes them migrate back to the cathode, and, as a consequence, analytes get concentrated on this pH boundary. As soon as the boundary gets neutralized, the stacked analytes migrate to the detector. The use of the inverse buffer system (base-acid-base) is also possible, albeit, less common. The technique gained popularity in the proteomics field, since even high-concentration acidic BGEs are MS-compatible. It is best suited for amphiprotic peptides but also weakly acidic or weakly basic ones since there is an ionization difference in loading buffer and BGE [20, 21]. Fine tuning of the buffers may increase the stacking effect even further. If the analyte of interest is well-studied and its isoelectric point is known, electrolytes can be selected for their buffering capacity; it should be negligible at the peptide's or protein's pI but sufficient in the basic and acidic regions. This will allow the pH junction to exist for a longer period and leads to even higher signal output [22]. As an example, Hasan et al. tested the applicability of pH junction concentration method (Fig. 3A) for four model proteins with different isoelectric point and reported 1000- to 10,000-fold enhancement without compromising the peak shape (Fig. 3B) [23].

Approaches like *field-amplified sample stacking (FASS)* and *field-enhanced sample injection (FESI)* can also be used to significantly improve the detection sensitivity. FASS and FESI events are achieved by creating a conductivity discontinuity in the CE capillary. Here the target analyte is dissolved in a low-concentration and low-conductivity loading buffer (LB) and injected in a high-conductivity and high-concentration background electrolyte (BGE). The difference between FASS and FESI is the way the sample is introduced, via hydrodynamic vs electrokinetic injection, respectively. When a high voltage is applied, the electrical field in the LB will be higher than the one in BGE. As a consequence, analytes will have a higher electrophoretic velocity in the sample plug as compared to the BGE. When analytes stumble upon sample plug/BGE interface, they will migrate slower and will be effectively stacked on this boundary. The anticipated sample concentration factor for FASS is in range of one to two orders of magnitude. To exemplify, integrating FASS in the CE-MS workflow allowed to quantitatively analyze single-cell proteome of *Xenopus laevis* embryo, reaching a 11 nmol/L LOD [24]. FESI typically delivers higher sensitivity enhancement factors. Monton and Terabe reported 3000-fold signal improvement reaching fmol/μL LODs when using FESI for peptide mapping of protein tryptic digests

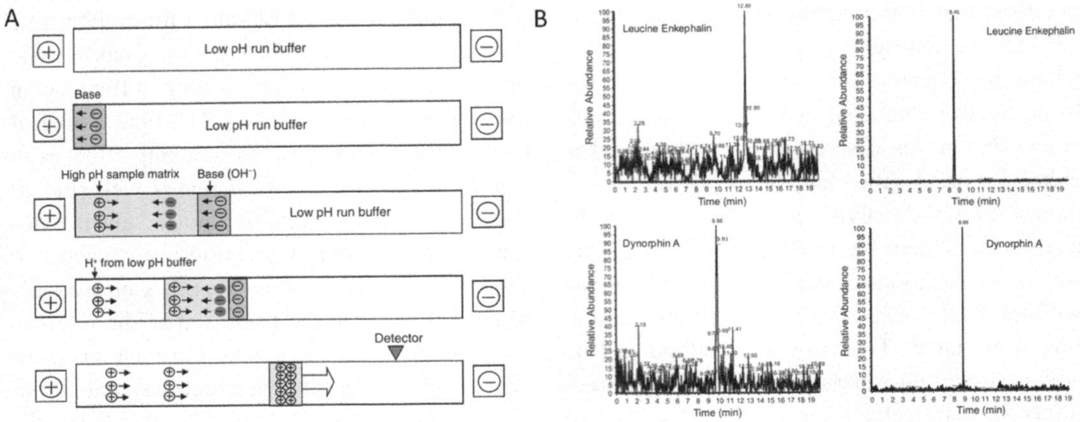

Fig. 3 Principle and application of pH-mediated stacking. (**a**) Schematic illustration of the pH-mediated sample stacking process. A short plug of strong base is injected into acidic BGE and is followed by a long hydrodynamic injection of a sample in alkaline matrix. Peptides gain positive charge from the protons in the BGE or loose charge to the hydroxyl ions in the basic plug and are consequently concentrated on the pH boundary upon application of the separation voltage. (**b**) Comparison of CE-MS electropherograms obtained for the analysis of two peptides by conventional pressure injection (left) and the pH junction stacking method (right). A clear increase in signal-to-noise ratio is observed when applying pH-mediated stacking [23]

[25]. They also mention FESI-bound acidic conditions as an additional benefit, since it lowers the peptide adsorption on the silica surface. Some years later, Pourhaghighi et al. used FESI for intact protein analysis and demonstrated 3200- and 4800-fold improvement in terms of peak height and peak area, respectively [26]. Despite FASS and FESI appear to be well-established preconcentration methods, the intrinsic medium to high conductivity of biological samples may demand prior purifications and sample pre-treatment before the techniques become applicable. Additionally, they also require high-conductivity BGEs, which might affect the electrospray ionization process and spray stability. Moreover, one should consider limited protein or peptide solubility in the loading buffer and still limited injection volumes [27]. On the other hand, the techniques appear promising due to their simplicity, not requiring careful design of electrolyte systems.

As mentioned above, biological samples are rarely low-conductive. To circumvent the extra sample pre-treatment steps, which often result in sample loss, the mechanism of *pH-mediated stacking* was suggested [28]. In this technique, a sample of high ionic strength is injected, and then it is directly followed by a plug of strong acid or base (for analyzing cations or anions, respectively). When the voltage is applied, the sample zone gets fully titrated, thus creating a low-conductivity region and initiating a FASS event. The zwitterionic properties of peptide sequences allow to effectively use the technique in proteomics. For CE-MS analysis of a protein digest, the concentration sensitivity of tens of fmol/mL can be reached without losing the separation efficiency [6]. The method was also successfully used for separation of site-specific phosphopeptide isomers, where sample concentrations are typically low [29].

Another way to use electrokinetic injection is exploited by the recently introduced *electrokinetic supercharging* (EKS). It is a combination of a long field-amplified EKI with a t-ITP refocusing step that combines the selectivity of t-ITP with concentration power of field-amplified sample injection resulting in several orders of magnitude enhancement in sensitivity. It gained notable popularity in microchip-based CE, since EKS exhibits flexibility not only towards different designs and geometries but also to various detectors and with many types of microchips [30, 31]. Currently, EKS-CE with MS detection is not widely used, mainly due to a limited amount of available MS-compatible electrolytes. However, initial tests have shown that sensitivity enhancement factors between 10 and 100 can be obtained. It was also reported to result in an improved separation when tested on a tryptic digest of beta-lactoglobulin [32]. EKS used in analysis of parathyroid hormone splice variants lowered limits of quantification (LOQs) to 10 pg/mL [33]. However, as for any stacking technique, buffer compatibility remains a challenge. That, along with relatively recent implementation of EKS in the field, may explain the limited amount of publications on the topic. However, the work of Busnel et al. demonstrated that efficient tryptic digest sensitivity with enhancement factors up to 10,000 can be obtained with full MS compatibility [32]. Unlike some other techniques, the EKS stacking mechanism is successfully explained with computer modelling approaches. Many experimental factors influencing the concentration efficiency can be integrated in mathematical models to ensure the best and the most controlled performance [34].

Injection manipulations may also be used to speed up the analysis, increase the throughput, and gain extra sample information. Comparing to RPLC for bottom-up proteomics applications, CZE is known to produce less peptide identifications per time unit. That is attributed to the dead time between the sample injection and the first compound reaching the detector that ranges around 25% of the total analysis time [35]. Multisegment injection (MSI) allows several analyses to be performed in a single run through introducing several sample plugs alternating with the BGE spacers. In that manner, the dead time for $n + 1th$ injection is used to detect peaks of the nth injection. Being initially developed for metabolite analysis, it also proved useful for proteomics applications. Dovichi's group demonstrated that by doing a triple injection of a yeast protein digest within a single injection time frame, the peptide identification rate doubled, and the separation profile remained the same for all the injections, as demonstrated in Fig. 4 [7]. MSI was used to evaluate B-type natriuretic peptide bioavailability after therapeutic infusion [36]. This peptide is a heart failure biomarker that gets catabolized by plasma proteases. Its dynamic generation and breakdown were monitored in time by means of MSI; up to seven injection plugs were used in this study.

2.1.2 Chromatographic Sample Concentration

Increase in sample loadability is effectively achieved by manipulating the injection process. However, chromatography-based techniques can be annexed to CE for the same purpose. An offline preconcentration, e.g., *solid-phase extraction (SPE)*, is sometimes considered as a separate step in the protocol; however, a CE-MS analysis often cannot go without it. Apart from increasing the analyte concentration and hence sensitivity of its detection, SPE is often necessary for sample pre-treatment and cleanup. Analytes in proteomics often come as complex mixtures with physical (viscosity) and chemical (pH, salts) properties unsuitable for a high-performance CE-MS analysis.

The offline preconcentration and pre-treatment methods are bound to a simple rule: separation of the analyte from other components and its elution in a suitable volume of the appropriate buffer. Accordingly, liquid-liquid extraction, centrifugation, precipitation, membrane-assisted filtration and dialysis, SPE, and immunoaffinity approaches may be used prior to CE-MS analysis of proteins and peptides. The latter two techniques have also developed towards in-line and on-line setups for time-efficient analysis with minimal sample loss. Integrating SPE directly in the CE-MS platform is often referred to as chromatographic sample concentration.

A SPE step can be coupled on-line to the CE-MS as a separate entity via a switching valve [37], or it can be introduced as a segment (in-line) of the capillary (Fig. 5) [38]. Various stationary phases (usually based on silica-based and polymer sorbents) and separation principles have been used in both setups, such as monolith-bracketed reversed-phase pre-columns [39], immunoaffinity chromatography using magnetic beads [40], reversed-phase [41, 42], and ion exchange [43]. For miniaturization of the setup,

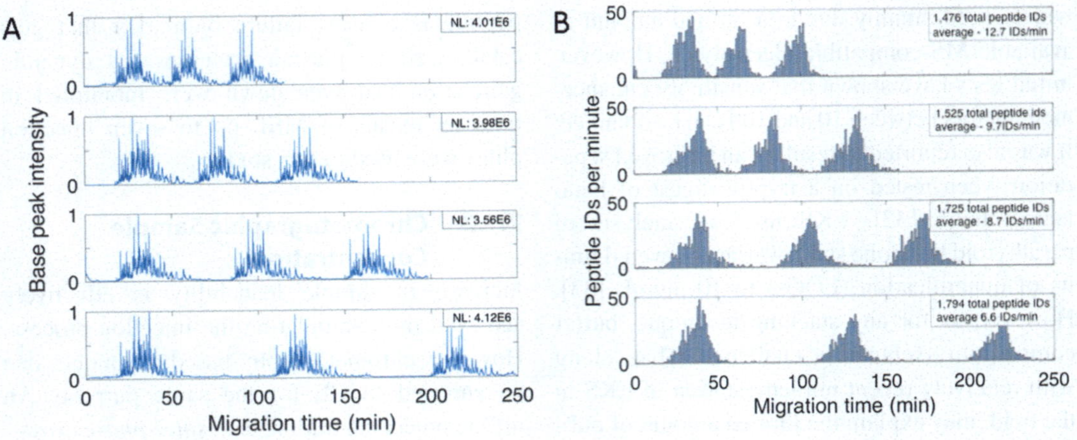

Fig. 4 Increase of CE-MS throughput by using multisegment injections as shown for a tryptic digest of yeast. (**a**) Three sequential injections are done with different separation time periods between them. Time between injections are 32.5 min (top), 47 min (2nd), 66.5 min (3rd), and 90 min (bottom). (**b**) Peptide identification rates obtained during the runs shown in (**a**). Average number of IDs is calculated by dividing the total number of IDs by the total separation time [7]

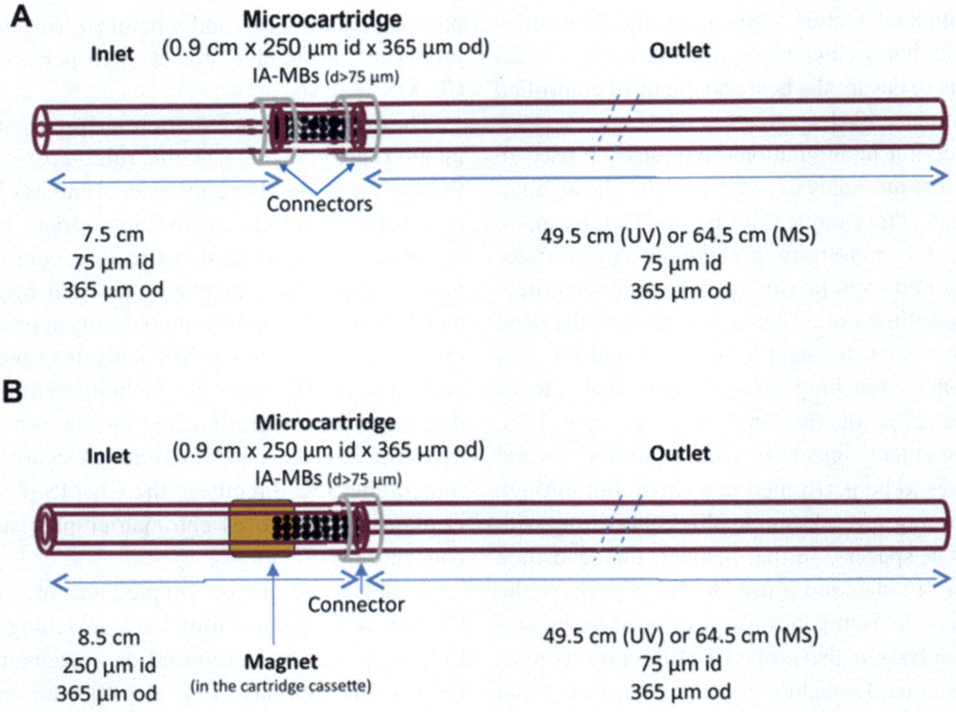

Fig. 5 Schematic representation of an integrated immunoaffinity SPE design for serum transthyretin analysis by CE-MS. Magnetic particles are either trapped in (**a**) a microcartridge body (using a large particle size) or are (**b**) retained by a magnet at the inlet of the capillary [38]

the standard capillary may be replaced by a microcartridge. This has been shown for the analysis of opioid peptides, where a C18 SPE cartridge was combined with a sheathless CZE setup, resulting in a 5000-fold decrease in LOD when compared to the same CE-MS setup without preconcentration [43]. The dead volume at the sleeve region is reported as a frequent limitation of that arrangement, but a seamless interface design solution has been already reported to fix the issue [44].

Immunoaffinity approaches often reinforce the conventional SPE setups discussed above with a target-specific enrichment. For that, antibodies or antibody fragments are immobilized on a solid support comprised of monoliths or particles. The technique has been largely applied by the group of Sanz-Nebot for the detection of erythropoietin [45], opioid peptides [46], serum transthyretin [46, 47], and other large proteins in complex biological samples [46]. The method was also successfully applied to detecting well-known cancer biomarkers: alpha-1-acid glycoprotein (AGP) [40] and C-reactive protein [48]. As for medical application, immunoaffinity CE separations are already used in clinical laboratories [49]; however, the addition of MS detection is still in development since the implementation costs of the instrument still remain rather high.

SPE does offer a great improvement for sample complexity reduction and preconcentration; however, average elution volumes of both in- and on-line SPE are larger than 1–2% of the capillary volume used for injection. Therefore, additional stacking techniques are often still required for appropriate sensitivity enhancement. To exemplify, the recent work of Dovichi's group shows possibility of doing large-scale proteomics analysis with an enrichment factor of 3000 compared to conventional electrokinetic injection [43]. They implemented an in-capillary cation-exchange monolith for solid-phase extraction with pH gradient elution. That allowed for integral sample stacking and pH junction events, since the elution buffers series had lower concentration but higher pH values than the BGE. This unique setup coupled to a high-resolution LTQ Orbitrap Velos allowed identification of 799 protein groups and 3381 peptides from only 50 ng of the digest.

To summarize, CE-MS supports numerous ways of signal enhancement, and the choice of the sample preconcentration methods would depend on both analyte and matrix properties. To ease the choice of the technique, we provide a decision flowchart (Fig. 6) [50] for the initial guidance.

2.2 Background Electrolyte

BGE pH is the key factor to the overall electromigration process, since both ionic/effective mobilities of the analytes and EOF strongly depend on pH and buffering capacity of the electrolyte. When selecting a suitable BGE, one should take into account the properties of analytes in a sample: the ideal BGE pH should lie in between anionic and cationic species pKas, maybe slightly shifted to the cationic value [51]. This will yield optimal and stable separation accompanied with good resolution due to a constant and stable ionization of the analyte. Tuning the pH of the BGE can increase the stacking effect of t-ITP. Furthermore, it is not uncommon to improve the separation efficiency and peak shape by increasing the ionic strength of the BGE. This results in analyte stacking and lowering of the EOF velocity. However, these approaches are rather limited by additional requirements for the electrolyte systems in CE-MS of proteins and peptides. The buffer components have to be volatile to avoid salt-induced signal suppression and high background signals.

These requirements limit the buffer systems options. Potential suitable ionic species with their ionic mobility and pKa values were evaluated in the review from Pantuckova and colleagues [52]. They also used a mathematical model of electrolyte migration in free solution using different BGE compositions [53]. The number of commonly used BGEs for CE-MS in proteomics is even smaller than this estimation [8]. Nonetheless, the set of volatile BGE systems often used in CZE covers almost the entire pH range and enables the direct coupling of CZE to MS using

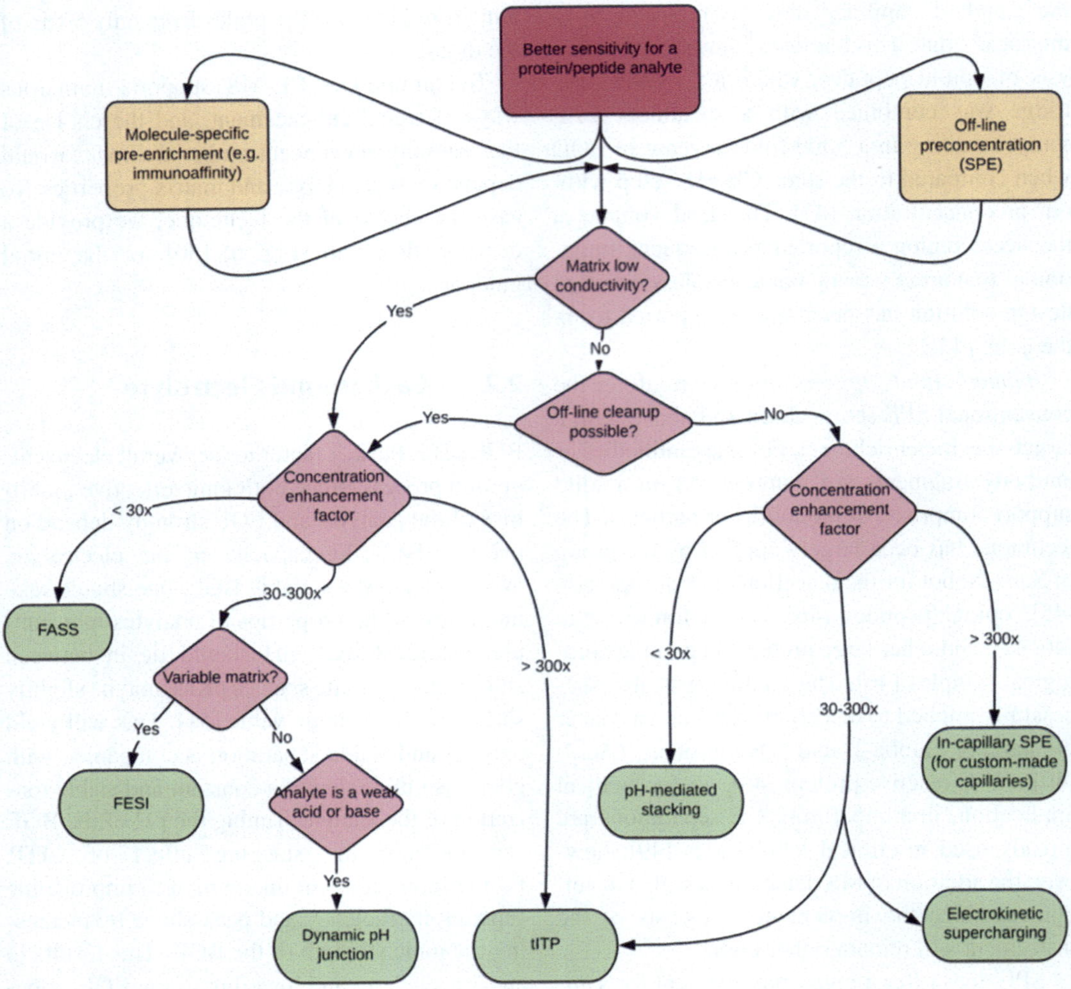

Fig. 6 Decision flowchart to select the most suitable CE sample preconcentration technique. (Modified and reproduced from Breadmore and Sänger-van de Griend [50])

ESI (Fig. 7) [52]. Free acetic acid [54], formic acid [55, 56], and the ammonium salts of both [57] are the first choice BGE buffers for peptides and protein separations in CE-MS. Sometimes, the use of carbonate as BGE component may be favorable, albeit it comes with a drawback of instability and gas bubbles formation, which can severely compromise the analysis. For particular analytes – highly hydrophobic peptides – a non-aqueous buffer system may be considered. Using a nonaqueous BGE proved to reliably allow the separation of temporin peptides in less than 12 min of analysis time, thus offering a good alternative to RPLC-MS systems [58].

Contemplating the complexity of protein sequence and structure, it is also important to ensure proper analyte solubility. Mathematically this was modelled by Ruckenstein and Shulgin, who focused on salts and organic additives affecting the protein solubilization in aqueous solvents [59]. The basic principles may be brought down as follows: proteins denature in high alcohol concentrations, high pH, and high salt concentration. Therefore, an investigator may want to be extremely cautious when considering, for example, isopropyl alcohol in the BGE or using untreated biological matrices as samples.

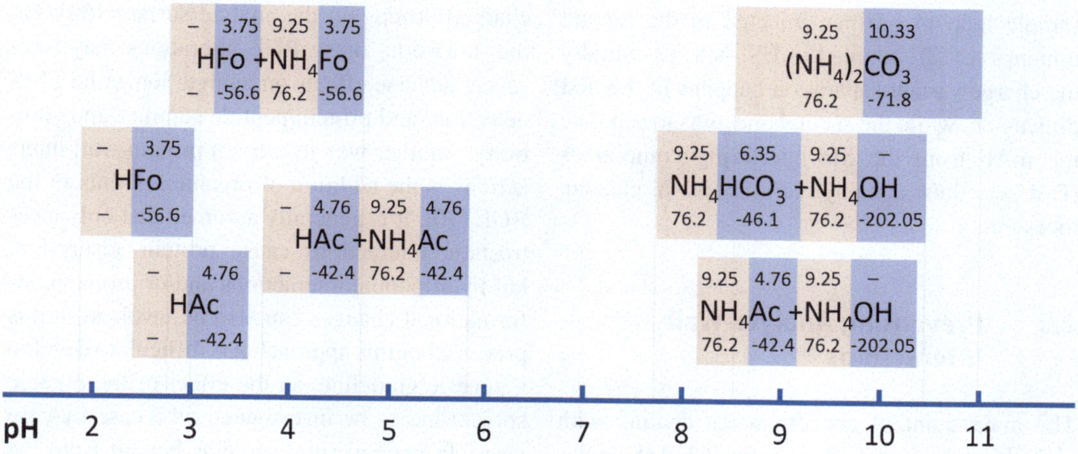

Fig. 7 Commonly used buffers and their operational pH regions recommended for on-line CZE–ESI-MS. Upper number represents pK_as and lower numbers represent ionic mobilities of the corresponding ionic species. (Modified and reproduced from Pantuckova et al. [52])

The use of nonvolatile BGEs in CE-MS is reported sporadically [60, 61]. Although this might work under specific choices of the system design, it is unlikely to become a commonly used arrangement. One of the main issues with nonvolatile components is the ion pairing phenomenon, where salts form complexes with analytes, leading to a decreased response in the MS. In case additives are used in the buffer – such as dynamic coatings, discussed later in this chapter – their compatibility with MS detection has to be ensured as well. Unfortunately, in general, BGE additives that effectively interact with proteins and peptides also result in a significant decrease of MS detection sensitivity. A way to circumvent it is to minimize the concentration of the MS-interfering substances or to dilute the unwanted compounds with the sheath liquid. This variant integrally comes as a sensitivity and/or CE performance trade-off, as Tang et al. has demonstrated for protein capillary isoelectric focusing coupled to ESI-MS [62] before the MS-friendly ampholytes were introduced [63]. One could also use alternatives to ESI as an ionization technique. The use of matrix-assisted laser desorption ionization (MALDI) has shown great tolerance to different CE additives as well as strong advantages for CE-MS analysis of proteomics samples (see corresponding paragraph on interfacing). However, since it is an offline hyphenation, it rarely performs as a high-throughput method.

Hyphenating incompatible methods is analytical challenging and requires novel solutions. Two-dimensional (2D) CE systems are widely used to tackle that problem. The ESI-interfering components are used in the first dimension to obtain optimal separation performance and are either removed from the fractions or separated from the analytes of interest in the second dimension. This approach was demonstrated by the group of Neusüß with a heart-cut 2D-CE separation system equipped with an isolated mechanical valve [64]. With the help of UV/Vis detection, in the first dimension, an analyte of interest was cut and sent for the second CE separation. The system feasibility was tested on a BSA digest, revealing minor peak broadening. Two years later they have published a review of the current trends in CE-CE-MS and their applications [65]. As an example, the ESI interference of ε-aminocaproic acid (EACA)-based BGE system (routinely used for pharmaceutical analysis of mAb charge variants by CZE [66]) was circumvented by using it as a first dimension in a CE-CE-MS setup [67]. When the target analytes reach the UV detector in the first separation system, analytes are being pressure-transferred into the mechanical valve sample loop to be then injected to the second dimension CZE coupled to ESI-MS. Essentially the charge variant separation happens in the first dimension, while the second one aims to separate the mAb from the MS-interfering compounds (EACA), thus actually serving as a cleanup measure.

2.3 Preventing Analyte-Wall Interactions

The main point of concern when dealing with high-throughput and comparative CE-MS analysis of biological samples is keeping the procedure repeatable, reproducible, and robust. To maintain that, the attention should be primarily paid to migration time stability and EOF consistency, and both can be easily compromised due to adverse interactions of analytes with the (bare-fused silica) capillary inner wall. Easily put, the silica surface bears a negative charge, which attracts basic components [68]. If an analyte is prone to interact with the silica, it may be retarded, and the migration time may be increased or stretched (thus promoting band broadening). In a worst-case scenario, adsorption may even be irreversible, preventing proper analyte quantitation and incapacitating the capillary for further use. Due to buffer composition inconsistency, the EOF stability also gets vulnerable; therefore sample matrix components should not interact with the wall as well.

To prevent protein adsorption, several approaches may be considered. The easiest one is to use BGEs with an extremely low or high pH, which ensures that the silica is not charged (low pH) or that both the silica and protein and peptides are fully negatively charged (high pH). However, proteins tend to be unstable at these pH values due to their complex structure. Furthermore, it is desirable to still have the possibility of tuning the BGE pH so that separations can be optimized in CZE mode. Increasing ionic strength of the BGE or adding surfactants and/or ion pairing reagents can also reduce the analyte-wall interactions by protonating, neutralizing, or creating steric hindrance towards the positively charged groups on the analyte surface [69]. On the downside, those BGE alterations may have severe adverse effects on analyte ionization, MS detection, and protein/peptide stability and solubility. Another way to prevent protein-wall interactions is the addition of organic solvents to the BGE [70]. It is generally assumed that only electrostatic interactions cause protein adsorption, but hydrophobic interactions and/or protein conformational changes can also be involved in this process. For this approach it is difficult to develop a generic guideline, so the effect of the organic solvent has to be investigated on a case-by-case basis. In general rigid proteins benefit from the addition of organic solvent to the BGE, whereas flexible proteins are negatively affected [70]. In other studies, organic solvents in the BGE have positively affected the separation efficiency of intact proteins [71, 72] and peptides [73]. Organic solvent addition to the BGE has other advantage, such as increasing separation selectivity, reducing Joule heating, shortening analysis time, and, in certain cases, enhancing the analyte solubility. However, some researchers argue whether these benefits are correctly estimated and relied upon [74].

The most successful and versatile strategy to prevent protein adsorption to the capillary wall in CE-MS is utilizing capillary coatings. Excellent reviews describing and discussing these coatings are available, and the reader is referred to those for more in-depth information [69, 75, 76]. Coatings may effectively protect the adverse interactions by creating either steric hindrance or surface inertness (preventing either electrostatic and/or hydrophobic interactions). Another way to achieve that is mounting opposite (positive or negative) charges on the analyte of interest and the capillary wall to create electrostatic repulsion.

The types of coatings and how they adhere to the surface could be split into three major groups. Dynamic coatings, composed by surfactant-like BGE additives, compete with the analyte for the silanol groups of the capillary wall. Since they are added in excess, they will nearly always win the competition. Structurally they are mostly polymers (polyamines or polysaccharides) and may be either cationic or neutral. These coatings are very easy to use, and they are frequently applied for CE analysis of proteins [75]. However, they may cause notable background noise and MS contamination and disturb the ESI process. Theoretically, there is a potential for their use in offline CE-MS, but that has not been reported so far. Capillaries can also be permanently coated with polymers, otherwise known as static coatings. They are attached covalently based on silane chemistry and polymerization reactions. Most of these coatings bear no charge (neutral coating) which results in little to absence of EOF. This can impair the ESI spray stability; however, it may be advantageous for increasing the separation window. These coatings have long-term stability but are impossible to regenerate if worn-out. The third way to craft a coating is by adsorption. Often the capillary is rinsed with a chemical agent solution prior to analysis and, if a refresher is necessary, in between the runs. These coatings may form neutral, cationic, or even anionic [77] layers on the capillary wall. Their adherence to the capillary wall is based on hydrogen bonding and hydrophobic interactions (neutral coatings) or ionic interactions (cationic and anionic coatings). Another interesting technique is to use a mixed interaction approach, where coating is first adsorbed to the capillary surface, but the electric field transforms it into a permanent one via silicate polymerization [78]. This kind of coating is relatively easy to apply (simple rinsing steps) and to strip off the capillary, and the reagent consumption is very moderate (microliter range). One disadvantage of adsorbed coatings is the bleeding effect that is sometimes observed at high BGE pH values [69] that often causes MS signal quenching.

While it is important to minimize the analyte migration distortion during CE-MS analysis in proteomics and protein analysis, coatings are also vital for the overall method development since they can modify the EOF strength and direction. As a short example, if a neutral coating is used, (almost) no EOF is generated, regardless of the BGE conditions. Thus, the CE polarity has to be set considering the net charge of the analyte of interest in the BGE, since it has to migrate towards the outlet. By making use of a charged coating, the direction and magnitude of the EOF may be tuned according to the separation purpose (see Fig. 8 [76] for the example of coating affecting the peptide separation and Fig. 9 [76] on glycopeptide separation). The advantages and limitations of various coatings for cation and anion analysis with respect to CE polarity and generated EOF were comprehensively reviewed by Huhn et al., and the reader is referred there for more details [69]. Besides dictating separation conditions, coatings also have an impact on the MS hyphenation by keeping the EOF stable. The stability of EOF is much more important for CE-MS than for conventional CE, since no outlet vial is present and sheath liquid and/or air can get in the capillary and ruin the separation.

A large variety of polymeric compounds are used as coatings, some being extensively used and commercialized, others being newly explored and shown as a proof of principle. The most popular coatings currently used for proteomics and intact protein analysis in CE-MS are shown in Table 1. Most of these coatings were initially developed with CE application in mind, and not all of them are straightforwardly compatible with MS ionization and detection. It must be also noted that very complex mixtures (real biological samples) may not be easily analyzed even with the use of appropriate coatings. Numerous reports showing high plate numbers and sharp peaks are reported only for a set of carefully selected, highly purified standards. However, when complex biological matrices come into play, the picture would most likely change, and therefore proper tests are to be conducted before any coating can be used on a newly generated sample.

2.4 CE-MS Interfacing

To hyphenate a liquid phase CE separation to MS, the analyte-carrying solvent needs to be transferred to the gas phase via an ionization source. Although MS systems can be equipped with a variety of ion sources [96], ESI holds absolute dominance. In ESI the analyte molecule is transferred directly to the MS and is prone to

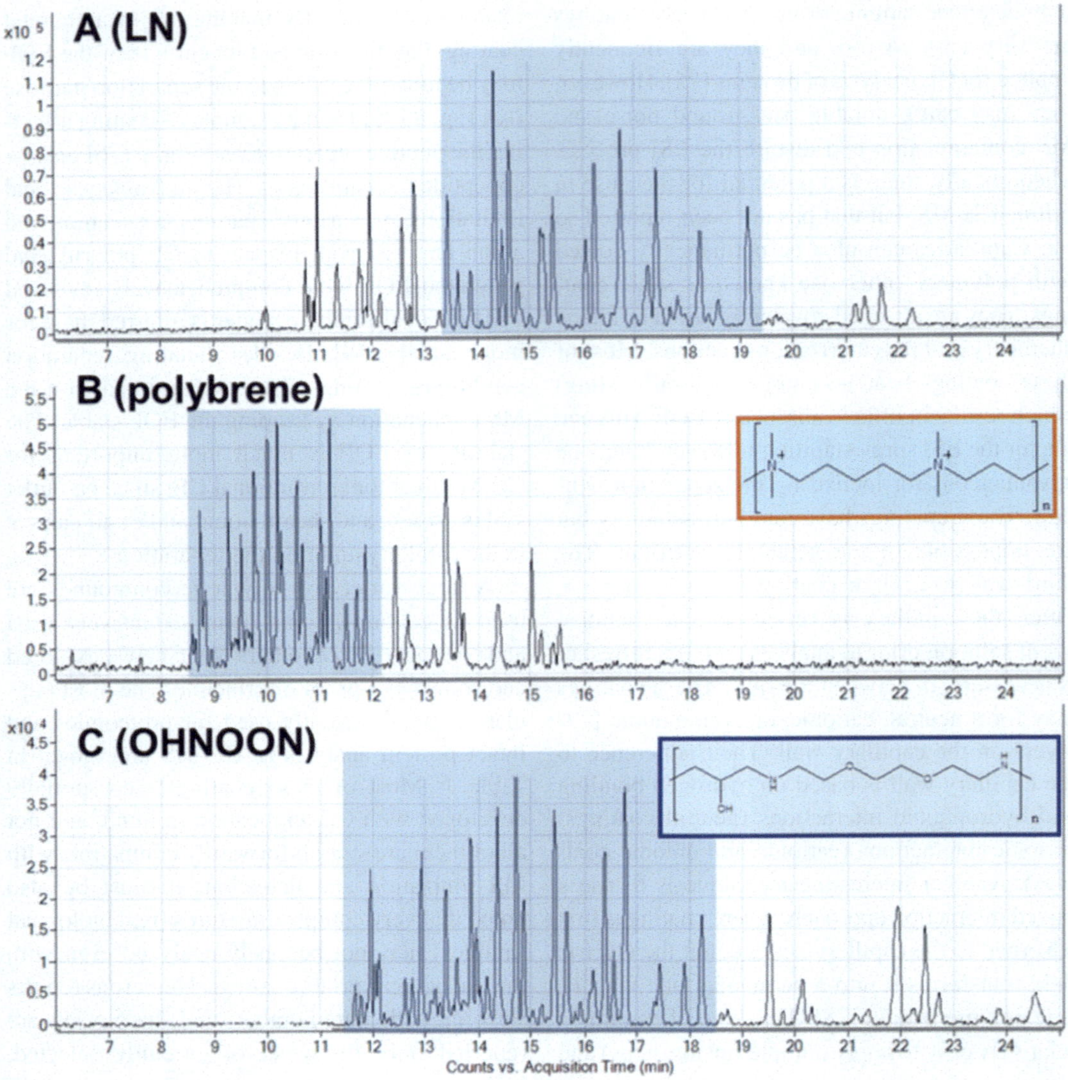

Fig. 8 CZE-ESI-MS analysis of a tryptic bovine serum albumin (BSA) digest using a capillary with (**a**) a neutral coating, (**b**) a high-density positively charged coating, and (**c**) a lower-density positively charged coating. Analytes were injected for 5 s at 50 mbar. The separation voltage was +30 kV for the neutral and −30 kV for the cationic coatings. Traces represent the base peak electropherogram constituted of the masses of all major BSA signals [76]

acquiring multiple positive or negative charges. Multiple charging is beneficial for proteomics applications, since it allows for detection of large peptides and full-size proteins, as well as give better results in protein/peptide fragmentation. The purpose of the ion source or interface is not only to facilitate ionization but also to complete and (preferably) decouple the electrical circuits of the CE and ESI-MS.

Generally, interfacing of a CE system with MS via an ESI source can be performed by a sheath-liquid or a sheathless approach (more details can be found in chapter "Capillary Electrophoresis-Mass Spectrometry for Metabolomics: Possibilities and Perspectives" of this book). The sheath liquid provides the electrical connection on the tip and outside of the separation capillary. The main advantage of this

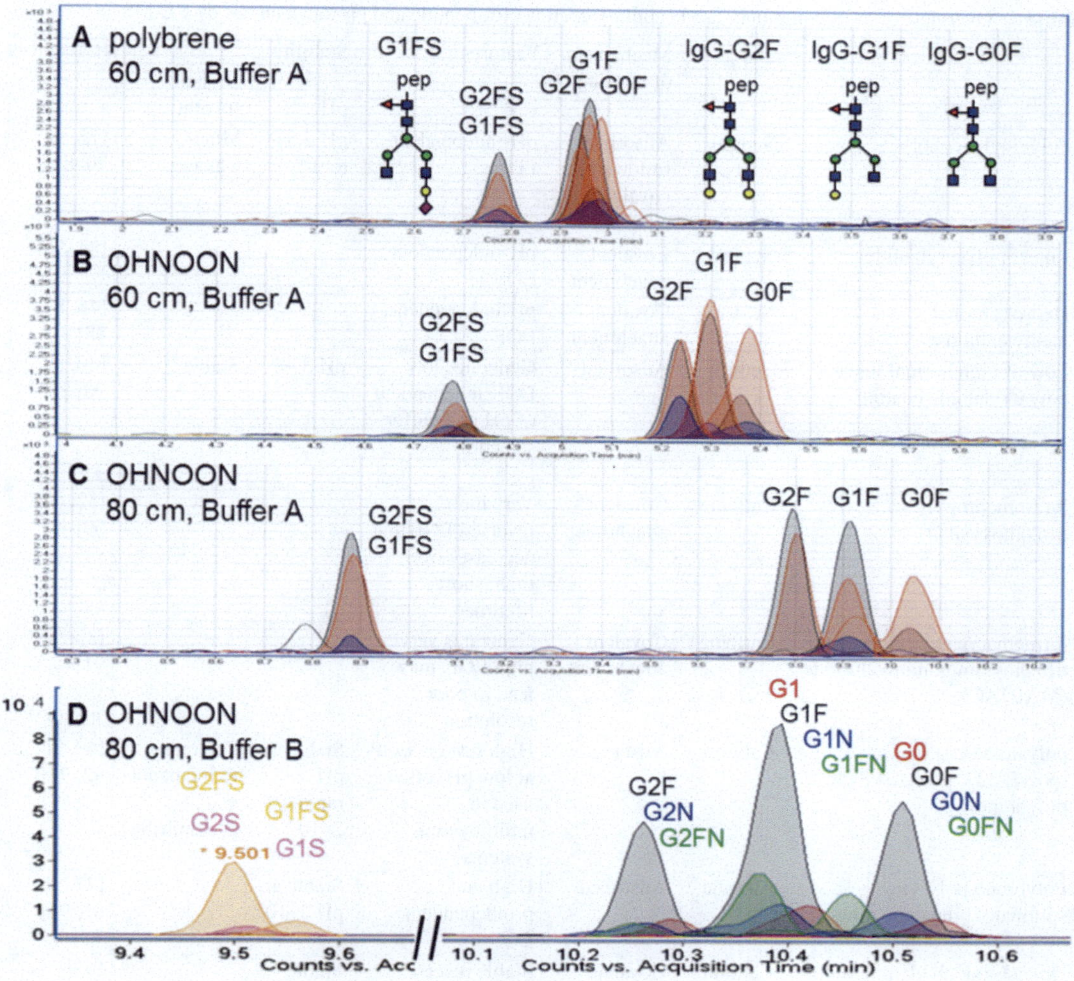

Fig. 9 CZE-ESI-MS separation of glycopeptides present in a human immunoglobulin G tryptic digest. Separations were performed on (**a**) a 60-cm capillary modified with a high-density positively charged coating, and (**b**) 60-cm or (**c**) 80-cm capillaries modified with a lower-density positively charged coating. Note the increase in migration time and resolution in the latter case. A further increase of separation efficiency could be achieved by switching from buffer A – 3:1 acetic acid/formic acid, 1 mol/L each – (A–C) to buffer B (**d**), which has the same composition but at 2 mol/L each [76]

approach is that the composition of the sheath liquid can be tuned to modify the ionization without changing CE selectivity and efficiency. For both peptide and protein analysis, the typical composition of a sheath liquid is a mixture of water and volatile organic modifier (methanol, acetonitrile, or isopropanol), often with addition of a volatile acid or base to improve the ionization efficiency. These parameters have shown to influence the signal intensity as well as the shape and position of the charge envelop. Therefore, pH, electrolyte concentration, and type and concentration of organic modifier used for the sheath liquid are utterly important variables. Additionally, unwanted CE buffer properties (described in the Sects. 2.2 and 2.3 of this chapter) could be balanced out by sheath liquid incorporation, although that results in sensitivity trade-off due to the sample dilution (CE effluent flows are in the nL/min, whereas sheath liquids are commonly applied in the μL/min range [97, 98]). Nonetheless, due to the robustness of the

Table 1 Overview of most commonly used capillary coatings for peptide and protein analysis by CE-MS

	Charge	Mechanism of crafting	Features	Stability		References
				pH	Organic solvent	
Polyvinyl alcohol	Neutral	Adsorbed followed by covalent attachment	pH-independent EOF	pH 2.5 to 9.5	Minor leakage	[75, 79–83]
Linear polyacrylamide	Neutral	Covalent attachment	pH-independent EOF			[84–86]
Hydroxypropyl methylcellulose	Neutral	Covalent attachment	pH-independent EOF			[46, 87, 88]
Low of high normal linear polyacrylamide coating	Neutral	Adsorbed	Generates low EOF, independent of pH and buffer selection	pH 3–9	Stable	[46, 89, 90]
3-(aminopropyl) trimethoxysilane	Cationic	Covalent attachment	Very high reversed EOF that decreases over time, short lifetime			[52, 64, 69]
3-(methacryloylaminopropyl) trimethylammonium chloride (MAPTAC)	Cationic	Covalent attachment	Generates very high EOF, may lead to poor resolution			[8, 69, 91]
polyamine coatings (polyE-323, poly-LA 313, polybrene)	Cationic	Adsorbed	High reverse EOF at low pH, often used in multilayered systems	Stable at pH range 2–10	Tolerance to methanol and acetonitrile	[69, 76, 92, 93]
Polybrene-poly(vinyl sulfonate) bilayer coating	Anionic	Adsorbed	High and p-independent EOF	Stable at pH 2–10		[37, 55, 81, 94]
Omega-iodoalkylammonium salts	Cationic	Covalent attachment	Stable reverse EOF	Stable up to pH 10		[95]

setup, it became widely acknowledged for CE-MS hyphenation. As an example, using this approach, LODs as low as 1 nmol/L were reached in protein digest analysis while having up to 12 times greater throughput than with nano-LC-MS [6]. Besides, extremely fast (<1 min) separations are easily hyphenated to MS using sheath-liquid interfacing without compromising the reproducibility and efficiency of the protein digest analysis [5].

The ability to dilute the undesirable compounds from CE effluent is not the only added value that sheath liquid interfacing can offer. Certain reagents could be added to the sheath liquid for promoting a particular chemical reaction upon mixing with the capillary effluent within the CE-MS interface. This process has been coined "sheath-flow chemistry." It may be used to enhance ionization rates for substances reluctant to acquire a charge. As a proof of concept, Bonvin et al. demonstrated possibility to manipulate intact proteins charge state by adding so-called supercharging reagents (3-nitrobenzyl alcohol and sulfolane) to the sheath liquid [99]. It reduces the *m/z* range required for protein analysis and (if applicable) facilitates further fragmentation of the molecule.

In liquid junction interface, the electrical connection is made through a small gap between the CE capillary and ESI spray needle. In some liquid-junction geometries, this gap also allows to add low volumes of a sheath liquid, thereby improving the CE effluent's MS compatibility. Expectedly, it also decreases the extent of the sample dilution; thus it is not surprising this interface has gained increased attention over the past 5 years [84, 100, 101]. Alongside this trend goes the development of a so-called electrokinetically pumped sheath-flow nanospray interface. It uses an electrokinetic flow to drive both the separation buffer and the sheath liquid, eliminating the necessity of a pump. It allows for low overall flow rates – nL/min range – and minimizes analyte dilution. The group of Dovichi is the main trailblazer for its development and application in proteomics: they demonstrate femtograms sensitivity for proteins and complex protein digests without compromising the robustness of the analysis [102–105].

The logical and, probably, final step for sensitivity gains is the direct coupling of the CE effluent to the ESI source and fully omitting the sheath liquid. In sheathless interfacing the CE terminating voltage is directly applied to the CE outlet. The required conductivity may be established by either metal coating of the capillary, by making the capillary tip porous or by adding a microelectrode in the CE system. In this case, the effluent flow is equal to the EOF one (ranges in tens of nL/min), and dilution effect is nonexistent. Moreover, since there are no connections – and thus dead volumes – separations are conserved in the most efficient manner. However, for all CE-MS separations with sheathless interfacing, the CE wet chemistry will have a direct influence on the spray performance (as described in Sect. 2.2), demanding a fine balance between optimal separation and efficient ionization. Many variants of sheathless interface designs were created and tested for both peptide and protein analysis [79, 106]. Even an "interface-free" approach – where the separation capillary and ESI tip are merged, requiring optimization of the CE current for obtaining a stable spray – have been suggested [107]. However, the most widely used and the only commercially available design is based on a porous tip design proposed by Moini in 2007 [98]. This interface quickly gained appreciation in the proteomics field for both sensitive analysis of intact proteins [108] and protein posttranslational modifications (PTMs), such as glycosylation [109], phosphorylation [110], and deamidation [19, 111]. This rather impressive performance in terms of sensitivity was mainly due to successful spray sustenance on very low CE effluent flow rates (below 20 nL/min), leading to low ionization suppression [112] and efficient sampling of the generated analyte vapor [113].

By design, spray properties in the sheathless interface cannot be tuned by an auxiliary liquid; therefore researchers have focused on gas phase solutions. For example, a dopant-enriched coaxial gas flow can be supplied to the ESI source [114]. By saturating the nitrogen gas with acetonitrile dopant flow, an enhancement of both ionization efficiency and spray stability is achieved. When that was integrated in a sheathless CE-MS approach for protein digest analysis, the signal intensity for peptides, and especially glycopeptides, increased twofold on average. The relative abundance of the analytes remained unaffected, showing that the effect was not analyte-dependent. Moreover, it decreased the noise and mass spectral interference levels and, concomitantly improved the measurement repeatability.

Few studies have focused on the direct comparison of the aforementioned interfaces for peptide and protein analysis. For example, Haselberg et al. pointed at a significant sensitivity gain (reaching sub-nmol/L LODs) in the analysis of intact proteins with a sheathless interface when compared to a sheath-liquid interface [108]. A direct and thorough comparison of the interface types was provided by the Neusüß's group [89]. They evaluated the performance of the nanoflow sheath liquid interface and compared it to a conventional sheath liquid setup and porous-tip sheathless interface in both positive and negative modes. Three types of analytes were used for the test, including tryptic peptides and a full-size monoclonal antibody in the reduced form. The nanoflow sheath liquid interface and the sheathless interface appeared to be comparable in sensitivity and both outperformed the sheath liquid interface. Independence of operating conditions for separation and electrospray ionization, that ensured full flexibility in method development, was mentioned as an added value of the nanoflow sheath liquid setup.

Undoubtedly, ESI is the most frequently used ion source when it comes to the CE-MS analysis of peptides and proteins. However, MALDI – another soft ionization technique – is also occasionally used in CE-MS-based proteomics. Since the ion source is physically separated from the CE separation, it allows for more flexibility in pH, ionic strength, and volatility of the BGE components choice [80, 115]. Additionally, offline detection does not press limits on the MS duty cycle and overall analysis time for data dependent mass analysis. The latter can be an issue as CE-ESI-MS is typically dealing with narrow electrophoretic peaks, and short duty cycles are preferred. On top of that, offline hyphenation gives room for additional sample treatment prior to MS analysis: for example, enzymatic digestion [116] or even sample enrichment by means of multiple analyte spotting [117]. Though, as any offline workflow, it may become labor-intensive and time-consuming; that predestined the further developments towards robotization. Ways to reduce sample deposition time include on-target fraction collection [118], using plates pre-coated with matrix, adding matrix to the sheath liquid used for deposition, and depositing analyte as a continuous flow instead of single droplets to avoid the CE resolution loss [119]. Overall, the developments towards online CE-MALDI-MS appear to be rather limited, albeit creative [120, 121]. A fascinating application of CE-MALDI-MS is reported by Rogowska and colleagues. They evaluated the changes of the electrophoretic mobility of whole yeast cells and subsequently deposited them on a target for a MALDI-MS analysis of the molecular profile using Biotyper platform. They showed that the electrophoretic separation mechanism relied on surface structural changes of *S. cerevisiae* subtypes [122].

2.5 Mass Analyzers and Fragmentation

In order to obtain specific mass spectrometric information from peptides and proteins, the choice of mass analyzer may be critical. Initially in CE-ESI-MS, quadrupole and ion trap mass analyzers were used. These mass analyzers typically can cover a mass range of up to *m/z* 4000, allowing detection of multiple charged analytes as obtained from ESI. However, the disadvantage of these mass analyzers is their relatively low resolution (about 0.5–1 mass unit) in the 100–2000 *m/z* range. This means that deconvolution of the mass spectra does not lead to a highly accurate mass determination and that it is impossible to distinguish compounds that are very close in molecular weight.

Over the last decade, high-resolution mass analyzers in CE-ESI-MS have been increasingly used. Currently, time-of-flight (TOF)-MS is most commonly applied for the analysis of intact proteins. Modern TOF-MS systems can achieve resolutions between 30,000 and 80,000. Combined with high mass accuracies (low ppm) provided by these instruments, this means that 30–80 kDa proteins can be analyzed with unit mass resolution allowing modifications as small as a deamidation (leading to a mass difference of 1 Da) to be confidently characterized. Fourier transform ion cyclotron resonance (FTICR) and Orbitrap mass analyzers have also been used occasionally for intact protein analysis by CE-ESI-MS. On the other hand, they are very efficient when it comes to bulk peptide analysis. Comigrating peptides with similar mass to charge ratios may not be precisely identified and/or quantified in the MS due to mass overlapping, but higher resolution and mass accuracy are capable of keeping the ambiguity off [123]. Furthermore, dynamic range and coverage of proteome analysis are a function of both the quality of the separation method and the MS system; thus poor MS performance may compromise even flawless CE separation [124].

Orbitrap and FTICR mass analyzers can provide extremely high resolutions (above 500,000 or even more). The downside is their longer duty cycles, which might compromise the achievable sensitivity of the system and fail to provide an adequate sampling of narrow CE peaks.

If analyte molecules get conditionally preselected and further fragmented in a mass spectrometer, the analysis is referred to as MS/MS or tandem MS. In proteomics applications it is mostly used for peptide sequencing and PTM mapping (more information on common PTMs is provided in Sect. 3.1 of this chapter). Currently almost all the known types of MS instruments can be coupled to a CE system, so in theory any type of fragmentation is possible. This encompasses "classical" collision-induced dissociation (CID), alongside with higher-energy collision dissociation (HCD) and electron transfer dissociation (ETD) that are shown to improve identification of long, highly charged, or strongly basic peptides, as well as ones bearing many PTMs [125]. Quality fragmentation (as well as overall MS performance) is especially crucial for sequencing of native peptides and full peptidome identification. Protein identification studies in classical proteomics have mass fingerprint data available for the tryptic digests, and reliable fragmentation patterns of two peptides may be enough for confident protein assignment. When dealing with peptidome, every peptide species has to undergo successful and full fragmentation and interpretation of MS/MS data. That motivated Mischak's group for a comparative performance study of different CE-coupled peptide sequencing platforms using urinary peptidome as a reference analyte [126]. CE-MALDI-TOF-TOF, CE-ESI-IT, CE-ESI-QTOF, and CE-ESI-LTQ-FTICR were compared, concluding FTICR instruments are to deliver the best fragmentation possibilities for the full and confident identification of peptide sequences. However, due to high instrument costs, MALDI-TOF-MS is the second best equipment by ion coverage when compared to the yields of the other CE-ESI-MS platforms tested.

However, intact protein fragmentation has long been a challenge. Over the last years, techniques that enable efficient intact protein fragmentation – like ETD and HCD – have been introduced. Over the last few years, a few studies have demonstrated that efficient CE separations can be combined with a top-down proteomics approach [127–132]. Besides evaluating this approach on protein standards, top-down proteomics has been applied to characterize unknown proteins in cell lysates [127, 130], a cell secretome [132], and culture filtrates [128], as well as to determine protein phosphorylation sites [130]. Whereas the top-down approach does allow identification of several dozen [127, 128, 131, 132] up to hundreds of proteins [130], unambiguously pinpointing site-specific modifications still proves troublesome [131]. It should be noted that the time to generate good-quality fragmentation spectrum is limited to the protein peak width (often around 0.5 min). Currently, top-down techniques still require long acquisition times (up to tens of minutes), so complete sequence coverage is often not obtained. Low-molecular-weight proteins are most confidently identified, although identifications of proteins with a molecular weight up to 80 kDa have already been reported [127]. To improve sequence coverage, a combination of several top-down fragmentation approaches might be used. For example, an ETD accompanied with HCD could improve intact protein identification [132] and was subsequently used to identify proteins in a cell secretome. Similar to the model proteins, an increase in sequence coverage was obtained, and a larger set of proteins was identified compared to separate ETD and HCD experiments.

3 Applications

Since the first introduction of CE-MS in the late 1980s of the previous century, many different peptide and protein-centered studies have been performed. This cannot all be discussed in this chapter. Therefore, for a more comprehensive overview of reported applications of proteomics and intact protein CE-MS, the reader is referred to relevant reviews [133–137]. In this chapter we only focus on a few important fields that have shown a significant body of work over the years, i.e., characterization of posttranslational modifications (PTMs), biopharmaceutical development, and biomarkers and clinical applications.

3.1 PTM Detection and Characterization

Since PTMs are an integral part of protein structure and are crucial for its stability and proper biological functioning, PTM characterization and profiling are an essential part of proteomics. The most common way to tackle the full PTM complexity is to generate a peptide pool by enzymatic digestion. As discussed previously in this chapter, the complexity of such samples is tremendous, which makes it hard to analyze them directly by MS. Therefore, methods to reduce complexity are vital for accessing as many modified peptides as possible while keeping sufficient signal intensities. In this setting, CE has many benefits to offer. Some PTMs are relatively bulky and increase the ionic radius, reducing the electrophoretic velocity of the peptide. Other PTMs alter both peptide mass and charge causing changes in the electrophoretic mobility. Even site-specific effects can shift the overall charge allowing CE separation of positional isomers [138]. Currently, CE-MS is an established and much-proven method to characterize PTMs like phosphorylation, deamidation, and glycosylation (and its nonenzymatic counterpart – glycation). Interestingly, this is not only true for the peptide level – they can also be relatively well-characterized on the intact protein level.

Phosphorylation affects the electrophoretic mobility of a peptide by adding a negative charge that is proportional to the number of phospho-groups present in the protein structure. Consequently, phospho-variants profile of the same peptide can be readily achieved. ESI-MS analysis of phosphorylated peptides is admittedly rather challenging, since their ionization efficiency appears to be lower than the non-modified counterparts [139]. On this issue, Heemskerk et al. presented a performance evaluation for ultralow-flow CE-ESI-MS system [110]. They report a significant increase in phosphopeptide ionization at flow rates around 15 nL/min. In the same work, they also demonstrate increase in sensitivity with introducing a t-ITP step, which allowed for injecting sample volumes up to 50% of the total capillary volume, while maintaining satisfactory resolution. Interestingly, in their work they used neutrally coated capillary to eliminate the EOF, which may be not the first choice for actual phosphopeptide identification. The work of Faserl et al. on capillary coatings and the overall CE-MS performance on PTM analysis named bare-fused silica the best option for multiply phosphorylated peptides [140]. They showed the added value of CE-MS analysis over nano-LC-MS for phosphopeptide identification in complex digests, since only 31% of identified peptides were detected by both methods. Furthermore, site-specific separation of phosphopeptide isomers can also be performed with CE-MS, as demonstrated by Dong and colleagues on the model panel of synthetic phosphopeptides [29]. CE-MS is also capable of performing quantitative phosphoproteomics on SILAC-labelled protein digests (Fig. 10) [138]. The analysis shown involved the pre-fractionation on RP-HPLC system with subsequent analysis of collected fractions on CE-MS, both of which are fully automated. To shift the peptides towards low-complexity regions of the chromatogram (i.e., lower their velocity and separate them from their unmodified peptide counterpart), an acetylation step was first performed. In that manner, sensitivity for low-abundance phosphopeptides was significantly increased. Low sample consumption, quick sample pre-treatment, and possibility to map extra modifications (acetylation, deamidation, oxidation) were listed as other benefits of the workflow. That methodology was reapplied in their following paper [140] with an extra step of phosphopeptide enrichment. Additionally, 2D-CE-MS coupled to an alkaline phosphatase microreactor was demonstrated to determine the peptide phosphorylation stoichiometry [141]. In the first dimension, unmodified peptides were separated from phosphorylated ones. The second dimension was subsequently used to separate and identify peptides after enzy-

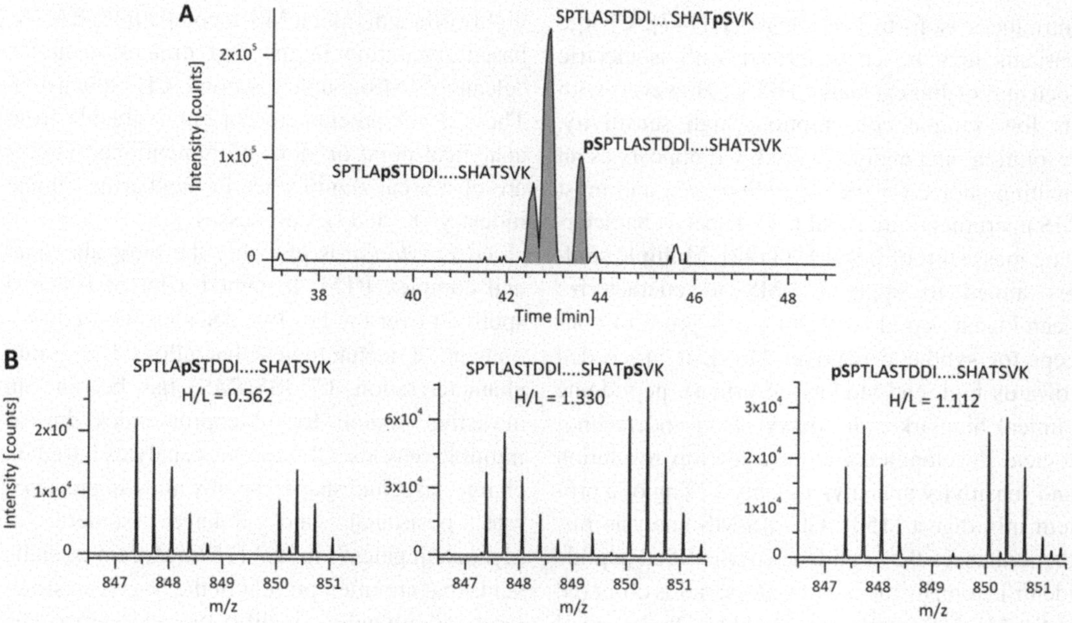

Fig. 10 CZE-ESI-MS analysis of phosphopeptide isomers. (**a**) Extracted ion electropherogram (m/z 850.2786) of the isobaric mono-phosphorylated peptide SPTLASTDDINSASASVNSHATSVK. (**b**) Using SILAC-labelled peptides (H) allowed quantification of the unlabelled (L) isobaric peptides. Reliable quantification of all three mono-phosphorylated peptides was achieved (ratios ranging from H/L = 0.562 to 1.330) [138]

matic pre-treatment. The ratio of the two fractions would tell the phosphorylation stoichiometry rates.

Promising attempts to elucidate phosphorylation pattern of the proteins are also made via top-down approach. The group of Norman Dovichi used CZE-ESI-MS/MS on a mixture of model proteins with deliberately added impurities, and all but one analyte were successfully profiled after HCD fragmentation [129]. Alongside with that, five modified phosphorylation sites of β-casein were identified using this method. Using similar methodology, Yates group has managed to go high scale in size and reported sensitive protein complex characterization with identifying processed forms of subunits and overall phosphorylation stoichiometry [131]. Summarizing the facts above, one may assert CE-MS systems to be a mighty and versatile tool to tackle peptide/protein phosphorylation, especially for discerning the phosphoisoforms.

Deamidation is a ubiquitous posttranslational modification, affecting L-asparagine (Asn) and to a lesser extent L-glutamine (Gln) residues of the protein sequence [142]. Deamidation appears to act as "molecular clock" for protein turnover and organism's aging processes [143, 144]. Simply put, proteins accumulating deamidated amino acid residues are likely to change their folding state and conformation, thereby will be predestined to utilization or destruction via proteasome degradation. Rates of deamidation vary for any individual amino acid residue and strongly depend on external factors (pH, temperature, ionic strength, and buffer ions) and protein 3D structure [145]. This PTM is largely studied for its functional role in biological systems, but, also, it is a point of concern in biopharmaceutics industry. Manufacturing and storage conditions of a therapeutic protein often lead to sufficient deamidation rates, compromising the efficiency and safety of the product [146, 147]. Thus, quite expectedly, the quality assurance process has to necessarily include deamidation characterization.

Deamidation of asparagine to aspartate, or isoaspartate, increases the negative charge of the protein or peptide, shifts its isoelectric point, and introduces isoform heterogeneity [147]. Charge variants may be characterized with isoelectric focusing or ion exchange HPLC. However, with its low sample consumption, high sensitivity, resolution, and analysis speed, CE appears as an inviting choice for the separation task, and most MS instruments are capable to detect a characteristic mass shift of 0.984 Da [142]. Multiple studies aimed to apply CE-MS to characterize deamidated peptides. Starting as a proof of concept for synthetic peptides [148], it proceeded towards high-end studies on urinary peptidome clinical biomarker discovery [149] and technological developments aiming for top resolution and sensitivity in analyzing only 129 ng of a protein mix digest [150]. CE-MS/MS analysis further enhances the confidence of modified peptide identification by the use of isotope ratios observed in the MS/MS fragment ions [111]. On the top of that, Gennaro et al. have presented a comparison of preparative RPLC-MS performance to CE-MS for the peptide mapping and deamidation analysis of protein digests, thereby asserting superior performance of the latter [151].

One should consider that the conditions of enzymatic digestion may actually induce deamidation processes, masking the initial state of the protein [142]. Therefore, it is not surprising to see a rising number of CE-MS-based deamidation studies on intact protein level, although they bear their own challenges. The deamidation mass shift is relatively small; hence MS detection has to be performed on high-resolution instruments (Rs > 40,000 for small proteins) offering high-quality isotope distributions [136]. Additionally, as for most CE separations, a capillary coating might be considered to prevent protein adsorption on silica (see Sect. 2.3 of this chapter). With these prerequisites, multiple deamidation studies of intact proteins with CE-MS(/MS) have been successfully conducted. Isoform and charge variant characterization (also encompassing glycosylation heterogeneity) were established for interferon-β1 [152], human growth hormone [153], human erythropoietin [154], ricin toxin [155], and last but not least monoclonal antibodies (mAbs) [150, 156]. Even a 2D CE-MS separation was reported for mAb charge variants with highly efficient (albeit MS-incompatible) EACA-based separation in the first dimension and a "cleanup" MS-coupled second CE dimension. These developments are not only valuable from analytical point of view: aforementioned targets are of a great significance in biopharmaceutical industry QC and QA processes.

Glycosylation is probably the most abundant and complex PTM. It gained a lot of research spotlight over the last two decades due to development of technologies that allowed its rapid characterization. CE-MS(/MS) has become an attractive system for glycoproteomics due to multiple reasons. CE separates analytes based on charge, size, and shape, thereby allowing to separate positional and linkage isomers of glycoconjugates (Fig. 11) [157]. Moreover, sialic acids that are often present in the N-glycan structures additionally modify the electrophoretic mobility of the molecules. CE is uniquely useful for monitoring small and hydrophilic glycopeptides which are troublesome to ionize in LC-MS-based proteomics [158]. A lot of work has been done in this area, which cannot all be covered in here. Therefore, the reader is referred to a recent comprehensive review of methods and applications for sialoglycosylation analysis [159]. The authors cover both glycopeptide and intact protein levels of analysis and highlight the added value of CE separation system. The majority of glycoprofiling studies were performed on total enzymatically released N-glycan pool of a certain sample, the so-called released glycan approach. In that case, the information on the microheterogeneity – location of a certain glycan subset on a certain asparagine residue – gets lost. That is often considered to be a vital piece of data, for example, in biomarker research [160], antibody quality control assurance [161], and structure-function relationships of glycoproteins [162]. Alongside with that, if the sample is complex (i.e., harbors more than one glycoprotein), released glycans cannot be correlated to a particular protein. Here, a glycopeptide-based approach comes of a great use, allowing to map glycan structures on a protein of interest. Since the physicochemical properties of a given glycopeptide depend on both glycan and peptide portion,

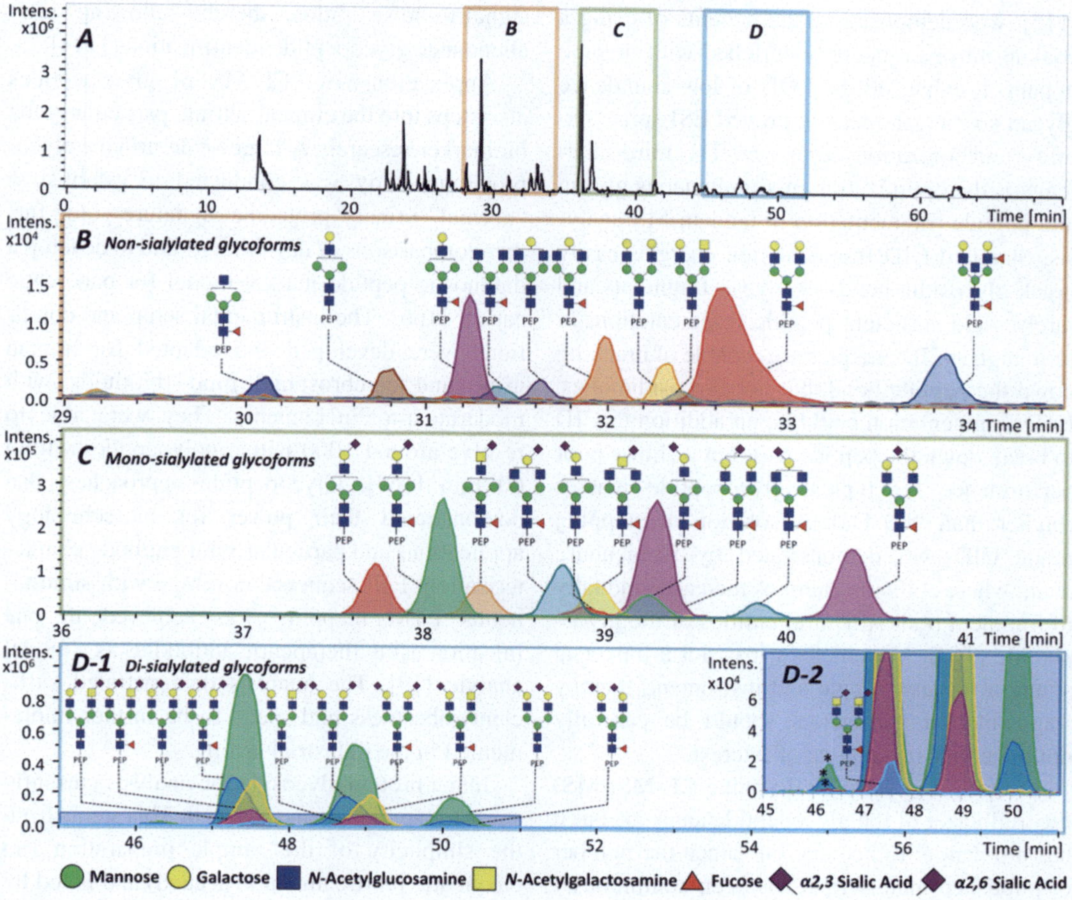

Fig. 11 CE-ESI-MS analysis of prostate-specific antigen (PSA) tryptic (glyco)peptides enabling the differentiation between α2,3- and α2,6-sialylated glycopeptides without prior derivatization. (**a**) Representative base peak electropherogram observed for a tryptic digest of PSA. (**b**) Separation of non-sialylated glycopeptides. (**c**) Separation of differentially linked mono-sialylated glycopeptides. (**d**) Separation of differentially linked di-sialylated glycopeptides. A total of 75 different glycopeptides and differentially linked variants were identified. The "PEP" label illustrates the tryptic peptide sequence N $_{69}$K to which the glycan is attached [157]

CE-MS is better than conventional reversed-phase LC-MS, which is common for proteomics analyses. Besides, the latter may result in broad peaks [81] and also requires larger sample amounts.

Glycopeptide analysis poses a challenge for three main reasons. First, it tremendously increases the complexity of the sample on separation and data analysis steps. The latter is based on the fact that the majority of proteomics software is peptide-centric and is not well-designed to annotate complex, sometimes chemically modified, PTM attachments. Only recently program solutions started to emerge, namely, GlycoQuest (Bruker Daltonics) and Byonic (Protein Metrics); however, manual curation is still a must for the majority of complex samples. For more information on current state-of-art strategies for interpretation of glycopeptide tandem mass spectral data, the reader is referred to a review article by Hu and colleagues [163]. Second, a glycopeptide approach results in more diversion of analyzed molecules, therefore complicating the identification of low-abundance species "hidden in the grass." Attempts to tackle this limitation involve increasing of CE-MS sensitivity and lowering the signal-to-noise ratio. A way to go about this issue was reported by Kammeijer and colleagues

[114], who demonstrated the benefits of using a coaxial nitrogen gas flow enriched with organic dopant. It enhanced the LOD of low-abundance glycan species and also improved ESI spray stability and ionization efficiency. The third challenge is the optimization of simultaneous glycan and peptide fragmentation in tandem MS analysis. Standard CID fragmentation energies easily break glycosidic bonds of glycan fragments and rarely yield sufficient peptide fragmentation. In turn, higher CID energies are capable of breaking down the peptide bond, but not glycosidic ones. ETD fragmentation could be an addition to CID to break down the peptide part, but it shows poor performance for typical glycopeptide masses (higher than 900 Da). A solution of stepping mode CID was demonstrated by Hinnenburg et al., where collision energies increase gradually to fragment both structural entities of the glycopeptide [164]. Nonetheless, it is not a universal solution for any sample and instrument, thereby fragmentation parameters should be carefully optimized for the analyte of interest.

Overall, it is yet early to name CE-MS(/MS) the trailblazer of the glycopeptide analysis, but it has a potential to become one since the number of published manuscripts has been continuously increasing every year. The standardized general conditions for the CZE-ESI-MS and CZE-MALDI-MS analysis of glycopeptides described in detail by Amon and colleagues paved the way for controlled and efficient separations, focusing on coatings influencing the EOF control [165]. The current state-of-the-art method is proposed by Dovichi's group, offering CZE-ESI-MS analysis with LODs in the low ng/μL range, minimal sample consumption, and impressively short analysis time of 9 min [105]. The developments for the in-depth analysis go alongside with better separation approaches. Where initially only main glycoforms were separated, now even positional glycan isomers can be separated on the glycopeptide level. Kammeijer et al. demonstrated a baseline separation method for α2,3- and α2,6-sialylated glycopeptides without prederivatization of the sample [157]. This method also provided increased sensitivity and lowered signal-to-noise ratios, thereby allowing low-abundance glycopeptide identifications [114].

Application-wise, CE-MS of glycopeptides also steps into the clinical setting, particularly the biomarker research. A large-scale urinary glycoproteome study was conducted to establish a "normal" urinary proteome signature – for further comparisons of deviations – and to develop a diagnostic peptide marker model for pancreatic cancer [166]. The instrumental setup and conditions were developed and adapted for human urine and cerebrospinal fluid (biofluids with moderate protein content). They were able to resolve around 1000 native polypeptides within 60 min [167]. Glycopeptide approaches also demonstrated their power for biotechnology applications and particularly for antibody characterization. Full sequence coverage with simultaneous PTM mapping was achieved in one injection using therapeutic antibodies as a model analyte [19]. The analysis demonstrated sufficient robustness and speed to be further implemented in the industrial setting.

Intact protein glycosylation analysis currently starts to rapidly advance as well. This stems from the simplicity of the sample preparation (no enzymatic pre-treatment is needed) and is fed by hardware technical developments of high-accuracy and high-resolution MS, as well as good electrophoretic resolution in CE. Glycotyping involves characterization of the number and relative abundancies of glycoforms, as well as characterization of associated oligosaccharide structures. Whereas the latter is better performed in glycopeptide-based manner, the former is better investigated at the intact glycoprotein level as it does not introduce the sample preparation bias. That is of a great importance for biopharmaceuticals, mainly for recombinant antibodies and antibody-based therapeutics (as will be discussed in the next section). For example, Han et al. presented a proof-of-concept CE-MS method to analyze degradation products and glycovariants of non-reduced antibodies, which is robust, efficient, and durable enough to be used in routine monitoring. In parallel, Haselberg with colleagues also delivered a low-flow sheathless

CE-MS method for stability and purity assessment of antibody-derived therapeutics. The method showed sufficient robustness (overall migration time RSDs below 2.2%), capacity to separate isomeric deamidated products, as well as reliably resolved sialylated glycoforms from their non-sialylated counterparts [168]. With increasing resolution and accuracy of the CE-MS instrumentation, intact protein level analysis of glycoprotein samples certainly has a potential to be the main technique for routine analyses, albeit yet is credible only for mildly glycosylated proteins.

3.2 Biopharmaceutical Development and QC

To develop a protein biopharmaceutical possessing certain physical, chemical, and biological properties and to ensure its stability in storage and use, one should first focus on the consistency of its amino acid sequence. That should be done in a quick and robust manner, and the possibility to get the separation within minutes, as would be feasible in CE, seems ideal. However, its use still remains limited, since more development was initially put into LC-MS systems, due to its initial coupling straightforwardness and better stability of operation. Nonetheless, more and more attempts towards CE-MS application in QC of biopharmaceuticals have been undertaken. As mentioned in the previous paragraph, alongside with resolving protein sequence consistency, CE-MS offers a possibility to simultaneously assess modifications, such as asparagine deamidation, methionine oxidation, C-terminal glutamic acid cyclization and aspartic acid isomerization. In each case, the modified peptides could be baseline separated from the intact counterpart, detecting modification levels as low as 2% [169]. The evaluation of these PTMs is essential, as they do influence the stability, half-life, and performance of the biopharmaceutical.

Erythropoietin (EPO) was the first biopharmaceutical extensively analyzed by CZE-ESI-MS. This protein is heavily glycosylated and therefore intrinsically hard to characterize. EOF suppression resulted in very high separation efficiencies of closely related recombinant human EPO glycoforms on the intact protein level [113, 154]. The separation was achieved mainly due to differences in amount of sialic acid residues. Differences in the hexose-N-acetyl-hexoseamine content also lead to small shifts in electrophoretic mobility and thus partial separation. Overall, more than 250 different isoforms, including glycosylation, oxidation, and acetylation products, could be distinguished in one CE-MS run of EPO [136]. The same protein has also been characterized on the peptide level, demonstrating the site-specific microheterogeneity of the glycosylation [170].

As just described the peptide-based approach still provides the most comprehensive, deep, and detailed analysis of a protein. The overall workflow time is significantly prolonged with sample pre-treatment steps but gives the reward of site-specific modification assignment. In that manner, Gahoual et al. have achieved full antibody peptide mapping simultaneously with microvariant characterization for four therapeutic monoclonal antibodies (mAbs) in a single injection [19]. To achieve that, they used a modified (surfactant-assisted) proteolytic scheme and analyzed the digest using sheathless CE-ESI-MS system with t-ITP preconcentration. This way they managed to lower down the required sample amount to 200 fmol of the digest. CE-MS of peptides also offers an advantage over LC-MS with the opportunity to fully separate isomers of aspartic acid (a marker of protein degradation) [82]. Without an efficient separation, these isomers could not have been discriminated since this modification does not change the net mass of the peptide.

A combination of an intact, middle-up, and bottom-up techniques shows great promise to characterize the total amount of modifications and its location concomitantly [171]. Such an approach is wholesome for biopharmaceuticals like antibody-drug conjugate (ADC) and serves as a requirement for their proper characterization [172]. The information about average molecular weight of ADC, drug to antibody ratio, and drug distribution is obtained time-efficiently by analyzing them on the intact level. Middle-up approaches establish the co-occurrence of drug attachment on a certain fragment. And last but not least, a bottom-up approach gives the information of the site-specific distribution alongside with peptide mapping of the protein backbone. This pipeline is essential for the development and quality assurance phases but might be too time and labor consuming for the routine quality control and monitoring. To gain throughput on that phase, the analysis of intact biopharmaceutical molecules comes into play. Alongside with the mass analysis to resolve truncated forms of the protein molecule [173, 174], intact mAbs may be resolved on the basics of their glycosylation profile and isomeric deamidation [168, 175]. Microfluidic integrated platforms also start to emerge to increase the throughput. The use of such a platform was demonstrated by Redman and colleagues to characterize lysine-linked ADCs [156]. The minimal sample preparation and short analysis time allowed drug load assessment, charge variant determination, and characterization of glycan heterogeneity.

To conclude, the intrinsic technical challenge of robust CE-MS coupling and analysis is being more and more addressed by analytical scientists, paving the CE-MS way to the biopharmaceutical industry. Nowadays, commercial solutions are being offered for antibody and ADC characterization, such as Thermo Scientific Q Exactive hybrid quadrupole-Orbitrap mass spectrometers with ZipChip system. The analytical method development, exploiting more and better hardware solutions, does not stop either, being mostly directed towards getting more valuable information from the intact level analysis. In light of all of this, a collaborative study on the robustness and portability of CE-MS can be mentioned as a notable undertaking. It was performed by an international team, consisting of 13 independent laboratories from both academia and industry [176]. All participants analyzed the samples from the same batch and used identical reagents and coated capillaries to run their assays; however, the analysis itself was performed on the equipment available in their laboratories. Migration time, peak height and peak area of ten representative target peptides of trypsin-digested bovine serum albumin were determined by every laboratory on two consecutive days. The results demonstrated that CE-MS is robust enough to allow a method transfer across multiple laboratories and should promote a more widespread use of peptide mapping and other CE-MS applications in biopharmaceutical analysis and related fields.

3.3 Biomarker Discovery and Clinical Application

As CE-MS is a powerful analytical technique, its appearance on the biomarker research scene is not surprising. It allows for identification of broader peptide range (if compared to the LC separations) and exhibits almost no sample carryover phenomena. Also, the reliability of the platform combined with worked-through data processing and mining methods, to date, makes CE-MS the most advanced technique for biomarker discovery of clinical significance. Most of the developments currently address the urinary proteome as a source of new biomarkers. Protein concentrations in urine are very moderate (when compared to serum); thus it does not require complex pre-purification steps and can be almost directly analyzed by CE-MS. Additionally, urine can be sampled in large quantities and in noninvasive manner, thus allowing enough material for hitting the concentration limit.

A big portion of studies focuses on the assessment of the urinary peptidome. Typical sample preparation involves a short pre-cleanup to eliminate large proteins, accompanied by buffer exchange. The remaining fraction is subsequently analyzed by CE-MS. Studies centering around kidney malfunctions are the most obvious area to find urinary biomarkers. For example, a multimarker model built to assess kidney injury and based on peptides assessed by CE-MS was proven viable by Metzger et al., revealing the intensity of only two peptides, which allowed for early and accurate prediction of acute kidney injury [177]. The same condition appeared to produce another set of peptidome signatures that were shown to predict kidney function improvement [178]. Urinary peptidome studies have also demonstrated their use for the diagnostics and patient stratification in other diseases, since 30% of the urinary proteins are derived from serum [179]. To exemplify, systematic lupus erythematosus [180] and various types of cancer [181] have been diagnosed using a peptide marker panel measured by CE-MS. One of the peptide biomarker panels, termed CKD273, has been successfully translated into the clinic and received an FDA approval for management of chronic kidney disease [182]. Despite being more challenging to handle, the urinary proteome also gained quite some attention as a protein biomarker source. The applications range from total protein fraction evaluation in patients with heart failure [183] to in-depth study of urinary PSA isoforms in prostate cancer [184].

The spectrum of other biofluids to be used as a protein/peptide biomarker source is very broad, namely, cerebrospinal fluid [185], saliva [186], seminal plasma [187], bile [188], and certainly blood plasma [38]. Some body fluids catalogued as "less-conventional" (cerumen, breast milk, and more) were also recently reviewed for their biomarker potential [189]. Typically, the CE-MS analysis itself remains relatively standardized for all the fluids, but the sample preparation differs dramatically. Plasma and serum pose the biggest challenge for their high dynamic range between high- and low-abundance proteins, which no contemporary instrument can handle directly. Ironically, the low-abundance proteins frequently hold the biggest biomarker potential. Therefore, strategies for abundant protein depletion are vital for plasma proteins [190] and other biofluids, alongside with other strategies of sample pretreatment. The latter issue has been recently protocolled by Mischak's research group, alongside with a roadmap and considerations for the analysis of proteomics and peptidomics biomarkers in biofluids [191]. For the plasma proteome biomarker studies, it is often important to combine background protein depletion with target protein enrichment. Here immunoaffinity and lectin enrichment SPE techniques may be engaged either in on-line manner [38] or performed offline [158].

4 Conclusions and Future Perspectives

CE-MS is masterfully climbing towards its rightful analytical niche in proteomics and intact protein analysis, being not a rival but a helping friend of LC-MS. Recalling CE advantageous features – high sensitivity and resolving power, low sample and solvent consumption, short analysis times, and many options to tune the separation – makes CE-MS a solid choice for protein samples characterization on multiple levels of complexity. CE-MS offers superb performance for PTM characterizations of therapeutic proteins, and it has already proven to be a promising tool in biomarker discovery and validation (as amply described in Sect. 3.3 of this chapter). The necessary step towards vast clinical implementation is the demonstration of robustness and validity of a certain analytical method, and CE-MS seems to already pass this milestone, being recognized for its reproducibility, stability, sensitivity and interlaboratory applicability and well-established data processing pipelines. We expect further developments of the system robustness, especially for complex setups with sample enrichment and/or cleanup and multidimensional separations. Ongoing implementation of microchip-based CE (MCE) separations presents a lucrative opportunity for commercialization, by lowering costs and improving user experience. Market-available MCE devices yet avoid MS coupling but already show potential in clinical setting. To bring an example, the 2100 Bioanalyzer MCGE system has been utilized for therapeutic protein analysis, including glycoprotein heterogeneity assessment [192]. The same platform has been implemented for C-reactive protein (CRP) detection as a sepsis marker on clinical serum samples [193]. The implementation of MS would help resolve issues with complex sample preparation or low accuracy of MW determination. Furthermore, researchers do highlight the problem of protein adsorption in CE/MCE systems and expect more research to be devoted towards novel coating materials development [194].

As a purely analytical tool for proteomics, CE-MS has plenty of high-end workflows, for example, those on intact protein (structural) analysis and single-cell "omics," that are yet presented only as a proof of concept with all the analyses performed on test proteins or peptides [117, 174, 195–197]. Given the impressive outcomes, we hope to see and possibly contribute to their development and use for real biological samples analyses.

References

1. Saraswathy N, Ramalingam P, Saraswathy N, Ramalingam P (2011) Mass spectrometry for proteomics. Concepts Tech Genomics Proteomics:171–183. https://doi.org/10.1533/9781908818058.171
2. Velghe S, Capiau S, Stove CP (2016) Opening the toolbox of alternative sampling strategies in clinical routine: a key-role for (LC-)MS/MS. TrAC Trends Anal Chem 84:61–73. https://doi.org/10.1016/J.TRAC.2016.01.030
3. Shi Y, Xiang R, Horváth C, Wilkins JA (2004) The role of liquid chromatography in proteomics. J Chromatogr A 1053:27–36. https://doi.org/10.1016/J.CHROMA.2004.07.044
4. Michalski A, Cox J, Mann M (2011) More than 100,000 detectable peptide species elute in single shotgun proteomics runs but the majority is inaccessible to data-dependent LC–MS/MS. J Proteome Res 10:1785–1793. https://doi.org/10.1021/pr101060v
5. Schiavone NM, Sarver SA, Sun L et al (2015) High speed capillary zone electrophoresis–mass spectrometry via an electrokinetically pumped sheath flow interface for rapid analysis of amino acids and a protein digest. J Chromatogr B 991:53–58. https://doi.org/10.1016/J.JCHROMB.2015.04.001
6. Neusüß C, Pelzing M, Macht M (2002) A robust approach for the analysis of peptides in the low femtomole range by capillary electrophoresis-tandem mass spectrometry. Electrophoresis 23:3149–3159. https://doi.org/10.1002/1522-2683(200209)23:18<3149::AID-ELPS3149>3.0.CO;2-8
7. Boley DA, Zhang Z, Dovichi NJ (2018) Multisegment injections improve peptide identification rates in capillary zone electrophoresis-based bottom-up proteomics. J Chromatogr A 1523:123–126. https://doi.org/10.1016/j.chroma.2017.07.022.Multisegment
8. Ramautar R, Heemskerk AAM, Hensbergen PJ et al (2012) CE-MS for proteomics: advances in interface development and application. J Proteomics 75:3814–3828. https://doi.org/10.1016/j.jprot.2012.04.050
9. Schmidt A, Karas M, Dülcks T (2003) Effect of different solution flow rates on analyte ion signals in nano-ESI MS, or: when does ESI turn into nano-ESI? J Am Soc Mass Spectrom 14:492–500. https://doi.org/10.1016/S1044-0305(03)00128-4
10. Wilm M, Mann M (1996) Analytical properties of the nanoelectrospray ion source. Anal Chem 68:1–8. https://doi.org/10.1021/ac9509519
11. Klein J, Papadopoulos T, Mischak H, Mullen W (2014) Comparison of CE-MS/MS and LC-MS/MS sequencing demonstrates significant complementarity in natural peptide identification in human urine. Electrophoresis 35:1060–1064. https://doi.org/10.1002/elps.201300327
12. Iadarola P, Fumagalli M, Bardoni AM et al (2016) Recent applications of CE- and HPLC-MS in the analysis of human fluids. Electrophoresis 37:212–230. https://doi.org/10.1002/elps.201500272
13. Di Venere M, Viglio S, Sassera D et al (2017) Do the complementarities of electrokinetic and chromatographic procedures represent the "Swiss knife" in proteomic investigation? An overview of the literature in the past decade. Electrophoresis 38:1538–1550. https://doi.org/10.1002/elps.201600504
14. Jooß K, Scholz N, Meixner J, Neusüß C (2019) Heart-cut nano-LC-CZE-MS for the characterization of proteins on the intact level. Electrophoresis:1–5. https://doi.org/10.1002/elps.201800411
15. Breadmore MC (2009) Electrokinetic and hydrodynamic injection: making the right choice for capillary electrophoresis. Bioanalysis 1:889–894. https://doi.org/10.4155/bio.09.73
16. Larsson M, Lutz ESM (2000) Transient isotachophoresis for sensitivity enhancement in capillary electrophoresis- mass spectrometry for peptide analysis CE and CEC. Electrophoresis 21:2859–2865
17. Mala Z, Bocek P, Gebauer P (2013) Recent progress in analytical capillary isotachophoresis. Electrophoresis 34:19–28. https://doi.org/10.1002/elps.201200323
18. Osbourn DM, Weiss DJ, Lunte CE (2008) On-line preconcentration methods for capillary electrophoresis. Electrophoresis 21:2768–2779. https://doi.org/10.1002/1522-2683(20000801)21
19. Gahoual R, Busnel JM, Beck A et al (2014) Full antibody primary structure and microvariant characterization in a single injection using transient isotachophoresis and sheathless capillary electrophoresis-tandem mass spectrometry. Anal Chem 86:9074–9081. https://doi.org/10.1021/ac502378e
20. Monton MRN, Imami K, Nakanishi M et al (2005) Dynamic pH junction technique for on-line preconcentration of peptides in capillary electrophoresis. J Chromatogr A 1079:266–273. https://doi.org/10.1016/j.chroma.2005.03.069
21. Aebersold R, Morrison HD (1990) Analysis of dilute peptide samples by capillary zone electrophoresis. J Chromatogr 516:79–88
22. Nesbitt CA, Lo JT, Yeung KK (2005) Over 1000-fold protein preconcentration for microliter-volume samples at a pH junction using capillary electropho-

resis. J Chromatogr A 1073:175–180. https://doi.org/10.1016/j.chroma.2004.09.081
23. Hasan N, Park SH, Oh E, Song EJ, Ban E, Yoo YS (2010) Sensitivity enhancement of CE and CE-MS for the analysis of peptides by a dynamic pH junction. J Sep Sci 33:3701–3709. https://doi.org/10.1002/jssc.201000020
24. Lombard-Banek C, Reddy S, Moody SA, Nemes P (2016) Label-free quantification of proteins in single embryonic cells with neural fate in the cleavage-stage frog (*Xenopus laevis*) embryo using capillary electrophoresis electrospray ionization high-resolution mass spectrometry (CE-ESI-HRMS). Mol Cell Proteomics 15:2756–2768. https://doi.org/10.1074/mcp.M115.057760
25. Monton MRN, Terabe S (2004) Field-enhanced sample injection for high-sensitivity analysis of peptides and proteins in capillary electrophoresis – mass spectrometry. J Chromatogr A 1032:203–211. https://doi.org/10.1016/j.chroma.2003.10.038
26. Pourhaghighi MR, Busnel J-M, Girault HH (2011) High-sensitive protein analysis by FESI-CE-MALDI-MS. Electrophoresis 32:1795–1803. https://doi.org/10.1002/elps.201100024
27. Zhu G, Sun L, Yan X, Dovichi NJ (2014) Bottom-up proteomics of *Escherichia coli* using dynamic pH junction preconcentration and capillary zone electrophoresis-electrospray ionization-tandem mass spectrometry. Anal Chem 86(13):6331–6336
28. Arnett SD, Lunte CE (2008) Investigation of the mechanism of pH-mediated stacking of anions for the analysis of physiological samples by capillary electrophoresis. Electrophoresis 24:1745–1752. https://doi.org/10.1002/elps.200305399. Investigation
29. Dong Y-M, Chien K, Chen J et al (2013) Site-specific separation and detection of phosphopeptide isomers with pH-mediated stacking capillary electrophoresis- electrospray ionization-tandem mass spectrometry. J Sep Sci 36:1582–1589. https://doi.org/10.1002/jssc.201300054
30. Dawod M, Chung DS (2011) High-sensitivity capillary and microchip electrophoresis using electrokinetic supercharging. J Sep Sci 34:2790–2799. https://doi.org/10.1002/jssc.201100384
31. Xu Z, Ando T, Nishine T et al (2003) Electrokinetic supercharging preconcentration and microchip gel electrophoretic separation of sodium dodecyl sulfate-protein complexes. Electrophoresis 24:3821–3827. https://doi.org/10.1002/elps.200305625
32. Busnel J-M, Niels L, Girault HH (2008) Electrokinetic supercharging for highly efficient peptide preconcentration in capillary zone electrophoresis. Electrophoresis 29:1565–1572. https://doi.org/10.1002/elps.200700643
33. Nyssen L, Fillet M, Cavalier E, Servais A-C (2019) Highly sensitive and selective separation of intact parathyroid hormone and variants by sheathless CE-ESI-MS/MS. Electrophoresis:1–25. https://doi.org/10.1002/elps.201800507
34. Xu Z, Timerbaev AR, Hirokawa T (2009) High-sensitivity capillary and microchip electrophoresis using electrokinetic supercharging preconcentration insight into the stacking mechanism via computer modeling. J Chromatogr A 1216:660–670. https://doi.org/10.1016/j.chroma.2008.10.077
35. Sun L, Hebert AS, Yan X et al (2014) Over 10 000 peptide identifications from the HeLa proteome by using single-shot capillary zone electrophoresis combined with tandem mass spectrometry. Angew Chem Int Ed 53:13931–13933. https://doi.org/10.1002/anie.201409075
36. Zhang S, Raedschelders K, Santos M, Eyk JEV (2017) Profiling B-type natriuretic peptide cleavage peptidoforms in human plasma by capillary electrophoresis with electrospray ionization mass spectrometry. J Proteome Res 16:4515–4522. https://doi.org/10.1021/acs.jproteome.7b00482
37. Tempels FWA, Underberg WJM, Somsen GW, de Jong GJ (2007) On-line coupling of SPE and CE-MS for peptide analysis. Electrophoresis 28:1319–1326. https://doi.org/10.1002/elps.200600403
38. Pero-Gascon R, Pont L, Benavente F et al (2016) Analysis of serum transthyretin by on-line immunoaffinity solid-phase extraction capillary electrophoresis mass spectrometry using magnetic beads. Electrophoresis 37:1220–1231. https://doi.org/10.1002/elps.201500495
39. Xia S, Yuan H, Liang Z et al (2015) Particulate capillary precolumns with double-end polymer monolithic frits for on-line peptide trapping and preconcentration. Chin Chem Lett 26:1068–1072. https://doi.org/10.1016/j.cclet.2015.05.042
40. Morales-Cid G, Diez-Masa JC, Frutos MD (2013) On-line immunoaffinity capillary electrophoresis based on magnetic beads for the determination of alpha-1 acid glycoprotein isoforms profile to facilitate its use as biomarker. Anal Chim Acta 773:89–96. https://doi.org/10.1016/j.aca.2013.02.037
41. Ortiz-Villanueva E, Benavente F, Giménez E et al (2014) Preparation and evaluation of open tubular C18-silica monolithic microcartridges for preconcentration of peptides by on-line solid phase extraction capillary electrophoresis. Anal Chim Acta 846:51–59. https://doi.org/10.1016/j.aca.2014.06.046
42. Benavente F, Vescina MC, Hernández E et al (2007) Lowering the concentration limits of detection by on-line solid-phase extraction-capillary electrophoresis-electrospray mass spectrometry. J Chromatogr A 1140:205–212. https://doi.org/10.1016/j.chroma.2006.11.092
43. Zhang Z, Sun L, Zhu G et al (2015) Integrated strong cation-exchange hybrid monolith coupled with capillary zone electrophoresis and simultaneous dynamic pH junction for large-volume proteomic analysis by mass spectrometry. Talanta 138:117–122. https://doi.org/10.1016/j.talanta.2015.01.040
44. Chen Y, Wang K, Yang H et al (2012) Synthesis of sulfo/vinyl biphasic silica hybrid monolithic cap-

illary column and its application to on-column preconcentration for capillary electrochromatography. J Chromatogr A 1233:91–99. https://doi.org/10.1016/j.chroma.2012.01.024
45. Benavente F, Hernández E, Guzman NA, Sanz-nebot V (2007) Determination of human erythropoietin by on-line immunoaffinity capillary electrophoresis: a preliminary report. Anal Bioanal Chem 387:2633–2639. https://doi.org/10.1007/s00216-007-1119-0
46. Medina-Casanellas S, Benavente F, Giménez E et al (2014) On-line immunoaffinity solid-phase extraction capillary electrophoresis mass spectrometry for the analysis of large biomolecules: a preliminary report. Electrophoresis 35:2130–2136. https://doi.org/10.1002/elps.201400119
47. Pont L, Benavente F, Barbosa J, Sanz-Nebot V (2017) On-line immunoaffinity solid-phase extraction capillary electrophoresis mass spectrometry using Fab'antibody fragments for the analysis of serum transthyretin. Talanta 170:224–232. https://doi.org/10.1016/j.talanta.2017.03.104
48. Peoples MC, Karnes HT (2008) Microfluidic capillary system for immunoaffinity separations of C-reactive protein in human serum and cerebrospinal fluid. Anal Chem 80:3853–3858
49. Amundsen LK, Siren H (2007) Immunoaffinity CE in clinical analysis of body fluids and tissues. Electrophoresis 28:99–113. https://doi.org/10.1002/elps.200500962
50. Breadmore MC, Sänger van de Griend CE (2016) In-capillary sample concentration in CE. LCGC N Am 32:174–186
51. Beckers JL, Bocek P (2003) The preparation of background electrolytes in capillary zone electrophoresis: golden rules and pitfalls. Electrophoresis 24:518–535
52. Pantuckova P, Gebauer P, Bocek P, Krivankova L (2009) Electrolyte systems for on-line CE – MS: detection requirements and separation possibilities. Electrophoresis 30:203–214. https://doi.org/10.1002/elps.200800262
53. Hruska V, Jaros M, Gas B (2006) Simul 5 – free dynamic simulator of electrophoresis. Electrophoresis 27:984–991. https://doi.org/10.1002/elps.200500756
54. Janini GM, Conrads TP, Wilkens KL et al (2003) A sheathless nanoflow electrospray interface for on-line capillary electrophoresis mass spectrometry. Anal Chem 75:1615–1619
55. Catai JR, Torano JS, de Jong GJ, Somsen GW (2006) Efficient and highly reproducible capillary electrophoresis – mass spectrometry of peptides using Polybrene-poly (vinyl sulfonate)-coated capillaries. Electrophoresis 27:2091–2099. https://doi.org/10.1002/elps.200500915
56. Liu T, Li J, Zeng R et al (2001) Capillary electrophoresis – electrospray mass spectrometry for the characterization of differential oxidation in glycoproteins by charge reversal and protease/glycosidase digestion. Anal Chem 73:5875–5885
57. Yang Y, Boysen RI, Matyska MT et al (2007) Open-tubular capillary electrochromatography coupled with electrospray ionization mass spectrometry for peptide analysis. Anal Chem 79:4942–4949
58. Cheng J, Chen DDY (2018) Nonaqueous capillary electrophoresis mass spectrometry method for determining highly hydrophobic peptides. Electrophoresis 39:1216–1221. https://doi.org/10.1002/elps.201700364
59. Ruckenstein E, Shulgin IL (2006) Effect of salts and organic additives on the solubility of proteins in aqueous solutions. Adv Colloid Interface Sci 126:97–103. https://doi.org/10.1016/j.cis.2006.05.018
60. Li F-A, Huang J-L, Her G-R (2008) Chip-CE/MS using a flat low-sheath-flow interface. Electrophoresis 29:4938–4943. https://doi.org/10.1002/elps.200800271
61. Li F-A, Huang J-L, Shen S-Y et al (2009) Development of a liquid-junction/low-flow interface for phosphate buffer capillary electrophoresis mass spectrometry. Anal Chem 81:2810–2814. https://doi.org/10.1021/ac802491y
62. Tang Q, Hamata AK, Lee G (1995) Capillary isoelectric focusing-electrospray mass spectrometry for protein analysis. Anal Chem 67:3515–3519
63. Dai J, Lamp J, Xia Q, Zhang Y (2018) Capillary isoelectric focusing-mass spectrometry method for the separation and online characterization of intact monoclonal antibody charge variants. Anal Chem 90:2246–2254. https://doi.org/10.1021/acs.analchem.7b04608
64. Kohl FJ, Montealegre C, Neusüß C (2016) On-line two-dimensional capillary electrophoresis with mass spectrometric detection using a fully electric isolated mechanical valve. Electrophoresis 37:954–958. https://doi.org/10.1002/elps.201500579
65. Schlecht J, Jooß K, Neusüß C (2018) Two-dimensional capillary electrophoresis-mass spectrometry (CE-CE-MS): coupling MS-interfering capillary electromigration methods with mass spectrometry. Anal Bioanal Chem 410:6353–6359
66. Moritz B, Schnaible V, Kiessig S et al (2015) Evaluation of capillary zone electrophoresis for charge heterogeneity testing of monoclonal antibodies. J Chromatogr B 983–984:101–110. https://doi.org/10.1016/j.jchromb.2014.12.024
67. Jooß K, Hühner J, Kiessig S et al (2017) Two-dimensional capillary zone electrophoresis – mass spectrometry for the characterization of intact monoclonal antibody charge variants, including deamidation products. Anal Bioanal Chem 409:6057–6067. https://doi.org/10.1007/s00216-017-0542-0
68. Cifuentes A, Santos JM, Frutos MD (1993) High-efficiency capillary electrophoretic separation of basic proteins using coated capillaries and cationic buffer additives evaluation of protein-capillary wall interactions. J Chromatogr 652:161–170
69. Huhn C, Ramautar R, Wuhrer M, Somsen GW (2010) Relevance and use of capillary coatings in capillary electrophoresis – mass spectrom-

etry. Anal Bioanal Chem 396:297–314. https://doi.org/10.1007/s00216-009-3193-y
70. Staub A, Comte S, Rudaz S et al (2010) Use of organic solvent to prevent protein adsorption in CE-MS experiments. Electrophoresis 31:3326–3333. https://doi.org/10.1002/elps.201000245
71. Moini M, Huang H (2004) Application of capillary electrophoresis/electrospray ionization-mass spectrometry to subcellular proteomics of Escherichia coli ribosomal proteins. Electrophoresis 25:1981–1987. https://doi.org/10.1002/elps.200305906
72. Simó C, Herrero M, Neusüß C et al (2005) Characterization of proteins from Spirulina platensis microalga using capillary electrophoresis-ion trap-mass spectrometry and capillary electrophoresis-time of flight-mass spectrometry. Electrophoresis 26:2674–2683. https://doi.org/10.1002/elps.200500055
73. Martin GB, Mansion F, Servais A-C et al (2009) CE-MS method development for peptides analysis, especially hepcidin, an iron metabolism marker. Electrophoresis 30:2624–2631. https://doi.org/10.1002/elps.200800794
74. Porras SP, Kenndler E (2005) Are the asserted advantages of organic solvents in capillary electrophoresis real? A critical discussion. Electrophoresis 26:3203–3220. https://doi.org/10.1002/elps.200500311
75. Lucy CA, Macdonald AM, Gulcev MD (2008) Non-covalent capillary coatings for protein separations in capillary electrophoresis. J Chromatogr 1184:81–105. https://doi.org/10.1016/j.chroma.2007.10.114
76. Pattky M, Barkovits K, Marcus K et al (2016) Statically adsorbed coatings for high separation efficiency and resolution in CE – MS peptide analysis: strategies and implementation. In: Capillary electrophoresis: methods and protocols. Humana Press, New York, NY, pp 53–75. https://doi.org/10.1007/978-1-4939-6403-1
77. Ciccone B (2001) A buffer system for capillary electrophoresis. Am Lab 33:30–33
78. Giorgio P, Gelfi C, Sebastiano R, Citterio A (2004) Surfing silica surfaces superciliously. J Chromatogr 1053:15–26. https://doi.org/10.1016/j.chroma.2004.05.073
79. Nguyen TTTN, Petersen NJ, Rand KD (2016) A simple sheathless CE-MS interface with a sub-micrometer electrical contact fracture for sensitive analysis of peptide and protein samples. Anal Chim Acta 936:157–167. https://doi.org/10.1016/j.aca.2016.07.002
80. Huck CW, Bakry R, Huber LA, Bonn GK (2006) Progress in capillary electrophoresis coupled to matrix-assisted laser desorption/ionization – time of flight mass spectrometry. Electrophoresis 27:2063–2074. https://doi.org/10.1002/elps.200600046
81. Zaia J (2014) Capillary electrophoresis-mass spectrometry of carbohydrates. Methods Mol Biol:13–25. https://doi.org/10.1007/978-1-62703-296-4
82. Gahoual R, Beck A, Leize-wagner E, Franc Y (2016) Cutting-edge capillary electrophoresis characterization of monoclonal antibodies and related products. J Chromatogr B 1032:61–78. https://doi.org/10.1016/j.jchromb.2016.05.028
83. Týčová A, Ledvina V, Klepárník K (2017) Recent advances in CE-MS coupling: instrumentation, methodology, and applications. Electrophoresis 38:115–134. https://doi.org/10.1002/elps.201600366
84. Krenkova J, Kleparnik K, Luksch J, Foret F (2019) Microfabricated liquid junction hybrid capillary electrophoresis-mass spectrometry interface for fully automated operation. Electrophoresis:1–20. https://doi.org/10.1002/elps.201900049
85. Bodnar J, Hajba L, Guttman A (2016) A fully automated linear polyacrylamide coating and regeneration method for capillary electrophoresis of proteins. Electrophoresis 37:3154–3159. https://doi.org/10.1002/elps.201600405
86. Zhang Z, Dovichi NJ (2018) Optimization of mass spectrometric parameters improve the identification performance of capillary zone electrophoresis for single-shot bottom-up proteomics analysis. Anal Chim Acta 1001:93–99. https://doi.org/10.1016/j.aca.2017.11.023
87. Stutz H (2009) Protein attachment onto silica surfaces – a survey of molecular fundamentals, resulting effects and novel preventive strategies in CE. Electrophoresis 30:2032–2061. https://doi.org/10.1002/elps.200900015
88. Volpi N, Maccari F (2009) Capillary electrophoresis of biomolecules: methods and protocols. Humana Press, New York, NY
89. Höcker O, Montealegre C, Neusüß C (2018) Characterization of a nanoflow sheath liquid interface and comparison to a sheath liquid and a sheathless porous-tip interface for CE-ESI-MS in positive and negative ionization. Anal Bioanal Chem 410:5265–5275. https://doi.org/10.1007/s00216-018-1179-3
90. Xia JQ (2016) Coated capillaries for CE-MS of therapeutic protein. In: Capillary electrophoresis–mass spectrometry. Springer, Cham, pp 7–12
91. Hernández-Borges J, Neusüß C, Cifuentes A, Pelzing M (2004) On-line capillary electrophoresis-mass spectrometry for the analysis of biomolecules. Electrophoresis 25:2257–2281. https://doi.org/10.1002/elps.200405954
92. Stutz H (2005) Advances in the analysis of proteins and peptides by capillary electrophoresis with matrix-assisted laser desorption/ionization and electrospray-mass spectrometry detection. Electrophoresis 26:1254–1290. https://doi.org/10.1002/elps.200410130
93. Khatri K, Klein JA, Haserick JR et al (2018) Microfluidic capillary electrophoresis–mass spectrometry for analysis of monosaccharides, oligosaccharides, and glycopeptides. Anal Chem 89:6645–6655. https://doi.org/10.1021/acs.anal--chem.7b00875.Detailed
94. Moreno-Gonz D, Laura G, Garc AM, Somsen GW (2013) Micellar electrokinetic chromatography – electrospray ionization mass spectrometry employ-

ing a volatile surfactant for the analysis of amino acids in human urine. Electrophoresis 34:2615–2622. https://doi.org/10.1002/elps.201300247
95. Elhamili A, Wetterhall M, Arvidsson B et al (2008) Rapid capillary electrophoresis time-of-flight mass spectrometry separations of peptides and proteins using a monoquaternarized piperazine compound (M7C4I) for capillary coatings. Electrophoresis 29:1619–1625. https://doi.org/10.1002/elps.200700737
96. Bhardwaj C, Hanley L (2014) Ion sources for mass spectrometric identification and imaging of molecular species. Nat Prod Rep 31:756–767. https://doi.org/10.1039/c3np70094a
97. Marginean I, Kelly RT, Prior DC et al (2008) Analytical characterization of the electrospray ion source in the nanoflow regime. Anal Chem 80:6573–6579
98. Moini M (2007) Simplifying CE – MS operation. 2. Interfacing low-flow separation techniques to mass spectrometry using a porous tip. Anal Chem 79:4241–4246
99. Bonvin G, Rudaz S, Schappler J (2014) In-spray supercharging of intact proteins by capillary electrophoresis – electrospray ionization – mass spectrometry using sheath liquid interface. Anal Chim Acta 813:97–105. https://doi.org/10.1016/j.aca.2013.12.043
100. Krenkova J, Kleparnik K, Grym J et al (2016) Self-aligning subatmospheric hybrid liquid junction electrospray interface for capillary electrophoresis. Electrophoresis 37:414–417. https://doi.org/10.1002/elps.201500357
101. Fanali C, Orazio GD, Fanali S (2012) Nano-liquid chromatography and capillary electrochromatography hyphenated with mass spectrometry for tryptic digest protein analysis: a comparison. Electrophoresis 33:2553–2560. https://doi.org/10.1002/elps.201200157
102. Wojcik R, Dada OO, Sadilek M, Dovichi NJ (2010) Simplified capillary electrophoresis nanospray sheath- flow interface for high efficiency and sensitive peptide analysis. Rapid Commun Mass Spectrom 24:2554–2560
103. Sun L, Zhu G, Zhao Y et al (2013) Ultrasensitive and fast bottom-up analysis of femtogram amounts of complex proteome digests. Angew Chemie 52:13661–13664. https://doi.org/10.1002/anie.201308139
104. Sun L, Zhu G, Zhang Z et al (2015) Third-generation electrokinetically pumped sheath-flow nanospray interface with improved stability and sensitivity for automated capillary zone electrophoresis – mass spectrometry analysis of complex proteome digests. J Proteome Res 14:2312–2321. https://doi.org/10.1021/acs.jproteome.5b00100
105. Qu Y, Sun L, Zhu G et al (2018) Sensitive and fast characterization of site-specific protein glycosylation with capillary electrophoresis coupled to mass spectrometry. Talanta 179:22–27. https://doi.org/10.1016/j.talanta.2017.10.015
106. Johnson RT, To NH, Stobaugh JF, Lunte CE (2017) The development of a Sheathless Interface for capillary electrophoresis electrospray ionization mass spectrometry using a cellulose acetate cast capillary. Chromatographia 80:1061–1067. https://doi.org/10.1007/s10337-017-3326-y
107. Tycova A, Foret F (2015) Capillary electrophoresis in an extended nanospray tip – electrospray as an electrophoretic column. J Chromatogr A 1388:274–279. https://doi.org/10.1016/j.chroma.2015.02.042
108. Haselberg R, Ratnayake CK, Jong GJD, Somsen GW (2010) Performance of a sheathless porous tip sprayer for capillary electrophoresis – electrospray ionization-mass spectrometry of intact proteins. J Chromatogr A 1217:7605–7611. https://doi.org/10.1016/j.chroma.2010.10.006
109. Heemskerk AAM, Deelder M, Mayboroda OA (2014) CE–ESI-MS for bottom-up proteomics: advances in separation, interfacing and applications. Mass Spectrom Rev 35:259–271. https://doi.org/10.1002/mas
110. Heemskerk AAM, Busnel J, Schoenmaker B et al (2012) Ultra-low flow electrospray ionization-mass spectrometry for improved ionization efficiency in phosphoproteomics. Anal Chem 84:4552–4559. https://doi.org/10.1021/ac300641x
111. Dominguez-Vega E, De Vijlder T, Romijn EP, Somsen GW (2017) Capillary electrophoresis-tandem mass spectrometry as a highly selective tool for the compositional and site-specific assessment of multiple peptide-deamidation. Anal Chim Acta 982:122–130. https://doi.org/10.1016/j.aca.2017.06.021
112. Jarvas G, Fonslow B, Iii JRY et al (2017) Characterization of a porous nano-electrospray capillary emitter at ultra-low flow rates. J Chromatogr Sci 55:47–51. https://doi.org/10.1093/chromsci/bmw148
113. Haselberg R, de Jong GJ, Somsen GW (2013) Low-flow sheathless capillary electrophoresis–mass spectrometry for sensitive glycoform profiling of intact pharmaceutical proteins. Anal Chem 85:2289–2296. https://doi.org/10.1021/ac303158f
114. Kammeijer GSM, Kohler I, Jansen BC et al (2016) Dopant enriched nitrogen gas combined with sheathless capillary electrophoresis – electrospray ionization-mass spectrometry for improved sensitivity and repeatability in glycopeptide analysis. Anal Chem 88:5849–5856. https://doi.org/10.1021/acs.analchem.6b00479
115. Busnel J, Josserand J, Lion N, Girault HH (2009) Iontophoretic fraction collection for coupling capillary zone electrophoresis with matrix-assisted laser desorption/ionization mass spectrometry. Anal Chem 81:3867–3872
116. Romson J, Jacksén J, Emmer Å (2019) An automated system for CE-MALDI and on-target digestion under a fluorocarbon lid applied on spermatophore proteins

from Pieris napi. J Chromatogr B 1104:228–233. https://doi.org/10.1016/j.jchromb.2018.11.021
117. Biacchi M, Said N, Beck A, Leize-wagner E (2017) Top-down and middle-down approach by fraction collection enrichment using off-line capillary electrophoresis – mass spectrometry coupling: application to monoclonal antibody F c/2 charge variants. J Chromatogr A 1498:120–127. https://doi.org/10.1016/j.chroma.2017.02.064
118. Johnson T, Bergquist J, Ekman R et al (2001) A CE-MALDI interface based on the use of prestructured sample supports. Anal Chem 73:2278–2283. https://doi.org/10.1021/ac0011888
119. Rejtar T, Hu P, Juhasz P et al (2002) Off-line coupling of high-resolution capillary electrophoresis to MALDI-TOF and TOF/TOF MS. J Proteome Res 1:171–179. https://doi.org/10.1021/pr015519o
120. Preisler J, Foret F, Karger BL (1998) On-line MALDI-TOF MS using a continuous vacuum deposition interface. Anal Chem 70:5278–5287. https://doi.org/10.1021/ac9807823
121. Musyimi HK, Narcisse DA, Zhang X et al (2004) Online CE – MALDI-TOF MS using a rotating ball interface. Anal Chem 76:5968–5973. https://doi.org/10.1021/ac0489723
122. Rogowska A, Pomastowski P, Złoch M, Viorica R (2018) The influence of different pH on the electrophoretic behaviour of Saccharomyces cerevisiae modified by calcium ions. Sci Rep 8:2–11. https://doi.org/10.1038/s41598-018-25024-4
123. Mann M, Kelleher NL (2008) Precision proteomics: the case for high resolution and high mass accuracy. Proc Natl Acad Sci U S A 105:18132–18138. https://doi.org/10.1073/pnas.0800788105
124. Shen Y, Jacobs JM, Camp DG I et al (2004) Ultra-high-efficiency strong cation exchange LC/RPLC/MS/MS for high dynamic range characterization of the human plasma proteome. Anal Chem 76(4):1134–1144. https://doi.org/10.1021/AC034869M
125. Guthals A, Bandeira N (2012) Peptide identification by tandem mass spectrometry with alternate fragmentation modes. Mol Cell Proteomics 11:550–557. https://doi.org/10.1074/mcp.R112.018556
126. Zürbig P, Renfrow MB, Schiffer E et al (2006) Biomarker discovery by CE-MS enables sequence analysis via MS/MS with platform-independent separation. Electrophoresis 27:2111–2125. https://doi.org/10.1002/elps.200500827
127. Li Y, Compton PD, Tran JC et al (2014) Optimizing capillary electrophoresis for top-down proteomics of 30–80 kDa proteins. Proteomics 14:1158–1164. https://doi.org/10.1002/pmic.201300381
128. Zhao Y, Sun L, Champion MM et al (2014) Capillary zone electrophoresis–electrospray ionization-tandem mass spectrometry for top-down characterization of the *Mycobacterium marinum* secretome. Anal Chem 86:4873–4878. https://doi.org/10.1021/ac500092q
129. Sun L, Knierman MD, Zhu G, Dovichi NJ (2013) Fast top-down intact protein characterization with capillary zone electrophoresis–electrospray ionization tandem mass spectrometry. Anal Chem 85:5989–5995. https://doi.org/10.1021/ac4008122
130. Han X, Wang Y, Aslanian A et al (2014) Sheathless capillary electrophoresis-tandem mass spectrometry for top-down characterization of *Pyrococcus furiosus* proteins on a proteome scale. Anal Chem 86:11006–11012. https://doi.org/10.1021/ac503439n
131. Han X, Wang Y, Aslanian A et al (2014) In-line separation by capillary electrophoresis prior to analysis by top-down mass spectrometry enables sensitive characterization of protein complexes. J Proteome Res 13:6078–6086. https://doi.org/10.1021/pr500971h
132. Zhao Y, Riley NM, Sun L et al (2015) Coupling capillary zone electrophoresis with electron transfer dissociation and activated ion electron transfer dissociation for top-down proteomics. Anal Chem 87:5422–5429. https://doi.org/10.1021/acs.analchem.5b00883
133. Pontillo C, Filip S, Borràs DM et al (2015) CE-MS-based proteomics in biomarker discovery and clinical application. Proteomics Clin Appl 9:322–334. https://doi.org/10.1002/prca.201400115
134. Robledo VR, Smyth WF (2014) Review of the CE-MS platform as a powerful alternative to conventional couplings in bio-omics and target-based applications. Electrophoresis 35:2292–2308. https://doi.org/10.1002/elps.201300561
135. Stolz A, Jooß K, Oliver H et al (2019) Recent advances in capillary electrophoresis- mass spectrometry: instrumentation, methodology and applications. Electrophoresis 40:79–112. https://doi.org/10.1002/elps.201800331
136. Haselberg R, de Jong GJ, Somsen GW (2013) CE-MS for the analysis of intact proteins 2010–2012. Electrophoresis:99–112. https://doi.org/10.1002/elps.201200439
137. Jiang Y, He MY, Zhang WJ et al (2017) Recent advances of capillary electrophoresis-mass spectrometry instrumentation and methodology. Chin Chem Lett 28:1640–1652. https://doi.org/10.1016/j.cclet.2017.05.008
138. Faserl K, Sarg B, Maurer V, Lindner HH (2017) Exploiting charge differences for the analysis of challenging post-translational modifications by capillary electrophoresis-mass spectrometry. J Chromatogr A 1498:215–223. https://doi.org/10.1016/j.chroma.2017.01.086
139. Steen H, Jebanathirajah JA, Rush J et al (2006) Phosphorylation analysis by mass spectrometry. Mol Cell Proteomics 5:172–181. https://doi.org/10.1074/mcp.M500135-MCP200
140. Faserl K, Sarg B, Gruber P, Lindner HH (2018) Investigating capillary electrophoresis-mass spectrometry for the analysis of common post-translational modifications. Electrophoresis:1–8. https://doi.org/10.1002/elps.201700437

141. Mou S, Sun L, Dovichi NJ (2014) Accurate determination of peptide phosphorylation stoichiometry via automated diagonal capillary electrophoresis coupled with mass spectrometry – proof of principle. Anal Chem 85:10692–10696. https://doi.org/10.1021/ac402858a.Accurate
142. Svozil J, Baerenfaller K (2017) A cautionary tale on the inclusion of variable posttranslational modifications in database-dependent searches of mass spectrometry data. Methods Enzymol 586:433–452. https://doi.org/10.1016/BS.MIE.2016.11.007
143. Robinson NE, Robinson AB (2001) Molecular clocks. Proc Natl Acad Sci 98:944–949. https://doi.org/10.1073/PNAS.98.3.944
144. Hains PG, Truscott RJW (2010) Age-dependent deamidation of lifelong proteins in the human lens. Invest Opthalmol Vis Sci 51:3107. https://doi.org/10.1167/iovs.09-4308
145. Pace AL, Wong RL, Zhang YT et al (2013) Asparagine deamidation dependence on buffer type, pH, and temperature. J Pharm Sci 102:1712–1723. https://doi.org/10.1002/jps.23529
146. Zheng JY, Janis LJ (2006) Influence of pH, buffer species, and storage temperature on physicochemical stability of a humanized monoclonal antibody LA298. Int J Pharm 308:46–51. https://doi.org/10.1016/J.IJPHARM.2005.10.024
147. Gervais D (2016) Protein deamidation in biopharmaceutical manufacture: understanding, control and impact. J Chem Technol Biotechnol 91:569–575. https://doi.org/10.1002/jctb.4850
148. Rosnack KJ, Stroh JG, Singleton DH et al (1994) Use of capillary electrophoresis-electrospray ionization mass spectrometry in the analysis of synthetic peptides. J Chromatogr A 675:219–225. https://doi.org/10.1016/0021-9673(94)85275-8
149. Coon JJ, Zürbig P, Dakna M et al (2008) CE-MS analysis of the human urinary proteome for biomarker discovery and disease diagnostics. Proteomics Clin Appl 2:964–973. https://doi.org/10.1002/prca.200800024
150. Gahoual R, Beck A, François Y-N, Leize-Wagner E (2016) Independent highly sensitive characterization of asparagine deamidation and aspartic acid isomerization by sheathless CZE-ESI-MS/MS. J Mass Spectrom 51:150–158. https://doi.org/10.1002/jms.3735
151. Gennaro LA, Salas-Solano O (2009) Characterization of deamidated peptide variants by micro-preparative capillary electrophoresis and mass spectrometry. J Chromatogr A 1216:4499–4503. https://doi.org/10.1016/j.chroma.2009.03.025
152. Bush DR, Zang L, Belov AM et al (2016) High resolution CZE-MS quantitative characterization of intact biopharmaceutical proteins: proteoforms of interferon-β1. Anal Chem 88:1138–1146. https://doi.org/10.1021/acs.analchem.5b03218
153. Staub A, Giraud S, Saugy M et al (2010) CE-ESI-TOF/MS for human growth hormone analysis. Electrophoresis 31:388–395. https://doi.org/10.1002/elps.200900315
154. Taichrib A, Pelzing M, Pellegrino C et al (2011) High resolution TOF MS coupled to CE for the analysis of isotopically resolved intact proteins. J Proteomics 74:958–966. https://doi.org/10.1016/J.JPROT.2011.01.006
155. Bergström T, Fredriksson S-Å, Nilsson C, Åstot C (2015) Deamidation in ricin studied by capillary zone electrophoresis- and liquid chromatography–mass spectrometry. J Chromatogr B 974:109–117. https://doi.org/10.1016/J.JCHROMB.2014.10.015
156. Redman EA, Mellors JS, Starkey JA, Ramsey JM (2016) Characterization of intact antibody drug conjugate variants using micro fluidic capillary electrophoresis – mass spectrometry. Anal Chem 88:2220–2226. https://doi.org/10.1021/acs.analchem.5b03866
157. Kammeijer GSM, Jansen BC, Kohler I et al (2017) Sialic acid linkage differentiation of glycopeptides using capillary electrophoresis – electrospray ionization – mass spectrometry. Sci Rep:1–10. https://doi.org/10.1038/s41598-017-03838-y
158. Amon S, Zamfir AD, Rizzi A (2008) Glycosylation analysis of glycoproteins and proteoglycans using capillary electrophoresis-mass spectrometry strategies. Electrophoresis 29:2485–2507. https://doi.org/10.1002/elps.200800105
159. Zhang Q, Li Z, Wang Y et al (2017) Mass spectrometry for protein sialoglycosylation. Mass Spectrom Rev:1–29. https://doi.org/10.1002/mas.21555
160. Pompach P, Brnakova Z, Sanda M et al (2013) Site-specific glycoforms of haptoglobin in liver cirrhosis and hepatocellular carcinoma. Mol Cell Proteomics:1281–1293. https://doi.org/10.1074/mcp.M112.023259
161. Yang N, Goonatilleke E, Park D et al (2016) Quantitation of site-specific glycosylation in manufactured recombinant monoclonal antibody drugs. Anal Chem 88:7091–7100. https://doi.org/10.1021/acs.analchem.6b00963
162. Nilsson J, Rüetschi U, Halim A et al (2009) Enrichment of glyco – peptides for glycan structure and attachment site identification. Nat Methods 6:809–811. https://doi.org/10.1038/nmeth.1392
163. Hu H, Khatri K, Klein J et al (2016) A review of methods for interpretation of glycopeptide tandem mass spectral data. Glycoconj J 33:285–296. https://doi.org/10.1007/s10719-015-9633-3
164. Hinneburg H, Stavenhagen K, Schweiger-hufnagel U et al (2015) The art of destruction: optimizing collision energies. J Am Soc Mass Spectrom 27:507–519. https://doi.org/10.1007/s13361-015-1308-6
165. Amon S, Plematl A, Rizzi A (2006) Capillary zone electrophoresis of glycopeptides under controlled electroosmotic flow conditions coupled to electrospray and matrix-assisted laser desorption/ionization mass spectrometry. Electrophoresis 27:1209–1219. https://doi.org/10.1002/elps.200500725

166. Schönemeier B, Metzger J, Klein J et al (2016) Urinary peptide analysis differentiates pancreatic cancer from chronic pancreatitis. Pancreas 00:1–9
167. Wittke S, Mischak H, Walden M et al (2005) Discovery of biomarkers in human urine and cerebrospinal fluid by capillary electrophoresis coupled to mass spectrometry: towards new. Electrophoresis 3:1476–1487. https://doi.org/10.1002/elps.200410140
168. Haselberg R, Vijlder TD, Heukers R et al (2018) Heterogeneity assessment of antibody-derived therapeutics at the intact and middle-up level by low-flow sheathless capillary electrophoresis-mass spectrometry. Anal Chim Acta 1044:181–190. https://doi.org/10.1016/j.aca.2018.08.024
169. Gahoual R, Houzé P, François YN (2018) Revealing the potential of capillary electrophoresis/mass spectrometry: the tipping point. 1–9 . https://doi.org/10.1002/rcm.8238
170. Giménez E, Ramos-Hernan R, Benavente F et al (2012) Analysis of recombinant human erythropoietin glycopeptides by capillary electrophoresis electrospray – time of flight-mass spectrometry. Anal Chim Acta 709:81–90. https://doi.org/10.1016/j.aca.2011.10.028
171. Said N, Gahoual R, Kuhn L et al (2016) Structural characterization of antibody drug conjugate by a combination of intact, middle-up and bottom-up techniques using sheathless capillary electrophoresis – tandem mass spectrometry as nanoESI infusion platform and separation method. Anal Chim Acta 918:50–59. https://doi.org/10.1016/j.aca.2016.03.006
172. Wakankar A, Chen Y, Gokarn Y, Jacobson FS (2011) Analytical methods for physicochemical characterization of antibody drug conjugates. MAbs 3:161–172. https://doi.org/10.4161/mabs.3.2.14960
173. Han M, Pearson JT, Wang Y et al (2017) Immunoaffinity capture coupled with capillary electrophoresis – mass spectrometry to study therapeutic protein stability in vivo. Anal Biochem 539:118–126. https://doi.org/10.1016/j.ab.2017.10.005
174. Han M, Rock BM, Pearson JT, Rock DA (2016) Intact mass analysis of monoclonal antibodies by capillary electrophoresis — mass spectrometry. J Chromatogr B 1011:24–32. https://doi.org/10.1016/j.jchromb.2015.12.045
175. Giorgetti J, Lechner A, Nero ED, Beck A (2018) Intact monoclonal antibodies separation and analysis by sheathless capillary electrophoresis-mass spectrometry. Eur J Mass Spectrom:1–9. https://doi.org/10.1177/1469066718807798
176. Wenz C, Barbas C, López-Gonzálvez Á et al (2015) Interlaboratory study to evaluate the robustness of capillary electrophoresis-mass spectrometry for peptide mapping. J Sep Sci 38:3262–3270. https://doi.org/10.1002/jssc.201500551
177. Metzger J, Mullen W, Husi H et al (2016) Acute kidney injury prediction in cardiac surgery patients by a urinary peptide pattern: a case-control validation study. Crit Care 20:157. https://doi.org/10.1186/s13054-016-1344-z
178. Markoska K, Pejchinovski M, Pontillo C et al (2018) Urinary peptide biomarker panel associated with an improvement in estimated glomerular filtration rate in chronic kidney disease patients. Nephrol Dial Transplant 33:751–759. https://doi.org/10.1093/ndt/gfx263
179. Albalat A, Husi H, Siwy J et al (2014) Capillary electrophoresis interfaced with a mass spectrometer (CE-MS): technical considerations and applicability for biomarker studies in animals. Curr Protein Pept Sci 15:23–35. https://doi.org/10.2174/1389203715666140221123920
180. Pejchinovski M, Siwy J, Mullen W et al (2018) Urine peptidomic biomarkers for diagnosis of patients with systematic lupus erythematosus. Lupus 27:6–16. https://doi.org/10.1177/0961203317707827
181. Belczacka I, Latosinska A, Siwy J et al (2018) Urinary CE-MS peptide marker pattern for detection of solid tumors. Sci Rep 8:1–11. https://doi.org/10.1038/s41598-018-23585-y
182. Pontillo C, Mischak H (2017) Urinary peptide-based classifier CKD273: towards clinical application in chronic kidney disease. Clin Kidney J 10:192–201. https://doi.org/10.1093/ckj/sfx002
183. Rossing K, Bosselmann HS, Gustafsson F et al (2016) Urinary proteomics pilot study for biomarker discovery and diagnosis in heart failure with reduced ejection fraction. PLoS One 11:1–15. https://doi.org/10.1371/journal.pone.0157167
184. Kammeijer GSM, Nouta J, De La Rosette JJMCH et al (2018) An in-depth glycosylation assay for urinary prostate-specific antigen. Anal Chem 90:4414–4421. https://doi.org/10.1021/acs.analchem.7b04281
185. Zürbig P, Jahn H (2012) Use of proteomic methods in the analysis of human body fluids in Alzheimer research. Electrophoresis 33:3617–3630. https://doi.org/10.1002/elps.201200360
186. Rossetti DV, Martelli C, Longhi R et al (2013) Quantitative analysis of thymosin β4 in whole saliva by capillary electrophoresis-mass spectrometry using multiple ions monitoring (CE-MIM-MS). Electrophoresis 34:2674–2682. https://doi.org/10.1002/elps.201300165
187. Neuhaus J, Schiffer E, von Wilcke P et al (2013) Seminal plasma as a source of prostate cancer peptide biomarker candidates for detection of indolent and advanced disease. PLoS One 8. https://doi.org/10.1371/journal.pone.0067514
188. Lankisch TO, Metzger J, Negm AA et al (2011) Bile proteomic profiles differentiate cholangiocarcinoma from primary sclerosing cholangitis and choledocholithiasis. Hepatology 53:875–884. https://doi.org/10.1002/hep.24103
189. Di Venere M, Viglio S, Cagnone M et al (2018) Advances in the analysis of "less-conventional" human body fluids: an overview of the CE- and HPLC-MS applications in the years 2015–

2017. Electrophoresis 39:160–178. https://doi.org/10.1002/elps.201700276

190. Tu C, Rudnick PA, Martinez MY et al (2010) Depletion of abundant plasma proteins and limitations of plasma proteomics. J Proteome Res 9:4982–4991. https://doi.org/10.1021/pr100646w
191. Stalmach A, Husi H, Mosbahi K, Albalat A, Mullen W, Mischak H (2015) Methods in capillary electrophoresis coupled to mass spectrometry for the identification of clinical proteomic/peptidomic biomarkers in biofluids. Methods Mol Biol 1243:187–205
192. Engel N, Weiss VU, Wenz C et al (2015) Challenges of glycoprotein analysis by microchip capillary gel electrophoresis. Electrophoresis 36:1754–1758. https://doi.org/10.1002/elps.201400510
193. Herwig E, Marchetti-Deschmann M, Wenz C et al (2015) Sensitive detection of C-reactive protein in serum by immunoprecipitation–microchip capillary gel electrophoresis. Anal Biochem 478:102–106. https://doi.org/10.1016/J.AB.2015.03.009
194. Dawod M, Kennedy RT, Arvin NE (2017) Recent advances in protein analysis by capillary and microchip electrophoresis. Analyst 142:1847–1866. https://doi.org/10.1039/c7an00198c
195. Bertoletti L, Schappler J, Colombo R et al (2016) Evaluation of capillary electrophoresis-mass spectrometry for the analysis of the conformational heterogeneity of intact proteins using beta 2 -microglobulin as model compound. Anal Chim Acta 945:102–109. https://doi.org/10.1016/j.aca.2016.10.010
196. Lombard-Banek C, Moody SA, Manzini MC, Nemes P (2019) Microsampling capillary electrophoresis mass spectrometry enables single-cell proteomics in complex tissues: developing cell clones in live Xenopus laevis and Zebrafish embryos. Anal Chem 91:4797–4805. https://doi.org/10.1021/acs.analchem.9b00345
197. He M, Luo P, Hong J et al (2019) Structural analysis of biomolecules through a combination of mobility capillary electrophoresis and mass spectrometry. ACS Omega 4:2377–2386. https://doi.org/10.1021/acsomega.8b03224

Peptidomics and Capillary Electrophoresis

Sille Štěpánová and Václav Kašička

Abstract

Peptides play a crucial role in many vitally important functions of living organisms. The goal of peptidomics is the identification of the "peptidome," the whole peptide content of a cell, organ, tissue, body fluid, or organism. In peptidomic or proteomic studies, capillary electrophoresis (CE) is an alternative technique for liquid chromatography. It is a highly efficient and fast separation method requiring extremely low amounts of sample. In peptidomic approaches, CE is commonly combined with mass spectrometric (MS) detection. Most often, CE is coupled with electrospray ionization MS and less frequently with matrix-assisted laser desorption/ionization MS. CE-MS has been employed in numerous studies dealing with determination of peptide biomarkers in different body fluids for various diseases, or in food peptidomic research for the analysis and identification of peptides with special biological activities. In addition to the above topics, sample preparation techniques commonly applied in peptidomics before CE separation and possibilities for peptide identification and quantification by CE-MS or CE-MS/MS methods are discussed in this chapter.

Keywords

Biomarkers · Bioactive peptides · Capillary electrophoresis · Mass spectrometry · Peptidomics

S. Štěpánová · V. Kašička (✉)
Institute of Organic Chemistry and Biochemistry, The Czech Academy of Sciences, Prague, Czechia
e-mail: kasicka@uochb.cas.cz

Abbreviations

AD	Alzheimer disease
BGE	background electrolyte
BLAST	basic local alignment search tool
BPE	base peak electropherogram
CE	capillary electrophoresis
CID	collision-induced dissociation
CKD	chronic kidney disease
CSF	cerebrospinal fluid
ECD	electron capture dissociation
EIE	extracted ion electropherogram
EOF	electroosmotic flow
ESI	electrospray ionization
ETD	electron transfer dissociation
FDR	false discovery rate
FT-ICR	Fourier-transform ion cyclotron resonance
HCD	high-energy collisional dissociation
IEF	isoelectric focusing
ITP	isotachophoresis

A. V. Colnaghi Simionato (ed.), *Separation Techniques Applied to Omics Sciences*, Advances in Experimental Medicine and Biology 1336, https://doi.org/10.1007/978-3-030-77252-9_5

IT	ion trap
LC	liquid chromatography
MALDI	matrix-assisted laser desorption ionization
MRM	multiple reaction monitoring
MS	mass spectrometry
MS/MS	tandem mass spectrometry
PTM	posttranslational modification
RP	reversed-phase
SDS	sodium dodecyl sulfate
SPE	solid-phase extraction
SPME	solid-phase microextraction
SRM	selected reaction monitoring
TOF	time of flight

1 Introduction

All human cells, tissues, and body fluids contain peptides that have many physiological functions, acting, e.g., as hormones, growth factors, antimicrobial agents, protease substrates and inhibitors, cytokines, and neurotransmitters. Peptides are involved in various biochemical processes, e.g., metabolism, immune response, and reproduction. Each peptide is characterized by its amino acid sequence and posttranslational modifications (PTMs). Liquid chromatography (LC) and capillary electrophoresis (CE) methods combined with mass spectrometry (MS) or other detection methods are widely utilized for the analysis of particular peptides [1–5] as well as for the peptidomic studies [6–9].

Peptidomics is defined as a comprehensive qualitative and quantitative analysis of all peptides of a given biological system (i.e., cell, organelle, tissue, body fluid, organ, or organism) at a defined time [10–18]. Peptidomics helps to clarify the role of endogenous peptides (generated in vivo within a biological system) in biochemical pathways. Peptidomic studies are dealing with peptide hormones and neuropeptides [19], antimicrobial peptides, peptide toxins, the discovery and analysis of new bioactive peptides [20], or peptide biomarkers [21–23]. Moreover, applications of peptidomics embrace studies of food protein digestion [24] and characterization of food processing-related proteolysis [25]. Peptidomic technologies are also applied in studies of different diseases [26, 27] and in drug research [28].

Besides many similarities [10], proteomic and peptidomic techniques differ in some important aspects. One of the goals of proteomics, particularly in its bottom-up approach, is to identify proteins using the mass and/or sequence information of different peptide fragments obtained by enzymatic protein digestion. Digestion is performed by various proteolytic enzymes, most frequently by trypsin, due to its high specificity and broad availability [29]. However, alternative proteases have been investigated and are used as well [30]. Peptidomics, on the other hand, aims not to identify the precursor protein but endogenous peptides according to their sequence information (see Fig. 1). Enzymatic digestion is not required for peptidomic analysis. Peptidomics generally embraces oligo- and polypeptides with $M_r \leq 20$ kDa.

Mass spectrometry (MS) combined with different LC and CE separation methods is extensively employed in proteomic and peptidomic studies [6–9, 31–35]. With LC-MS methods, higher sensitivity than with CE-MS techniques is achieved, mostly because of higher loading capacity of LC column. In addition, flow rates in LC systems can be decreased to achieve a wider separation window, which results in a higher number of generated MS/MS spectra and thus in higher amount of peptide identifications. Nevertheless, in recent CE-MS studies, application of dynamic pH junction sample stacking method enabled to remarkably enlarge sample loading capacity and to widen the separation window [36, 37]. A limitation of LC columns can be sample carryover from previous injections [38]. CE-MS methods show several advantages, such as short analysis times, high resolution, low mass limits of detection, and excellent mass accuracy. Proteomic and peptidomic studies, where the two methods were compared, have clearly shown that CE-MS/MS and LC-MS/MS are highly complementary, and application of both techniques substantially increases sequence coverage [39–42].

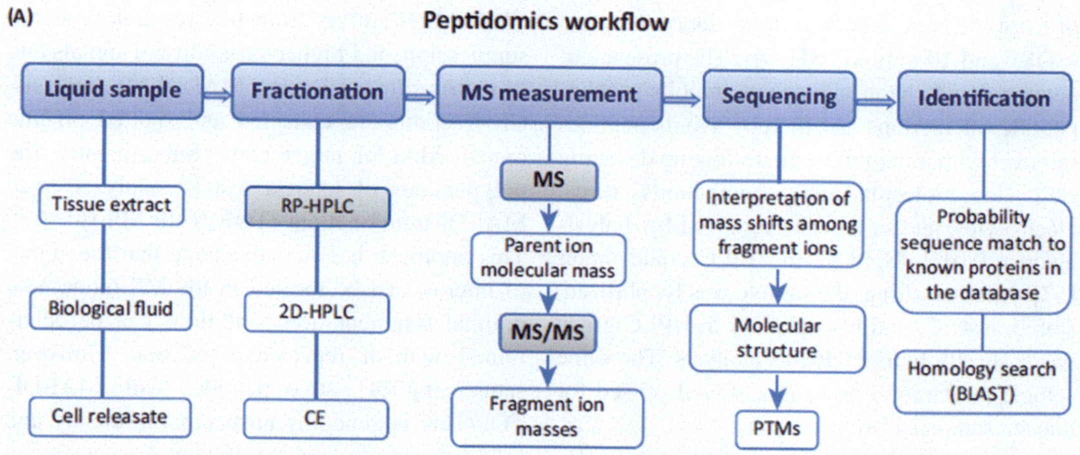

Fig. 1 A simplified flow chart showing general peptidomic workflow. (Adapted with permission from [19])

Large part of peptidomic research focuses on the discovery of peptides that display certain functions or properties, e.g., being biomarkers for disease [21] or having special biological activity [43].

2 Sample Preparation

In peptidomics, it is recommended to analyze a crude, unprocessed sample [44–46]. However, in most cases, sample preparation cannot be completely avoided. Nevertheless, it should be robust, highly reproducible, and kept to a minimum. During sample preparation, minimal loss of peptides should be ensured. For biological samples, where proteinases or peptidases are active, it is essential to inactivate them by adding protease/peptidase inhibitors, applying low or high pH, utilizing chaotropic substances, or denaturing solvents to avoid proteome/peptidome degradation. When tissues or cell lines are being analyzed, first they have to be disintegrated by mechanical homogenization, sonication, and chemical or enzymatic digestion [47, 48]. Then, the peptides/proteins need to be solubilized, usually by treating the homogenized tissues and cell lines with urea or detergents. In body fluids, the available peptides and proteins are already dissolved. However, body fluids contain many different molecules with a wide range of polarity, hydrophobicity, charge, and size. Thus, in peptidomics, due to the high complexity of samples, it is necessary to selectively enrich the compounds of interest. Also, various types of samples or peptides of interest need different sample preparation methods [49, 50]. Ultrafiltration; selective precipitation of proteins with organic solvents, acids, or chaotropic agents; solid-phase extraction (SPE); solid-phase microextraction (SPME); and magnetic beads are the most frequently utilized methods for peptide extraction in peptidomics [51]. Magnetic nanoparticles functionalized with different active moieties including enzymes, ligands, metal oxides, and other species have shown to be efficient for protein digestion and enrichment of low-abundance peptides or proteins [52]. Additionally, electromigration-based sample pretreatment techniques such as free-flow electrophoresis, isoelectric focusing (IEF), isotachophoresis (ITP), electrodialysis, electromembrane extraction, and electroextraction have been suggested for peptidomic and metabolomic studies [53, 54].

Peptides and proteins are present in biological samples in a large concentration range of 10–12 orders of magnitude [55]. Thus, high-abundance proteins (e.g., albumin) tend to interfere in the analysis of low-abundance peptides. For example, in urine samples, the high albumin content caused poor resolution and clogging of the CE capillary [56]. Before urine CE-MS analysis, Mischak et al. suggested to employ ultrafiltration (20 kDa molecular mass cutoff) in the presence of 2 mol/L urea, 0.02% sodium dodecyl sulfate (SDS), and 10 mmol/L NH_4OH. The presence of detergent and chaotropic agents inhibits protein-peptide interactions and thereby avoids peptides removal with albumin. In the following desalting step, low-molecular-mass compounds (urea, electrolytes, and salts) were removed by dialysis against 0.01% NH_4OH in HPLC-grade water [57]. After desalting, the sample was lyophilized, stored at 4 °C, and resuspended in HPLC-grade water shortly before CE-MS analysis. The same sample preparation procedure was also used for plasma samples [58].

For the cerebrospinal fluid (CSF), an analogous method was used [59]. CSF was centrifuged immediately after collection (at 4 °C), frozen, and stored at −80 °C. After thawing, alkaline buffer (pH 10.5) containing urea, NaCl, and NH_4OH was added to the CSF, followed by ultrafiltration and desalting procedures similarly as in urine and plasma samples.

3 CE-MS

3.1 Basic Characteristics

MS is the most broadly employed method for proteomic and peptidomic studies [60–62]. MS-based peptidomic methods are able to detect high number of peptides in a single experiment and identify their PTMs. However, in biological samples, relative levels of different peptides vary by many orders of magnitude, and low-abundance peptides are not present in sufficiently high levels to provide reasonable signals in the MS/MS analysis that is required for identification. In addition, the most abundant peptides can suppress the signals of the less abundant ones. Thus, it is frequently necessary to apply pre-separation techniques such as LC or CE before electrospray ionization (ESI)-MS or matrix-assisted laser desorption ionization (MALDI)-MS analysis. Commonly, CE system is coupled on-line to ESI ion source [60, 62] and off-line to MALDI-MS [63, 64]. The advantage of off-line regime is better tolerance to salts and possibility to store separated samples for long time. However, CE-MALDI suffers from low resolution, signal suppression, and higher variability of signals due to matrix effects. During MALDI-MS analysis, CE fractions are collected and spotted off-line onto a MALDI target plate. Subsequently, the polypeptides of interest can be analyzed with MALDI-time of flight (TOF)/TOF MS [65–67]. This approach has the advantage that the signal of interest can be located in the MS mode, and optimal fragmentation conditions can be determined without repeated separation. However, sequencing of native peptides with MALDI-TOF/TOF is generally unsuccessful, mostly due to low sensitivity and insufficient mass accuracy [68].

CE allows fast separation with high resolution and high efficiency, requires small sample amounts (advantage for rare and volume-restricted samples) and solvent volumes, and provides stable constant flow [69]. Disadvantages of the method include nonspecific adsorption of peptides and proteins onto the inner surface of the fused silica capillary, difficulties with analysis of diluted biological samples due to small injection volume (typically few to several nanoliters), and relatively short separation window. Nonspecific adsorption of peptides and proteins can be avoided by use of coated capillaries that also allow regulation of electroosmotic flow (EOF). Based on the mode of attachment of the coating material, the capillary coatings are divided to permanent, semipermanent, and dynamic coatings [70]. Dynamic coating agents (e.g., amines, anionic and cationic surfactants, neutral hydrophilic polymers, etc.) are added to the background electrolyte (BGE). In CE-MS, the use of dynamic coating agents can give rise to intense background signals, suppress ionization, and contaminate the ion source [71]. Permanent coating materials are irreversibly attached to the inner capillary wall either by physical adsorption or by covalent bonding. These coatings are usually highly stable over a wide pH range and do not interfere on the separation and/or detection system. The suitability of different capillary coatings (i.e., neutrally coated, positively coated, and bare-fused silica) was evaluated for the analysis of 70 synthetic peptides modified with common

PTMs (i.e., phosphorylation, acetylation, methylation, and nitration) [72]. For singly modified peptides, the neutrally coated capillary resulted in the highest signal intensity, but for multi-phosphorylated peptides, bare-fused silica with a stronger EOF transporting even multiply negatively charged peptides to cathodic capillary end yielded the best results. Sensitivity of CE analysis can be improved by application of different preconcentration techniques [73, 74] such as sample stacking [75], transient-isotachophoresis (t-ITP) [76, 77], dynamic pH junction [78], SPE [79], and SPME [80].

Over the years, various CE-ESI-MS interfaces have been developed [81–83], some of which are currently commercially available such as coaxial sheath-flow interface at G1607A analyzer (Agilent Technologies, USA), porous tip interface at CESI 8000 apparatus (AB SCIEX, USA), and electrospray emitter with EOF-driven sheath liquid at EMASS II device (Prince Technologies, The Netherlands). Two types of coupling, sheathless and sheath-flow interfaces, are currently being employed. In a recent study, the analytical performance of a nanoflow sheath liquid interface for CE-ESI-MS was compared to standard triple-tube sheath liquid and sheathless porous tip interfaces [84]. Both nanoflow interfaces gave similar results that were better than the results achieved with triple-tube interface. Nevertheless, sheath liquid systems offer higher flexibility in method development because of less dependence on separation and electrospray conditions.

3.2 Peptide Identification

For peptide identification, the MS instrument should enable accurate mass measurement and provide tandem mass fragmentation to give additional information for peptide sequencing [16, 51]. Fragmentation techniques most often used in peptidomics are collision-induced dissociation (CID), electron transfer dissociation (ETD), and high-energy collisional dissociation (HCD). Shen et al. demonstrated the complementarity of these three fragmentation methods and also the dependence of peptide identification rates on their combination with different identification methods [85]. The combination of all three abovementioned fragmentation methods enabled also to identify larger amount of endogenous tear peptides [86]. Another fragmentation technology, electron capture dissociation (ECD) with Fourier-transform ion cyclotron resonance (FT-ICR) MS, enables localization of even labile PTMs, such as glycosylation [87].

In peptidomic analysis, MS/MS data are interpreted with bioinformatics tools. These tools can identify peptides by database search engines, spectral matching algorithms, de novo sequencing, or hybrid approaches [88]. The commonly applied strategy for peptide identification is to employ sequence database searching where the obtained ion spectra are compared to theoretical spectra predicted for each protein or peptide contained in the sequence database. Software programs available for peptide identification of MS/MS data were originally developed for the identification of proteins. Thus, the majority of data in spectral libraries are generated from site-specific enzyme digestion. As the endogenous peptides are not produced by cleavage of one specific protease, their identification by sequence database searching is more difficult than identification of proteins in tryptic digests. Most proteomic software adjustable to peptidomic analysis can search for simple PTMs such as phosphorylation, deamidation, and oxidation. Few software platforms can identify endogenous proteins with more complex modifications such as glycosylation [16]. The most known database search software are Mascot [89], SeQuest [90], X!Tandem [91], OMSSA [92], and MS-Fit [93]. Majority of the software tools require the user to specify the expected PTM types. For peptide identification with unspecified PTMs, the software tool PeaksPTM has been developed [94].

Peptides can be also identified by spectral matching using databases such as SwePep [95], Erop-Moscow [96], PeptideDB [97], and Peptidome [98]. The databases can be searched using different peptide characteristics, e.g., peptide monoisotopic mass with(out) PTMs, length, and amino acid sequence. The SwePep [95] database was later extended with CID MS/MS spectra of endogenous peptides, which enables its use for validation of other experimentally derived spectra or for studying fragmentation patterns of peptides without specific enzymatic cleavage sites [99].

Sometimes, due to the lack of available amino acid sequences, the peptides can only be identified applying de novo sequencing methods. De novo sequencing needs an algorithm to search the optimal peptide out of enormous pool of peptides. Computational algorithms try to relate the peptide fragment masses to a series of amino acids in order to derive the peptide sequence [51, 100]. The method requires very good quality spectra. Most widely used programs for de novo sequencing in proteomics are Peaks [101], PepNovo [102–105], Sherenga [106], DirecTag [107], MS-Tag [108], and UStags [109]. Partial sequences (sequence tags) are easier to find than the full sequence out of fragmentation spectra. The tags found by de novo sequencing program are used to search a protein database. If a peptide is identified from the fragmentation spectrum with different approaches (database searches, spectral matching, and de novo sequencing), the chances for correct hit are higher (fewer false positive and negative hits). Combining of the different approaches allows also larger number of peptides to be identified.

It is necessary to statistically validate the results by means of identification probabilities and false discovery rates (FDRs). Some search engines contain their own statistical scoring mechanism (e.g., Mascot [89], X!Tandem [91]). Additional confirmation can be attained by taking into account fragment ion intensities, fragmentation patterns, and retention/migration times [51, 110–112]. In CE separations of biological fluids performed at acidic pH, "streaks" of peptides are observed when migration time is plotted against mass (see Fig. 2) [113]. Thus, according to the migration time in CE separation and according to their mass, the net charge of the peptide can be estimated.

Due to the limited amount of sample that can be loaded into CE capillary, often low-intensity peaks are obtained in MS, which complicate subsequent MS/MS analysis. Comparison of CE-MS/MS and LC-MS/MS sequencing applied for the determination of natural peptides in human urine resulted in higher amount of peptide sequences identified by LC-MS/MS (50%) than with CE-MS/MS (20%); with both methods 30% peptide sequences were identified [41]. At mass range < 1000 Da and between 5001 Da and 15,000 Da, more peptides in urine samples were detected with CE than LC [38]; LC was more effective in the region between 2000 Da and 4000 Da (see Fig. 3). To achieve higher identification rates, both CE-MS/MS and LC-MS/MS approaches were combined in the recent study comparing urine and plasma peptidomes [58].

3.3 Peptide Quantification

Besides identification purposes, MS-based methods can be used for the quantitation of endogenous peptides [114]. Absolute analyte quantification is performed by adding stable isotope labelled peptides to the sample that serve as internal standards [115]. The concentration can be calculated by comparing the ratio of intensity of the labelled analogue and the naturally occurring peptide. However, synthetic marker peptides can be used only in cases where exact sequence information (including posttranslational modifications) of the analytes to be quantified is available. In addition, this approach requires synthesis of the compound and will add further complexity to an already highly complex sample. It is also possible to examine the relative level of a peptide in two different samples by labelling the peptides in one sample with the light stable isotope and in the other sample with the heavy stable isotope; the two samples are then combined and analyzed together. The relative abundance of the peptide in the two samples can be calculated from the ratio of the peak intensity or peak area of the two isotopic forms.

Additionally, label-free ion counting methods are being utilized in peptidomics. These approaches include peak intensity (area under the curve), spectral counting, and selected reaction monitoring (SRM) [116, 117]. Peak intensity method involves the measurement of ion current

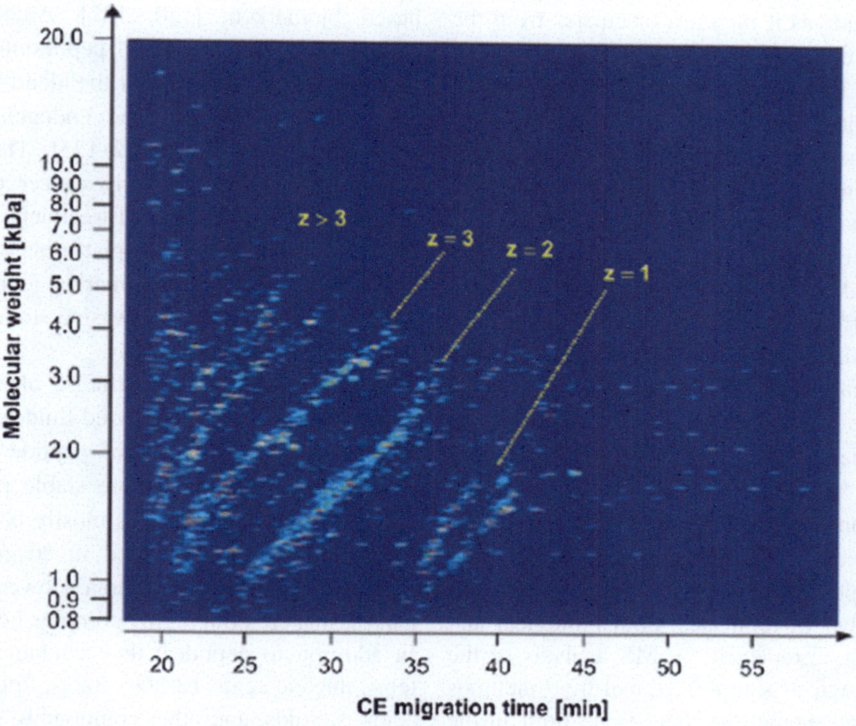

Fig. 2 Contour plot showing the organization of peptides in the CE-MS spectrogram in distinct charge-specific lines. The membership to a certain charge line allows reliable prediction of the number of basic amino acids. (Reprinted with permission from [113])

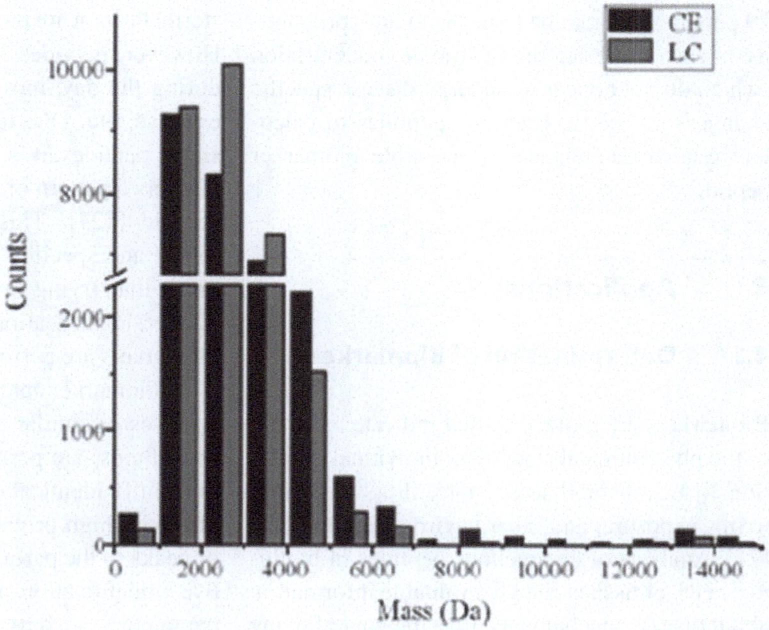

Fig. 3 Histogram representing the mass ranges of the peptide peaks detected in the CE-MS and LC-MS analysis of human urine samples. (Reprinted with permission from [38])

of the peptide as it migrates or eluates from the CE or LC columns into the MS detector, either integrating the signal over the entire migration/elution period or taking the peak value. Spectral counting method counts how many times a peptide is identified with MS/MS analysis. More abundant peptides are sampled more frequently; larger coverage of lower abundance peptides can be achieved by enhancing the number of experiments performed with sample [118]. SRM (referred also as multiple reaction monitoring, MRM) quantitation is implemented on triple quadrupole mass spectrometers and requires a synthetic peptide standard for each peptide to be quantified. This method enables to correctly quantify large as well as small changes of peptides between different sample groups [117]. For label-free peptidomics, it is important to include large number of replicates, both biological and technical (i.e., repeated CE-MS analysis of the same biological sample). Label-free methods have the advantage that there is no limit in the number of samples that can be compared; also smaller sample volumes can be employed.

Jantos-Siwy et al. [119] have suggested a special approach for the relative quantification of peptides of interest in urine samples. They chose 29 peptides that can be found with high probability in each urine sample in similar concentration (which do not appear to undergo disease specific changes) as "house-keeping" peptides to calculate relative abundances of possible biomarker peptides.

4 Applications

4.1 Determination of Biomarkers

Biomarkers are molecules that indicate a change of the physiological state of an individual in relation to health or disease state, drug treatment, toxins exposure, and other environmental effects [21]. Analysis of endogenous peptides in biofluids, cells, or tissues can give valuable information about disease mechanisms; thus the goal of many peptidomic studies is the search of molecules that correlate with disease states, mostly peptide-based biomarkers [120, 121]. Among other approaches, proteomics and peptidomics represent promising strategies for the identification of cancer-specific proteins and endogenous peptides, respectively [23, 122–125]. These techniques may help with early cancer detection, prognosis, and prediction of treatment response. Proteomic and peptidomic platforms employing biological fluids allow biomarker identification with soft-invasive or noninvasive strategy without the need for tissue biopsy.

Among various body fluids, blood serum [126] is the most often utilized fluid for clinical diagnosis. In the context of peptide profiling, blood plasma provides more stable peptidome readout than serum. This is mostly because the proteolytic activity involved in coagulation is higher in serum than in plasma. However, serum and plasma are both highly complex body fluids. In addition to peptides, they contain also proteins, nucleic acids, carbohydrates, lipids, amino acids, steroids, and other compounds. Urine has some advantages among different body fluids. It is easily collected in large quantities in a noninvasive manner, and its composition reflects health status of the kidney, the bladder, the prostate, and the vascular system. Urine samples collected in a sterile fashion are more stable compared to blood. However, peptides in urine change remarkably during the day, most likely because of the diet, exercise, etc. Due to the high variability of the single peptides, it is recommended to use as biomarkers a pattern of peptides rather than a single peptide [127]. This makes the method more robust and specific. Therefore, in many studies, rather than trying to identify a unique diagnostic marker, identification and validation of panels of biomarkers are performed [113, 128].

In biomarker analysis, frequently comparative analyses of similar samples (e.g., urine or other body fluids) are performed. Thus, it is important to identify identical compounds in different samples with high probability of correct assignment of peaks to the particular compounds. This tentative identification can be achieved utilizing parameters, such as mass, migration time, effective mobility, and ion counting as measure of relative abundance (or any other unique measure)

[129]. Because of some variability of migration times in CE analysis, it is suggested to calibrate migration times using internal standards (e.g., peptides frequently present in the sample). It is assumed that the relative migration time in correlation to the other peptides present in the sample does not change considerably. Softwares that automatically pick peaks based on parameters such as signal-to-noise ratio or appearance in several consecutive spectra have been developed (e.g., Mosaiques Visu [130], MSight [131], DeCyder MS [132]). An important feature of such softwares is the ability to perform charge deconvolution with a low error rate and combine peaks of the same mass but different charge states and to carry out an efficient normalization of migration times and amplitudes to compensate for any differences between individual measurements [87].

It is important that biomarker discovery studies use validated pre-analytical procedures and analytical platforms, appropriate statistics, relatively large patient groups for discovery, and independent groups (diseased patients, healthy individuals) for validation [113, 133]. The amino acid sequence of a peptide biomarker is not always necessary for its clinical or diagnostic use. However, the sequence can provide valuable information, which can be used in therapeutic development.

CE-MS is employed for the identification of biomarkers of disease as well as assessment of biomarkers in clinical diagnosis [133, 134]. The method has been also suggested for the evaluation of disease progression or the effects of therapy [127]. Most frequently, CE-MS has been applied in the studies of human urinary peptidome [135, 136], but the method has been also utilized for the analysis of other body fluids (e.g., CSF, plasma, bile) [137]. Peptide pattern for possible early diagnosis of Alzheimer disease in CSF based on 12 discriminatory peptides was developed by a CE-MS method (see Fig. 4) [59]. The human urinary peptidome was analyzed by CE-MS, and the pattern of peptide urinary biomarkers for the detection of chronic kidney disease (CKD) was described [138]. A set of 273 urinary peptides that were remarkably altered in the urine of 230 patients with CKD was identified, when compared with 379 healthy subjects. Additionally, urinary biomarkers for diabetes, diabetic nephropathy, and nondiabetic proteinuric renal diseases were defined and validated in blinded data sets using CE-MS [139]. Forty biomarkers that differentiated between patients with diabetes and healthy persons were identified. Moreover, the determined biomarker patterns enabled to distinguish patients with diabetic nephropathy from patients with other chronic renal diseases. In other studies, CE-MS was utilized to identify 22 potential polypeptide markers of urothelial carcinoma [140] or determine the panel of twelve new biomarkers of prostate cancer from the first-void urine [141]. In addition, utilizing CE-MS method, a characteristic biomarker panel for coronary artery disease composed of fifteen peptides was defined [142].

Peptidomic studies also helped to clarify behavioral observations of animals [143] or to develop peptide biomarker panels for diagnosis of various animal diseases (e.g., bovine mastitis) [144, 145].

4.2 Food Peptidomics

Food or nutritional peptidomics as a part of foodomics [146–148] focuses on the composition, interactions, and properties of bioactive peptides present in different food matrices [20, 149]. The food peptidome can be defined as the whole peptide pool present in food products or raw materials or obtained during processing and storage [150]. Food peptidomic studies are dealing with authenticity, origin, and history of food products but also with functional and sensory properties, allergenicity, and biological activity of peptides found in foods or raw materials. Food-originating peptides possessing particular biological activity (e.g., antihypertensive, antithrombotic, antimicrobial, immunomodulating) can be released by gastrointestinal digestion or during food processing and storage, or by in vitro hydrolysis with proteolytic enzymes [151, 152].

Dairy products that are considered to be a valuable source of biologically active peptides

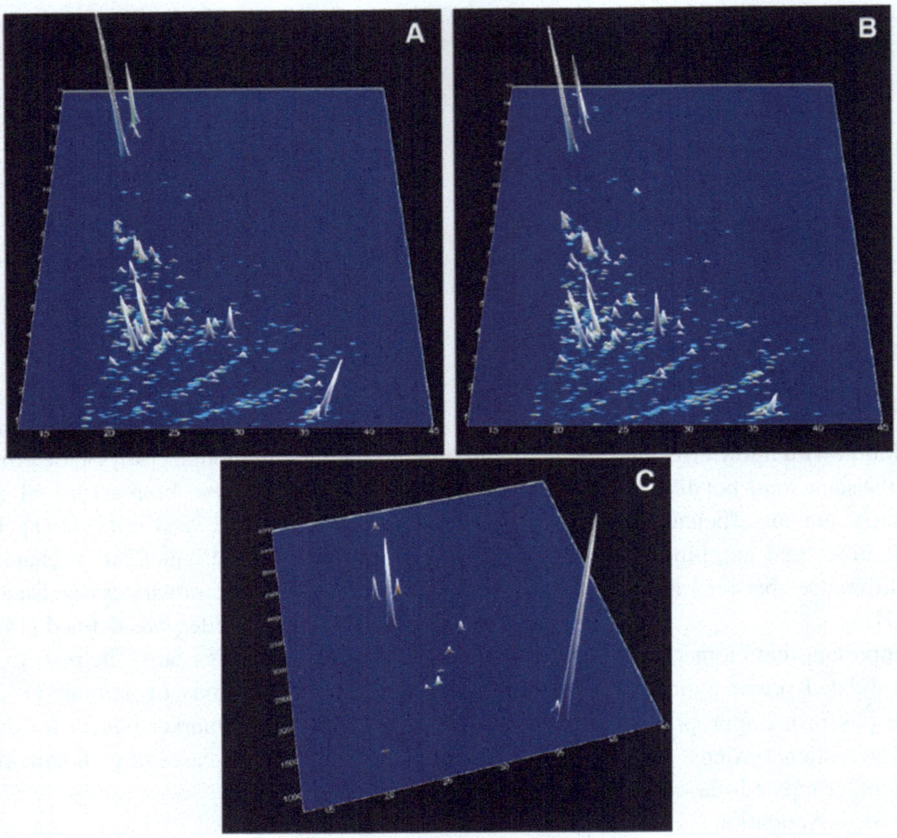

Fig. 4 Contour plots of the training set. The normalized CE-migration time (in min) is plotted on the x-axis and the relative molecular mass (in kDa) on the y-axis. As a third dimension, the signal intensity is color-coded (blue lowest and white highest signal intensity). Each dot represents one peptide. (**a**) Compiled 3D protein contour plot from CSF samples of 34 patients with Alzheimer's disease (AD). (**b**) Compiled 3D protein contour plot for healthy controls (n = 17). (**c**) Discriminative biomarker pattern for subjects suffering from AD (n = 34). Depicted is a 3D plot of 12 peptides that serve as specific biomarkers for AD brain damage. (Reprinted with permission from [59])

have become the main subject of many peptidomic studies [149, 153, 154]. CE-MS combined with semipreparative reversed-phase (RP)-HPLC separation enabled to determine four potential angiotensin-converting enzyme-inhibitory peptides from raw ovine milk [155]. The active sequences could be obtained by enzymatic hydrolysis either of the individual κ-casein fraction or the total casein fraction from ovine milk.

Peptide profile of commercial Prato cheese was characterized by MALDI-MS and CE [156]. CE revealed a characteristic pattern of hydrolysis, with formation of para-κ-casein, hydrolysis of αs1-casein at the Phe23-Phe24 bond, and hydrolysis of β-casein. By MALDI-MS, 44 previously reported peptides were identified.

In a nutraceutical derived from a bovine milk protein hydrolysate, 17 antihypertensive peptides were identified using CE-MS in combination with chemometric tools. The identity of these peptides was confirmed by CE-MS/MS [157].

For the separation and identification of bioactive peptides in hypoallergenic infant milk formulas, CE-ion trap (IT)-MS and CE-TOF-MS were employed (see Fig. 5) [158]. Additionally, classical semiempirical relation between electrophoretic mobility and charge-to-mass ratio (m_e vs. q/M^{α}, $\alpha = 1/2$ for the classical polymer model) was used

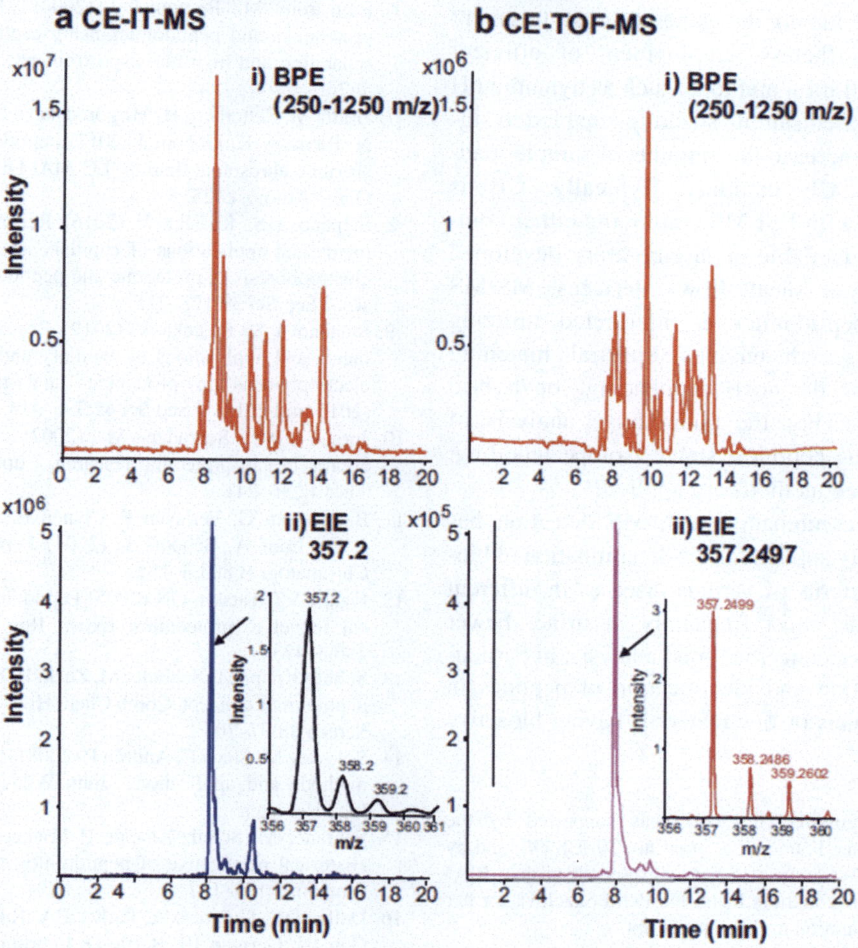

Fig. 5 Analysis of infant milk formulae by (**a**) CE-IT-MS and (**b**) CE-TOF-MS. (i) Base peak electropherograms (BPE) and (ii) extracted ion electropherograms (EIE); CE-IT-MS 357.2 ± 0.5 *m/z* and CE-TOF-MS 357.2497 ± 0.0005 *m/z*. A piece of the mass spectra of the electrophoretic peak corresponding to the bioactive peptide is also shown. (Reprinted with permission from [158])

to describe the migration behavior of detected bioactive peptides [111]. This approach helped with the identification of the peptides and improved the reliability of the results. Mostly, the angiotensin-converting enzyme inhibitors and antihypertensive peptides were identified but peptides with other biological activities were also found.

5 Conclusions

CE is an alternative and complementary method for LC in peptidomic studies. Advantages of CE are high separation efficiency, short analysis time, requirement of small sample amounts, low consumption of chemicals and solvents, and tolerance to interfering compounds that can be separated from the compounds of interest. Since lower sample amounts can be loaded into CE capillary than to LC column in peptidomic studies with CE-MS methods, generally less peptides are identified than with LC-MS techniques. However, even though CE has limited loading capacity, in typical urine or plasma sample, several thousand peptides and proteins can be routinely detected using CE-MS [133]. Low loading capacity is more challenging in CE-MS/MS applications, where more material is needed due to the generation of fragment ions. Nevertheless, employment of efficient preconcentration methods, such as dynamic pH junction-based sample stacking, enabled to significantly increase the amounts of sample loadable into CE capillary. Typically, CE is combined with ESI MS employing either commercially available or in laboratory developed sheathless or sheath-flow interfaces. MS/MS data in peptidomics is interpreted utilizing database search engines, spectral matching algorithms, de novo sequencing, or hybrid approaches. For the quantitative analysis of endogenous peptides, stable isotope labelling, or label-free methods are applied.

CE in combination with MS detection has been mainly applied for the determination of biomarker patterns of various diseases in different body fluids, most frequently in urine. Fewer reports have concerned food analysis, in particular separation and identification of peptides in food products or raw materials having bioactive properties.

Acknowledgments The work was supported by the Czech Science Foundation, grant no. 20-03899S, and by the Czech Academy of Sciences, research project RVO 61388963. The authors thank Dr. Petra Sázelová for her help in preparation of this manuscript.

Conflict of Interest Statement The authors declare no conflict of interest.

References

1. Saz JM, Marina ML (2008) Application of micro- and nano-HPLC to the determination and characterization of bioactive and biomarker peptides. J Sep Sci 31:446–458
2. Khalaf R, Baur D, Pfister D (2015) Optimization of reversed-phase chromatography methods for peptide analytics. J Chromatogr A 1425:198–203
3. Le Maux S, Nongonierma AB, FitzGerald RJ (2015) Improved short peptide identification using HILIC-MS/MS: retention time prediction model based on the impact of amino acid position in the peptide sequence. Food Chem 173:847–854
4. Kašička V (2018) Recent developments in capillary and microchip electroseparations of peptides (2015-mid 2017). Electrophoresis 39:209–234
5. Mikšík I (2019) Coupling of CE-MS for protein and peptide analysis. J Sep Sci 42:385–397
6. Ramstrom M, Bergquist J (2004) Miniaturized proteomics and peptidomics using capillary liquid separation and high mass spectrometry. FEBS Lett 567:92–95
7. Holtta M, Zetterberg H, Mirgorodskaya E, Mattsson N, Blennow K, Gobom J (2012) Peptidome analysis of cerebrospinal fluid by LC-MALDI MS. PLoS One 7:Art. no. e42555
8. Štěpánová S, Kašička V (2016) Recent developments and applications of capillary and microchip electrophoresis in proteomic and peptidomic analyses. J Sep Sci 39:198–211
9. Štěpánová S, Kašička V (2019) Recent developments and applications of capillary and microchip electrophoresis in proteomics and peptidomics (2015-mid 2018). J Sep Sci 42:398–414
10. Jurgens M, Schrader M (2002) Peptidomic approaches in proteomic research. Curr Opin Mol Ther 4:236–241
11. Baggerman G, Verleyen P, Clynen E, Huybrechts J, De Loof A, Schoofs L (2004) Peptidomics. J Chromatogr B 803:3–16
12. Ivanov VT, Yatskin ON (2005) Peptidomics: a logical sequel to proteomics. Expert Rev Proteomics 2:463–473
13. Schulz-Knappe P, Schrader M, Zucht HD (2005) The peptidomics concept. Comb Chem High Throughput Screen 8:697–704
14. Soloviev M, Shaw C, Andrén P (2008) Peptidomics: methods and applications. John Wiley and Sons, Inc., Hoboken
15. Schrader M, Schulz-Knappe P, Fricker LD (2014) Historical perspective of peptidomics. EuPA Open Proteom 3:171–182
16. Dallas DC, Guerrero A, Parker EA, Robinson RC, Gan JN, German JB, Barile D, Lebrilla CB (2015) Current peptidomics: applications, purification, identification, quantification, and functional analysis. Proteomics 15:1026–1038
17. Mahboob S, Mohamedali A, Ahn SB, Schulz-Knappe P, Nice E, Baker MS (2015) Is isolation of comprehensive human plasma peptidomes an achievable quest? J Proteome 127:300–309
18. Schrader M (2018) Origins, technological development, and applications of peptidomics. Methods Mol Biol 1719:3–39
19. Romanova EV, Sweedler JV (2015) Peptidomics for the discovery and characterization of neuropeptides and hormones. Trends Pharmacol Sci 36:579–586
20. Agyei D, Tsopmo A, Udenigwe CC (2018) Bioinformatics and peptidomics approaches to the discovery and analysis of food-derived bioactive peptides. Anal Bioanal Chem 410:3463–3472
21. Schrader M, Selle H (2006) The process chain for peptidomic biomarker discovery. Dis Markers 22:27–37
22. Bauca JM, Martinez-Morillo E, Diamandis EP (2014) Peptidomics of urine and other biofluids for cancer diagnostics. Clin Chem 60:1052–1061

23. Di Meo A, Pasic MD, Yousef GM (2016) Proteomics and peptidomics: moving toward precision medicine in urological malignancies. Oncotarget 7:52460–52474
24. Picariello G, Mamone G, Nitride C, Addeo F, Ferranti P (2013) Protein digestomics: integrated platforms to study food-protein digestion and derived functional and active peptides. Trends Anal Chem 52:120–134
25. Toldra F, Reig M, Aristoy MC, Mora L (2018) Generation of bioactive peptides during food processing. Food Chem 267:395–404
26. Crameri R (2005) The potential of proteomics and peptidomics for allergy and asthma research. Allergy 60:1227–1237
27. Krochmal M, Schanstra JP, Mischak H (2018) Urinary peptidomics in kidney disease and drug research. Expert Opin Drug Discovery 13:259–268
28. Clynen E, Baggerman G, Husson SJ, Landuyt B, Schoofs L (2008) Peptidomics in drug research. Expert Opin Drug Discovery 3:425–440
29. Tsiatsiani L, Heck AJR (2015) Proteomics beyond trypsin. FEBS J 282:2612–2626
30. Giansanti P, Tsiatsiani L, Low TY, Heck AJR (2016) Six alternative proteases for mass spectrometry-based proteomics beyond trypsin. Nat Protoc 11:993–1006
31. Tinoco AD, Saghatelian A (2011) Investigating endogenous peptides and peptidases using peptidomics. Biochemistry 50:7447–7461
32. Sun LL, Zhu GJ, Yan XJ, Zhang ZB, Wojcik R, Champion MM, Dovichi NJ (2016) Capillary zone electrophoresis for bottom-up analysis of complex proteomes. Proteomics 16:188–196
33. Vitorino R (2018) Digging deep into peptidomics applied to body fluids. Proteomics 18:Art. No. 1700401. https://doi.org/10.1002/pmic.201700401
34. Dams M, Dores-Sousa JL, Lamers RJ, Treumann A, Eeltink S (2019) High-resolution nano-liquid chromatography with tandem mass spectrometric detection for the bottom-up analysis of complex proteomic samples. Chromatographia 82:101–110
35. Gan JN, Robinson RC, Wang JQ, Krishnakumar N, Manning CJ, Lor Y, Breck M, Barile D, German JB (2019) Peptidomic profiling of human milk with LC-MS/MS reveals pH-specific proteolysis of milk proteins. Food Chem 274:766–774
36. Chen DY, Shen XJ, Sun LL (2017) Capillary zone electrophoresis-mass spectrometry with microliter-scale loading capacity, 140 min separation window and high peak capacity for bottom-up proteomics. Analyst 142:2118–2127
37. Lubeckyj RA, McCool EN, Shen XJ, Kou Q, Liu XW, Sun LL (2017) Single-shot top-down proteomics with capillary zone electrophoresis-electrospray ionization-tandem mass spectrometry for identification of nearly 600 Escherichia coli proteoforms. Anal Chem 89:12059–12067
38. Mullen W, Albalat A, Gonzalez J, Zerefos P, Siwy J, Franke J, Mischak H (2012) Performance of different separation methods interfaced in the same MS-reflection TOF detector: a comparison of performance between CE versus HPLC for biomarker analysis. Electrophoresis 33:567–574
39. Li YH, Champion MM, Sun LL, Champion PAD, Wojcik R, Dovichi NJ (2012) Capillary zone electrophoresis-electrospray ionization-tandem mass spectrometry as an alternative proteomics platform to ultraperformance liquid chromatography-electrospray ionization-tandem mass spectrometry for samples of intermediate complexity. Anal Chem 84:1617–1622
40. Sarg B, Faserl K, Kremser L, Halfinger B, Sebastiano R, Lindner HH (2013) Comparing and combining capillary electrophoresis electrospray ionization mass spectrometry and nano-liquid chromatography electrospray ionization mass spectrometry for the characterization of post-translationally modified histones. Mol Cell Proteomics 12:2640–2656
41. Klein J, Papadopoulos T, Mischak H, Mullen W (2014) Comparison of CE-MS/MS and LC-MS/MS sequencing demonstrates significant complementarity in natural peptide identification in human urine. Electrophoresis 35:1060–1064
42. Chen DY, Shen XJ, Sun LL (2018) Strong cation exchange-reversed phase liquid chromatography-capillary zone electrophoresis-tandem mass spectrometry platform with high peak capacity for deep bottom-up proteomics. Anal Chim Acta 1012:1–9
43. Boonen K, Creemers JW, Schoofs L (2009) Bioactive peptides, networks and systems biology. BioEssays 31:300–314
44. Bodzon-Kulakowska A, Bierczynska-Krzysik A, Dylag T, Drabik A, Suder P, Noga M, Jarzebinska J, Silberring J (2007) Methods for samples preparation in proteomic research. J Chromatogr B 849:1–31
45. Finoulst I, Pinkse M, Van Dongen W, Verhaert P (2011) Sample preparation techniques for the untargeted LC-MS-based discovery of peptides in complex biological matrices. J Biomed Biotechnol:Article No. 245291. https://doi.org/10.1155/2011/245291
46. Niu ZL, Zhang WW, Yu CW, Zhang J, Wen YY (2018) Recent advances in biological sample preparation methods coupled with chromatography, spectrometry and electrochemistry analysis techniques. Trends Anal Chem 102:123–146
47. Xue YJ, Gao H, Ji QC, Lam Z, Fang XP, Lin ZP, Hoffman M, Schulz-Jander D, Weng ND (2012) Bioanalysis of drug in tissue: current status and challenges. Bioanalysis 4:2637–2653
48. Smith KM, Xu Y (2012) Tissue sample preparation in bioanalytical assays. Bioanalysis 4:741–749
49. Aristoteli LP, Molloy MP, Baker MS (2007) Evaluation of endogenous plasma peptide extraction methods for mass spectrometric biomarker discovery. J Proteome Res 6:571–581
50. Vitorino R, Barros AS, Caseiro A, Ferreira R, Amado F (2012) Evaluation of different extraction procedures for salivary peptide analysis. Talanta 94:209–215

51. Maes E, Oeyen E, Boonen K, Schildermans K, Mertens I, Pauwels P, Valkenborg D, Baggerman G (2018) The challenges of peptidomics in complementing proteomics in a clinical context. Mass Spectrom Rev. https://doi.org/10.1002/mas.21581
52. Li Y, Zhang XM, Deng CH (2013) Functionalized magnetic nanoparticles for sample preparation in proteomics and peptidomics analysis. Chem Soc Rev 42:8517–8539
53. Kamphorst JJ, Tjaden UR, van der Heijden R, DeGroot J, Van der Greef J, Hankemeier T (2009) Feasibility of electrodialysis as a fast and selective sample preparation method for the profiling of low-abundant peptides in biofluids. Electrophoresis 30:2284–2292
54. Lindenburg PW, Ramautar R, Hankemeier T (2013) The potential of electrophoretic sample pretreatment techniques and new instrumentation for bioanalysis, with a focus on peptidomics and metabolomics. Bioanalysis 5:2785–2801
55. Anderson NL, Anderson NG (2002) The human plasma proteome – history, character, and diagnostic prospects. Mol Cell Proteomics 1:845–867
56. Theodorescu D, Fliser D, Wittke S, Mischak H, Krebs R, Walden M, Ross M, Eltze E, Bettendorf O, Wulfing C, Semjonow A (2005) Pilot study of capillary electrophoresis coupled to mass spectrometry as a tool to define potential prostate cancer biomarkers in urine. Electrophoresis 26:2797–2808
57. Mischak H, Vlahou A, Ioannidis JPA (2013) Technical aspects and inter-laboratory variability in native peptide profiling: the CE-MS experience. Clin Biochem 46:432–443
58. Magalhaes P, Pontillo C, Pejchinovski M, Siwy J, Krochmal M, Makridakis M, Carrick E, Klein J, Mullen W, Jankowski J, Vlahou A, Mischak H, Schanstra JP, Zurbig P, Pape L (2018) Comparison of urine and plasma peptidome indicates selectivity in renal peptide handling. Proteomics Clin Appl 12:Article Number: 1700163. https://doi.org/10.1002/prca.201700163
59. Jahn H, Wittke S, Zurbig P, Raedler TJ, Arlt S, Kellmann M, Mullen W, Eichenlaub M, Mischak H, Wiedemann K (2011) Peptide fingerprinting of Alzheimer's disease in cerebrospinal fluid: identification and prospective evaluation of new synaptic biomarkers. PLoS One 6:Art. No.: e26540. https://doi.org/10.1371/journal.pone.0026540
60. Herrero M, Ibanez E, Cifuentes A (2008) Capillary electrophoresis-electrospray-mass spectrometry in peptide analysis and peptidomics. Electrophoresis 29:2148–2160
61. Simo C, Cifuentes A, Kašička V (2013) Capillary electrophoresis-mass spectrometry for peptide analysis: target-based approaches and proteomics/peptidomics strategies. In: Volpi N, Maccari F (eds) Capillary electrophoresis of biomolecules. Methods and protocols. Humana Press (Springer), New York, pp 139–151
62. Robledo VR, Smyth WF (2014) Review of the CE-MS platform as a powerful alternative to conventional couplings in bio-omics and target-based applications. Electrophoresis 35:2292–2308
63. Wang JH, Jiang XY, Sturm RM, Li LJ (2009) Combining tissue extraction and off-line capillary electrophoresis matrix-assisted laser desorption/ionization Fourier transform mass spectrometry for neuropeptide analysis in individual neuronal organs using 2,5-dihydroxybenzoic acid as a multifunctional agent. J Chromatogr A 1216:8283–8288
64. Wang JH, Zhang YZ, Xiang F, Zhang ZC, Li LJ (2010) Combining capillary electrophoresis matrix-assisted laser desorption/ionization mass spectrometry and stable isotopic labeling techniques for comparative crustacean peptidomics. J Chromatogr A 1217:4463–4470
65. Rejtar T, Hu P, Juhasz P, Campbell JM, Vestal ML, Preisler J, Karger BL (2002) Off-line coupling of high-resolution capillary electrophoresis to MALDI-TOF and TOF/TOF MS. J Proteome Res 1:171–179
66. Mischak H, Kaiser T, Walden M, Hillmann M, Wittke S, Herrmann A, Knueppel S, Haller H, Fliser D (2004) Proteomic analysis for the assessment of diabetic renal damage in humans. Clin Sci 107:485–495
67. Wittke S, Mischak H, Walden M, Kolch W, Radler T, Wiedemann K (2005) Discovery of biomarkers in human urine and cerebrospinal fluid by capillary electrophoresis coupled to mass spectrometry: towards new diagnostic and therapeutic approaches. Electrophoresis 26:1476–1487
68. Zurbig P, Renfrow MB, Schiffer E, Novak J, Walden M, Wittke S, Just I, Pelzing M, Neusüss C, Theodorescu D, Root KE, Ross MM, Mischak H (2006) Biomarker discovery by CE-MS enables sequence analysis via MS/MS with platform-independent separation. Electrophoresis 27:2111–2125
69. Voeten RLC, Ventouri IK, Haselberg R, Somsen GW (2018) Capillary electrophoresis: trends and recent advances. Anal Chem 90:1464–1481
70. Hajba L, Guttman A (2017) Recent advances in column coatings for capillary electrophoresis of proteins. Trends Anal Chem 90:38–44
71. Huhn C, Ramautar R, Wuhrer M, Somsen GW (2010) Relevance and use of capillary coatings in capillary electrophoresis-mass spectrometry. Anal Bioanal Chem 396:297–314
72. Faserl K, Sarg B, Gruber P, Lindner HH (2018) Investigating capillary electrophoresis-mass spectrometry for the analysis of common post-translational modifications. Electrophoresis 39:1208–1215
73. Kitagawa F, Otsuka K (2014) Recent applications of on-line sample preconcentration techniques in capillary electrophoresis. J Chromatogr A 1335:43–60
74. Breadmore MC, Grochocki W, Kalsoom U, Alves MN, Phung SC, Rokh MT, Cabot JM, Ghiasvand A, Li F, Shallan AI, Keyon ASA, Alhusban AA,

See HH, Wüthrich A, Dawod M, Quirino JP (2019) Recent advances in enhancing the sensitivity of electrophoresis and electrochromatography in capillaries and microchips (2016-2018). Electrophoresis 40:17–39
75. Šlampová A, Malá Z, Gebauer P (2019) Recent progress of sample stacking in capillary electrophoresis (2016-2018). Electrophoresis 40:40–54
76. Guo XJ, Fillmore TL, Gao YQ, Tang KQ (2016) Capillary electrophoresis-nanoelectrospray ionization-selected reaction monitoring mass spectrometry via a true sheathless metal-coated emitter interface for robust and high-sensitivity sample quantification. Anal Chem 88:4418–4425
77. Malá Z, Gebauer P (2019) Recent progress in analytical capillary isotachophoresis. Electrophoresis 40:55–64
78. Zhu GJ, Sun LL, Dovichi NJ (2016) Dynamic pH junction preconcentration in capillary electrophoresis-electrospray ionization-mass spectrometry for proteomics analysis. Analyst 141:5216–5220
79. Ramautar R, Somsen GW, de Jong GJ (2016) Developments in coupled solid-phase extraction-capillary electrophoresis 2013-2015. Electrophoresis 37:35–44
80. Kohler I, Schappler J, Rudaz S (2013) Microextraction techniques combined with capillary electrophoresis in bioanalysis. Anal Bioanal Chem 405:125–141
81. Lindenburg PW, Haselberg R, Rozing G, Ramautar R (2015) Developments in interfacing designs for CE-MS: towards enabling tools for proteomics and metabolomics. Chromatographia 78:367–377
82. Týčová A, Ledvina V, Klepárník K (2017) Recent advances in CE-MS coupling: instrumentation, methodology, and applications. Electrophoresis 38:115–134
83. Stolz A, Jooss K, Hocker O, Romer J, Schlecht J, Neusüss C (2019) Recent advances in capillary electrophoresis-mass spectrometry: instrumentation, methodology and applications. Electrophoresis 40:79–112
84. Hocker O, Montealegre C, Neusüss C (2018) Characterization of a nanoflow sheath liquid interface and comparison to a sheath liquid and a sheathless porous-tip interface for CE-ESI-MS in positive and negative ionization. Anal Bioanal Chem 410:5265–5275
85. Shen YF, Tolic N, Xie F, Zhao R, Purvine SO, Schepmoes AA, Ronald JM, Anderson GA, Smith RD (2011) Effectiveness of CID, HCD, and ETD with FT MS/MS for degradomic-peptidomic analysis: comparison of peptide identification methods. J Proteome Res 10:3929–3943
86. Azkargorta M, Soria J, Ojeda C, Guzman F, Acera A, Iloro I, Suarez T, Elortza F (2015) Human basal tear peptidome characterization by CID, HCD, and ETD followed by in silico and in vitro analyses for antimicrobial peptide identification. J Proteome Res 14:2649–2658
87. Mischak H, Coon JJ, Novak J, Weissinger EM, Schanstra JP, Dominiczak AF (2009) Capillary electrophoresis-mass spectrometry as a powerful tool in biomarker discovery and clinical diagnosis: an update of recent developments. Mass Spectrom Rev 28:703–724
88. Menschaert G, Van de Kerckhove TTM, Baggerman G, Schoofs L, Luyten W, Van Criekinge W (2010) Peptidomics coming of age: a review of contributions from a bioinformatics angle. J Proteome Res 9:2051–2061
89. Perkins DN, Pappin DJC, Creasy DM, Cottrell JS (1999) Probability-based protein identification by searching sequence databases using mass spectrometry data. Electrophoresis 20:3551–3567
90. Eng JK, McCormack AL, Yates JR (1994) An approach to correlate tandem mass-spectral data of peptides with amino-acid-sequences in a protein database. J Am Soc Mass Spectrom 5:976–989
91. Craig R, Beavis RC (2004) TANDEM: matching proteins with tandem mass spectra. Bioinformatics 20:1466–1467
92. Geer LY, Markey SP, Kowalak JA, Wagner L, Xu M, Maynard DM, Yang XY, Shi WY, Bryant SH (2004) Open mass spectrometry search algorithm. J Proteome Res 3:958–964
93. Jimenez CR, Huang L, Qiu Y, Burlingame AL (1998) Searching sequence databases over the internet: protein identification using MS-fit. Curr Protoc Protein Sci 14:unit 16.5
94. Han X, He L, Xin L, Shan BZ, Ma B (2011) PeaksPTM: mass spectrometry-based identification of peptides with unspecified modifications. J Proteome Res 10:2930–2936
95. Falth M, Skold K, Norrman M, Svensson M, Fenyo D, Andren PE (2006) SwePep, a database designed for endogenous peptides and mass spectrometry. Mol Cell Proteomics 5:998–1005
96. Zamyatnin AA, Borchikov AS, Vladimirov MG, Voronina OL (2006) The EROP-Moscow oligopeptide database. Nucleic Acids Res 34:D261–D266
97. Liu F, Baggerman G, Schoofs L, Wets G (2008) The construction of a bioactive peptide database in Metazoa. J Proteome Res 7:4119–4131
98. Minamino N (2001) Peptidome: the fact-database for endogenous peptides. Tanpakushitsu Kakusan Koso 46:1510–1517
99. Falth M, Svensson M, Nilsson A, Skold K, Fenyo D, Andren PE (2008) Validation of endogenous peptide identifications using a database of tandem mass spectra. J Proteome Res 7:3049–3053
100. Boonen K, Landuyt B, Baggerman G, Husson SJ, Huybrechts J, Schoofs L (2008) Peptidomics: the integrated approach of MS, hyphenated techniques and bioinformatics for neuropeptide analysis. J Sep Sci 31:427–445
101. Ma B, Lajoie G (2009) De novo interpretation of tandem mass spectra. Curr Protoc

Bioinformatics 25:Chapter 13, Unit 10. https://doi.org/10.1002/0471250953.bi1310s25
102. Frank A, Pevzner P (2005) PepNovo: De novo peptide sequencing via probabilistic network modeling. Anal Chem 77:964–973
103. Frank AM, Savitski MM, Nielsen ML, Zubarev RA, Pevzner PA (2007) De novo peptide sequencing and identification with precision mass spectrometry. J Proteome Res 6:114–123
104. Frank AM (2009) A ranking-based scoring function for peptide-spectrum matches. J Proteome Res 8:2241–2252
105. Frank AM (2009) Predicting intensity ranks of peptide fragment ions. J Proteome Res 8:2226–2240
106. Dancik V, Addona TA, Clauser KR, Vath JE, Pevzner PA (1999) De novo peptide sequencing via tandem mass spectrometry. J Comput Biol 6:327–342
107. Tabb DL, Ma ZQ, Martin DB, Ham AJL, Chambers MC (2008) DirecTag: accurate sequence tags from peptide MS/MS through statistical scoring. J Proteome Res 7:3838–3846
108. Jimenez CR, Huang L, Qiu Y, Burlingame AL (1998) Searching sequence databases over the internet: protein identification using MS-tag. Curr Protoc Protein Sci 14:Chapter 16, unit 16.6. https://doi.org/10.1002/0471140864.ps1606s14
109. Shen YF, Tolic N, Hixson KK, Purvine SO, Pasa-Tolic L, Qian WJ, Adkins JN, Moore RJ, Smith RD (2008) Proteome-wide identification of proteins and their modifications with decreased ambiguities and improved false discovery rates using unique sequence tags. Anal Chem 80:1871–1882
110. Simo C, Gonzalez R, Barbas C, Cifuentes A (2005) Combining peptide modeling and capillary electrophoresis mass spectrometry for characterization of enzymes cleavage patterns: recombinant versus natural bovine pepsin A. Anal Chem 77:7709–7716
111. Catala-Clariana S, Benavente F, Gimenez E, Barbosa J, Sanz-Nebot V (2013) Identification of bioactive peptides in hypoallergenic infant milk formulas by CE-TOF-MS assisted by semiempirical model of electromigration behavior. Electrophoresis 34:1886–1894
112. Barroso A, Gimenez E, Benavente F, Barbosa J, Sanz-Nebot V (2015) Modelling the electrophoretic migration behaviour of peptides and glycopeptides from glycoprotein digests in capillary electrophoresis-mass spectrometry. Anal Chim Acta 854:169–177
113. Metzger J, Luppa PB, Good DM, Mischak H (2009) Adapting mass spectrometry-based platforms for clinical proteomics applications: the capillary electrophoresis coupled mass spectrometry paradigm. Crit Rev Clin Lab Sci 46:129–152
114. Romanova EV, Dowd SE, Sweedler JV (2013) Quantitation of endogenous peptides using mass spectrometry based methods. Curr Opin Chem Biol 17:801–808
115. Fricker LD, Lim JY, Pan H, Che FY (2006) Peptidomics: identification and quantification of endogenous peptides in neuroendocrine tissues. Mass Spectrom Rev 25:327–344
116. Neilson KA, Ali NA, Muralidharan S, Mirzaei M, Mariani M, Assadourian G, Lee A, van Sluyter SC, Haynes PA (2011) Less label, more free: approaches in label-free quantitative mass spectrometry. Proteomics 11:535–553
117. Fricker L (2018) Quantitative peptidomics: general considerations. Methods Mol Biol 1719:121–140
118. Liu HB, Sadygov RG, Yates JR (2004) A model for random sampling and estimation of relative protein abundance in shotgun proteomics. Anal Chem 76:4193–4201
119. Jantos-Siwy J, Schiffer E, Brand K, Schumann G, Rossing K, Delles C, Mischak H, Metzger J (2009) Quantitative urinary proteome analysis for biomarker evaluation in chronic kidney disease. J Proteome Res 8:268–281
120. Rodriguez-Ortiz ME, Pontillo C, Rodriguez M, Zurbig P, Mischak H, Ortiz A (2018) Novel urinary biomarkers for improved prediction of progressive eGFR loss in early chronic kidney disease stages and in high risk individuals without chronic kidney disease. Sci Rep 8:Article number 15940. https://doi.org/10.1038/s41598-018-34386-8
121. Pejchinovski M, Siwy J, Metzger J, Dakna M, Mischak H, Klein J, Jankowski V, Bae KT, Chapman AB, Kistler AD (2017) Urine peptidome analysis predicts risk of end-stage renal disease and reveals proteolytic pathways involved in autosomal dominant polycystic kidney disease progression. Nephrol Dial Transplant 32:487–497
122. Simionato AVC, Carrilho E, Tavares MFM (2010) CE-MS and related techniques as a valuable tool in tumor biomarkers research. Electrophoresis 31:1214–1226
123. Latosinska A, Frantzi M, Vlahou A, Mischak H (2013) Clinical applications of capillary electrophoresis coupled to mass spectrometry in biomarker discovery: focus on bladder cancer. Proteomics Clin Appl 7:779–793
124. Frantzi M, van Kessel KE, Zwarthoff EC, Marquez M, Rava M, Malats N, Merseburger AS, Katafigiotis I, Stravodimos K, Mullen W, Zoidakis J, Makridakis M, Pejchinovski M, Critselis E, Lichtinghagen R, Brand K, Dakna M, Roubelakis MG, Theodorescu D, Vlahou A, Mischak H, Anagnou NP (2016) Development and validation of urine-based peptide biomarker panels for detecting bladder cancer in a multi-center study. Clin Cancer Res 22:4077–4086
125. Belczacka I, Latosinska A, Siwy J, Metzger J, Merseburger AS, Mischak H, Vlahou A, Frantzi M, Jankowski V (2018) Urinary CE-MS peptide marker pattern for detection of solid tumors. Sci Rep 8:Article Number: 5227. https://doi.org/10.1038/s41598-018-23585-y

126. Gao Y, Lin L, Huang ZZ, Chen YJ, Hang W (2011) Peptidome workflow of serum and urine samples for biomarker discovery. Anal Methods 3:773–779
127. Kolch W, Neusüss C, Pelzing M, Mischak H (2005) Capillary electrophoresis - mass spectrometry as a powerful tool in clinical diagnosis and biomarker discovery. Mass Spectrom Rev 24:959–977
128. Albalat A, Mischak H, Mullen W (2011) Clinical application of urinary proteomics/peptidomics. Expert Rev Proteomics 8:615–629
129. Mischak H, Julian BA, Novak J (2007) High-resolution proteome/peptidome analysis of peptides and low-molecular-weight proteins in urine. Proteomics Clin Appl 1:792–804
130. Neuhoff NV, Kaiser T, Wittke S, Krebs R, Pitt A, Burchard A, Sundmacher A, Schlegelberger B, Kolch W, Mischak H (2004) Mass spectrometry for the detection of differentially expressed proteins: a comparison of surface-enhanced laser desorption/ionization and capillary electrophoresis/mass spectrometry. Rapid Commun Mass Spectrom 18:149–156
131. Palagi PM, Walther D, Quadroni M, Catherinet S, Burgess J, Zimmermann-Ivol CG, Sanchez JC, Binz PA, Hochstrasser DF, Appel RD (2005) MSight: an image analysis software for liquid chromatography-mass spectrometry. Proteomics 5:2381–2384
132. Kaplan A, Soderstrom M, Fenyo D, Nilsson A, Falth M, Skold K, Svensson M, Pettersen H, Lindqvist S, Svenningsson P, Andren PE, Bjorkesten L (2007) An automated method for scanning LC-MS data sets for significant peptides and proteins, including quantitative profiling and interactive confirmation. J Proteome Res 6:2888–2895
133. Mischak H, Schanstra JP (2011) CE-MS in biomarker discovery, validation, and clinical application. Proteomics Clin Appl 5:9–23
134. Pontillo C, Filip S, Borras DM, Mullen W, Vlahou A, Mischak H (2015) CE-MS-based proteomics in biomarker discovery and clinical application. Proteomics Clin Appl 9:322–334
135. Pejchinovski M, Siwy J, Mullen W, Mischak H, Petri MA, Burkly LC, Wei R (2018) Urine peptidomic biomarkers for diagnosis of patients with systematic lupus erythematosus. Lupus 27:6–16
136. Carleo A, Chorostowska-Wynimko J, Koeck T, Mischak H, Czajkowska-Malinowska M, Rozy A, Welte T, Janciauskiene S (2017) Does urinary peptide content differ between COPD patients with and without inherited alpha-1 antitrypsin deficiency? Int J Chron Obstruct Pulmon Dis 12:829–837
137. Schiffer E, Mischak H, Novak J (2006) High resolution proteome/peptidome analysis of body fluids by capillary electrophoresis coupled with MS. Proteomics 6:5615–5627
138. Good DM, Zurbig P, Argiles A, Bauer HW, Behrens G, Coon JJ, Dakna M, Decramer S, Delles C, Dominiczak AF, Ehrich JHH, Eitner F, Fliser D, Frommberger M, Ganser A, Girolami MA, Golovko I, Gwinner W, Haubitz M, Herget-Rosenthal S, Jankowski J, Jahn H, Jerums G, Julian BA, Kellmann M, Kliem V, Kolch W, Krolewski AS, Luppi M, Massy Z, Melter M, Neusüss C, Novak J, Peter K, Rossing K, Rupprecht H, Schanstra JP, Schiffer E, Stolzenburg JU, Tarnow L, Theodorescu D, Thongboonkerd V, Vanholder R, Weissinger EM, Mischak H, Schmitt-Kopplin P (2010) Naturally occurring human urinary peptides for use in diagnosis of chronic kidney disease. Mol Cell Proteomics 9:2424–2437
139. Rossing K, Mischak H, Dakna M, Zurbig P, Novak J, Julian BA, Good DM, Coon JJ, Tarnow L, Rossing P (2008) Urinary proteomics in diabetes and CKD. J Am Soc Nephrol 19:1283–1290
140. Theodorescu D, Wittke S, Ross MM, Walden M, Conaway M, Just I, Mischak H, Frierson HF (2006) Discovery and validation of new protein biomarkers for 4 urothelial cancer: a prospective analysis. Lancet Oncol 7:230–240
141. Theodorescu D, Schiffer E, Bauer HW, Douwes F, Eichhorn F, Polley R, Schmidt T, Schofer W, Zurbig P, Good DM, Coon JJ, Mischak H (2008) Discovery and validation of urinary biomarkers for prostate cancer. Proteomics Clin Appl 2:556–570
142. Zimmerli LU, Schiffer E, Zurbig P, Good DM, Kellmann M, Mouls L, Pitt AR, Coon JJ, Schmieder RE, Peter KH, Mischak H, Kolch W, Delles C, Dominiczak AF (2008) Urinary proteomic biomarkers on coronary artery disease. Mol Cell Proteomics 7:290–298
143. Torati LS, Migaud H, Doherty MK, Siwy J, Mullen W, Mesquita PEC, Albalat A (2017) Comparative proteome and peptidome analysis of the cephalic fluid secreted by Arapaima gigas (Teleostei: Osteoglossidae) during and outside parental care. PLoS One 12:art.no. e0186692
144. Mansor R, Mullen W, Albalat A, Zerefos P, Mischak H, Barrett DC, Biggs A, Eckersall PD (2013) A peptidomic approach to biomarker discovery for bovine mastitis. J Proteome 85:89–98
145. Thomas FC, Mullen W, Tassi R, Ramirez-Torres A, Mudaliar M, McNeilly TN, Zadoks RN, Burchmore R, Eckersall PD (2016) Mastitomics, the integrated omics of bovine milk in an experimental model of Streptococcus uberis mastitis: 1. High abundance proteins, acute phase proteins and peptidomics. Mol BioSyst 12:2735–2747
146. Garcia-Canas V, Simo C, Herrero M, Ibanez E, Cifuentes A (2012) Present and future challenges in food analysis: foodomics. Anal Chem 84:10150–10159
147. Ibanez C, Simo C, Garcia-Canas V, Cifuentes A, Castro-Puyana M (2013) Metabolomics, peptidomics and proteomics applications of capillary electrophoresis-mass spectrometry in Foodomics: a review. Anal Chim Acta 802:1–13
148. Alvarez G, Montero L, Llorens L, Castro-Puyana M, Cifuentes A (2018) Recent advances in the application of capillary electromigration methods for food analysis and foodomics. Electrophoresis 39:136–159

149. Capriotti AL, Cavaliere C, Piovesana S, Samperi R, Lagana A (2016) Recent trends in the analysis of bioactive peptides in milk and dairy products. Anal Bioanal Chem 408:2677–2685
150. Minkiewicz P, Dziuba J, Darewicz M, Iwaniak A, Dziuba M, Nalecz D (2008) Food peptidomics. Food Technol Biotechnol 46:1–10
151. Carrasco-Castilla J, Hernandez-Alvarez AJ, Jimenez-Martinez C, Gutierrez-Lopez GF, Vila-Ortiz G (2012) Use of proteomics and peptidomics methods in food bioactive peptide science and engineering. Food Eng Rev 4:224–243
152. Saavedra L, Hebert EM, Minahk C, Ferranti P (2013) An overview of "omic" analytical methods applied in bioactive peptide studies. Food Res Int 54:925–934
153. Sanchez-Rivera L, Martinez-Maqueda D, Cruz-Huerta E, Miralles B, Recio I (2014) Peptidomics for discovery, bioavailability and monitoring of dairy bioactive peptides. Food Res Int 63:170–181
154. Giacometti J, Buretic-Tomljanovic A (2017) Peptidomics as a tool for characterizing bioactive milk peptides. Food Chem 230:91–98
155. Gomez-Ruiz JA, Ramos M, Recio I (2007) Identification of novel angiotensin-converting enzyme-inhibitory peptides from ovine milk proteins by CE-MS and chromatographic techniques. Electrophoresis 28:4202–4211
156. Baptista DP, Araujo FDD, Eberlin MN, Gigante ML (2017) A survey of the peptide profile in prato cheese as measured by MALDI-MS and capillary electrophoresis. J Food Sci 82:386–393
157. Benavente F, Pero-Gascon R, Pont L, Jaumot J, Barbosa J, Sanz-Nebot V (2018) Identification of antihypertensive peptides in nutraceuticals by capillary electrophoresis-mass spectrometry. J Chromatogr A 1579:129–137
158. Catala-Clariana S, Benavente F, Gimenez E, Barbosa J, Sanz-Nebot V (2010) Identification of bioactive peptides in hypoallergenic infant milk formulas by capillary electrophoresis-mass spectrometry. Anal Chim Acta 683:119–125

Discovery of Native Protein Complexes by Liquid Chromatography Followed by Quantitative Mass Spectrometry

Wasim Aftab and Axel Imhof

Abstract

Discovering protein complexes in vivo is of vital importance to understand the evolution and function of biological systems. Proteomics analysis has evolved as a state-of-the-art technique in elucidating the above information. A combination of liquid chromatography (LC) and liquid chromatography coupled to shotgun mass spectrometry (LC-MS) provides the most exhaustive information in this regard. However, a significant amount of computational effort is required for the meaningful interpretation of the generated datasets. In this chapter we describe in detail the state-of-the-art pipeline to discover soluble protein complexes and provide practical advice focusing on typical situations a biologist faces while analyzing such proteomics datasets. Furthermore, we briefly describe two commonly used software packages to solve the described problem: *Weka* for training protein-protein interactions (PPIs) using machine learning (ML) and *Cytoscape* for clustering the interaction network.

Keywords

Protein-protein interactions · Protein networks · Liquid chromatography · Mass spectrometry · Shotgun proteomics · Label-based protein quantification · Label-free protein quantification · Data-independent data acquisition (DIA) · Spectral libraries · Weighted cross correlation · Machine learning · Denoising protein-protein interaction matrices · Network analysis · Network display

W. Aftab
Biomedical Center, Protein Analysis Unit, Faculty of Medicine, Ludwig-Maximilians-Universität München, Munich, Germany

Graduate School for Quantitative Biosciences (QBM), Ludwig-Maximilians-Universität Munich, Munich, Germany

A. Imhof (✉)
Biomedical Center, Protein Analysis Unit, Faculty of Medicine, Ludwig-Maximilians-Universität München, Munich, Germany
e-mail: imhof@lmu.de

1 Introduction

Proteins execute almost all cellular functions. They seldom exist as single subunits but rather interact to form larger protein assemblies or complexes. These complexes are responsible for a wide range of functions within cells including formation of cytoskeleton, transportation of cargo, metabolism of substrates for the production of energy, replication of DNA, protection and maintenance of the genome, transcription and translation of genes to gene

A. V. Colnaghi Simionato (ed.), *Separation Techniques Applied to Omics Sciences*, Advances in Experimental Medicine and Biology 1336, https://doi.org/10.1007/978-3-030-77252-9_6

products, maintenance of protein turnover, and protection of cells from internal and external damaging agents [1]. Therefore, determining the composition of protein complexes is an essential step toward understanding the cell as an integrated system [2].

All cellular systems contain a densely connected network of proteins/protein complexes (see Fig. 1). If a tiny mutation occurs somewhere in these highly connected networks, the effect(s) will diffuse throughout the entire network via multiple PPIs. Therefore, when a gene is mutated such that the corresponding change in amino acid perturbs its interaction with one or multiple partners, (a phenomenon which is commonly called edgetic perturbation [3]) this mutation could lead to various diseases [4]. Due to the high connectivity of the network, such perturbations can combine to affect a large part of the protein network and hence the physiology of the cell [5]. One way to investigate and predict the effect of such crucial changes in the cell is to create a map of protein complexes and then compare the network maps of wild type and the disturbed (mutant) system. This can potentially facilitate the understanding of the molecular biology of protein interaction networks. Moreover, the MS-based protein correlation profiling that we review in this chapter can also increase our knowledge about hypothetical proteins of unknown functions [6–8].

Here we present an integration of experimental and computational techniques state of the art to discover protein complexes in a cell extract. The method consists of fractionation of native protein complexes by orthogonal chromatography methods like size exclusion chromatography (SEC) and ion exchange chromatography (IEX) followed by a tryptic digest and quantitative reversed-phase chromatography (RPC) coupled to mass spectrometry to quantify proteins complexes and sophisticated computational methods to extract information from complex sets of data (workflow is depicted in Fig. 2).

This chapter mainly consists of two parts: experimental and computational. The goal of the chapter is to introduce readers to the exciting field of global protein complex discovery with a

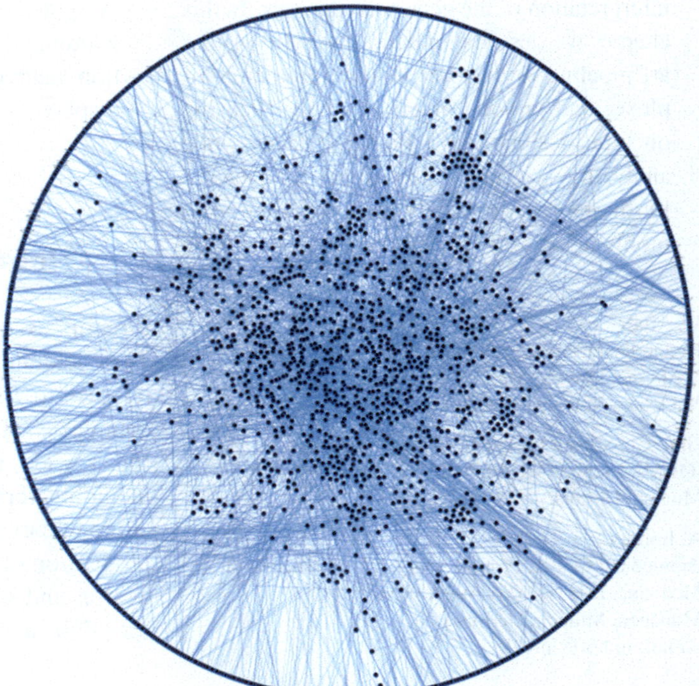

Fig. 1 Cells contain highly connected protein networks; the tiny blue dots represent protein complexes, and the connections among them are shown by light blue lines

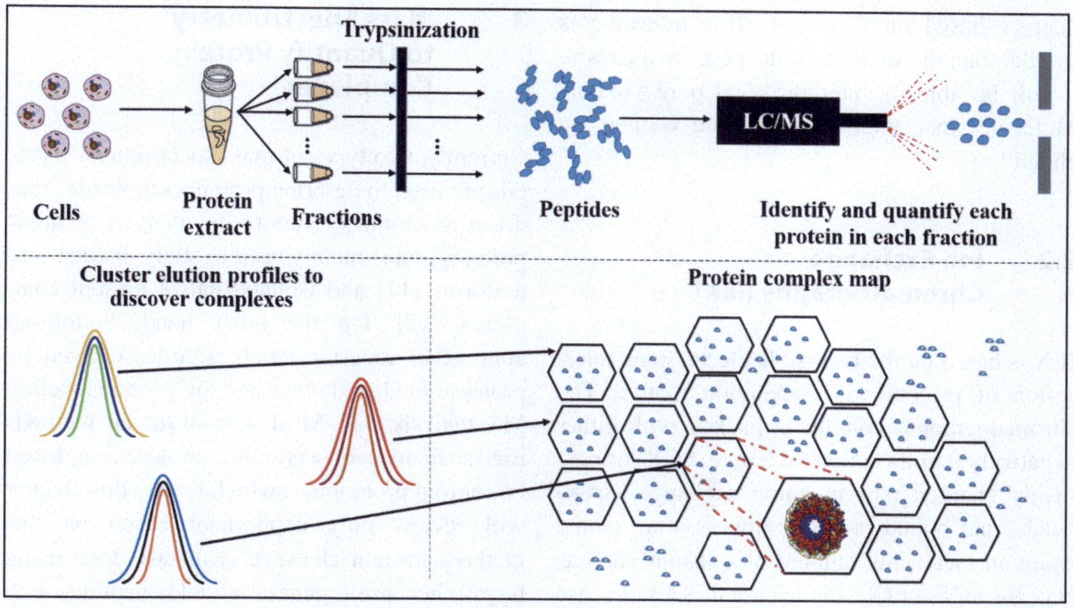

Fig. 2 Workflow to detect protein complexes in a cell extract; the protein extract is obtained from a population of cells and is further fractionated using liquid chromatography in order to separate protein complexes from each other

major focus on state-of-the-art computational methods. Though a description of the method experimental details is not a goal of this chapter, we will introduce necessary concepts to facilitate the better comprehension of the entire workflow.

2 Liquid Chromatography to Fractionate Native Protein Complexes

In order to discover protein complexes successfully from complex mixtures, the quality of the extracts is of utmost importance. To do this the complexity of the mixture should be investigated using a classical data-dependent acquisition (DDA) shotgun proteomics experiment. If the number of proteins thus identified is on a par with the expert knowledge (usually 5000–10,000 proteins can be identified from whole cell extracts isolated from higher eukaryotes), then the extract can be further fractionated. One of the biggest challenges in this procedure is the limited dynamic range of many MS-based experiments. This leads to the detection, of only highly abundant protein complexes or subunits thereof. In order to deal with it, Havugimana et al. [8] has proposed very deep fractionation by using multiple orthogonal modalities. Popularly size and charge-based chromatographic separation techniques have been employed to solve this problem [6–8].

2.1 Size Exclusion Chromatography (SEC)

SEC is a chromatographic technique in which protein molecules are separated by size differences when they elute through a SEC column. The column is packed with a matrix of spherical particles carrying defined pore sizes. Protein molecules bigger than the pore sizes cannot diffuse into the beads, thereby eluting first. Molecules that range in size between the very big and very small can penetrate the pores to varying degrees based on their size. If a molecule is smaller than the smallest of the pores in the resin, it will be able to enter the total pore volume. Molecules that enter the total pore volume are eluted last.

2.2 Ion Exchange Chromatography (IEX)

IEX is based on the reversible electrostatic interaction of protein with separation matrix. The chromatographic column is packed with either negatively (strong cation exchange, SCX) or positively charged (strong anion exchange, SAX) beads, and bound protein complexes are eluted using an increasing amount of salt ions such as Na^+, for SCX, or Cl^-, in the case of SAX. In case of anionic exchange chromatography, protein complexes with a higher negative charge require a higher concentration of anions; hence eluting later than the proteins with lesser net negative charge and vice versa is true for cationic exchange chromatography.

The resolution of a native liquid chromatography column (i.e., how well can very similar protein complexes be separated) is determined by a number of parameters such as the number of theoretical plates, the peak capacity of the resin, the nature of the analyte, and several other parameters [9, 10]. In very complex samples such as whole cell lysates, the limited resolution inevitably results in the co-elution of unrelated protein complexes. An increase in resolution can only be achieved at the cost of a substantial increase of measurement time, which quickly becomes impractical. The compromise between the high resolution and the ability to be able to measure different physiological conditions may be achieved by performing replicate experiments on different modalities and subsequently checking how consistently the apexes are aligned for a pair of proteins in all the replicates. This will result in a co-apex score (defined later) that can be computed for every pair of elution and can be used to penalize/boost them based on the score during data analysis.

3 Mass Spectrometry to Quantify Protein Complexes

Currently, two types of mass spectrometry methods are used to describe protein complexes. Top-down proteomics refers to the analysis of intact proteins and is mainly used to study distinct proteoforms [11] and isolated native protein complexes [12]. On the other hand, bottom-up approaches measure small peptides derived by protease-mediated cleavage of proteins before MS analysis [13]. Most state-of-the-art methods used so far to resolve protein complexes employed bottom-up proteomics, which is why this chapter will review only experiment based on this method. Protein cleavage is mostly done using trypsin because it generally yields peptides with 6–25 amino acids long with a defined positively charged C-terminus (K or R) creating ideal analytes for the mass spectrometer. A typical LC-MS workflow is shown in Fig. 3. After tryptic digest the mixture of peptides is still very complex. For example a simple protein mixture, such as an *E. coli* cell lysate, contains approximately 2.5–5 k proteins [13]. If one considers that 2.5 k proteins are expressed in *E. coli* with an average of 20 peptides per protein, then it will result in 50 k peptides. Therefore, the peptide sample needs to be further separated using RPC coupled directly to the mass spectrometer. RPC separates peptides mostly based on their hydrophobicity. After that, the sample is ionized and injected in the mass spectrometer. Most commonly used ionization sources are matrix-assisted laser desorption/ionization (MALDI) and electrospray ionization (ESI). MALDI yields singly charged precursor ions and ESI produces precursor ions with two or more charges. Within the mass spectrometer, the ions are separated based on their mass to charge ration (*m/z*) by the mass analyzer and detected.

A mass spectrum is generated when the ions hit the detector and the corresponding data is recorded in files having proprietary format defined by the vendor of the MS instrument. Then MS data processing software programs are used to identify the peptides in the sample, which

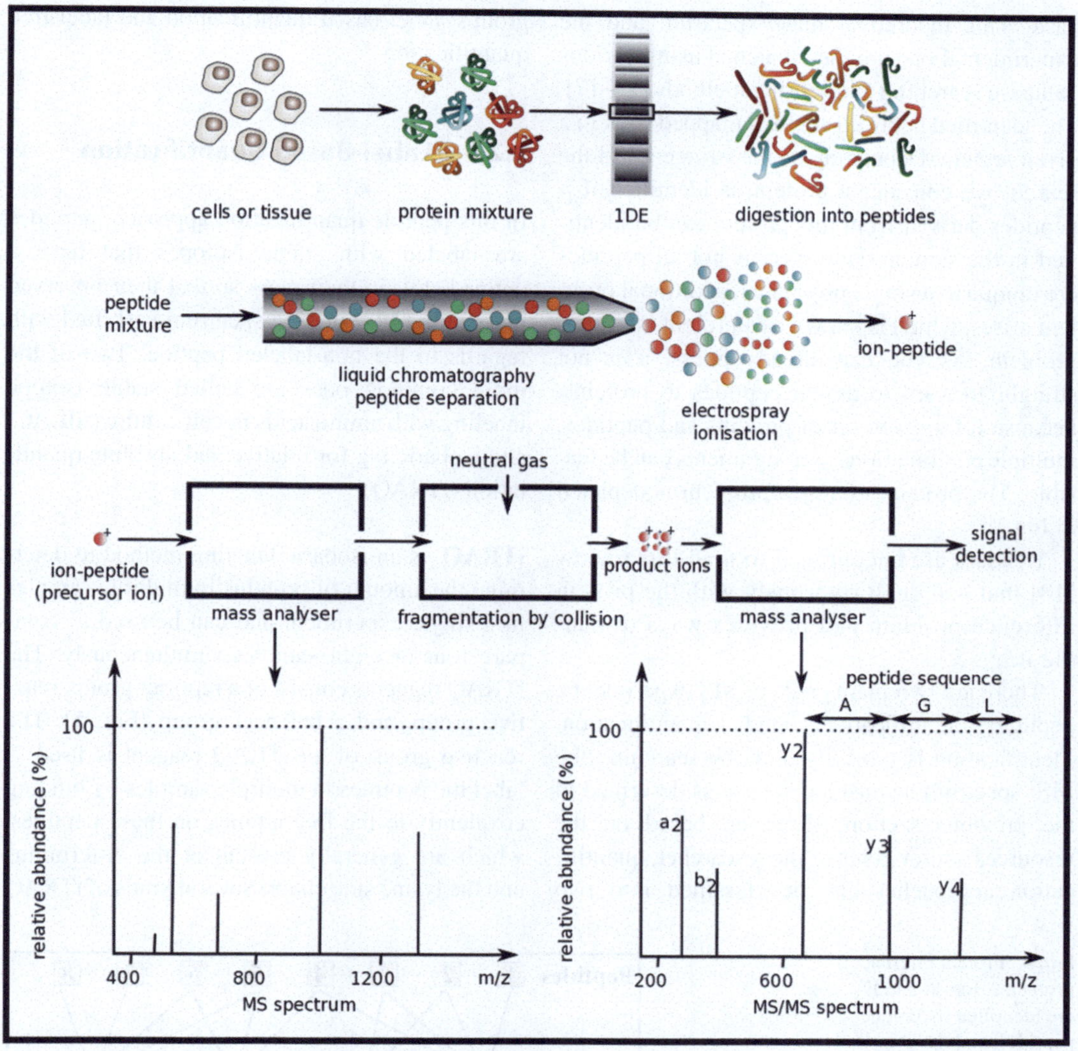

Fig. 3 A typical workflow of shotgun proteomics [14]

can be mapped to the corresponding proteins. This peptide-protein mapping is one of the difficult challenges in proteomics. We briefly talk about peptide identification and protein inference in the next section.

3.1 The Protein Inference Problem

Let's say we have a cell/tissue extract as shown in Fig. 3 and want to know what proteins are present in that sample. In other words we ask: Which proteins can we identify? First the process starts with digestion: the proteins are broken into smaller parts called peptides. Let's call them experimental peptides. RPC is employed on these experimental peptides before feeding them into a mass spectrometer. The mass spectrometer records the mass of these peptides, and, upon collision-induced fragmentation, a signature or fragment mass spectrum for each of them is included in a file. Then based on our expert knowledge, we decide upon a database, which should contain sequence and summary description of the proteins in the sample. After that we perform an in silico digestion of the proteins in the database, i.e., the proteins in the database will be broken into peptides to create a peptide database. This theoretical mass spectrum and the experimental one are then matched using various database searching and scoring methods [15–17]. The identified peptides are then mapped back to a given protein. Depending on the stringency of the search, we consider a protein as identified if *n* peptides derived from this protein can be identified in the sample. However, as not all peptides are unique to a single protein, an additional problem arises, which is known as a protein inference problem. As you can already guess, it is not straight forward to assign peptides to proteins, because for a given set of proteins and peptides, multiple combinations of assignments can be feasible. The protein inference problem is depicted in Fig. 4.

Readers are encouraged to read this review [18] that had dealt rigorously with the protein inference problem and provides ways to handle it.

There are two main goals of MS experiments: peptide identification and quantification. Identification is typically done by searching the MS^2 spectrum against a database as described in the previous section. However, based on the resources and/or goals of the researcher, quantification approaches can be classified into two groups: label-based quantification and label-free quantification.

3.2 Label-Based Quantification

In this peptide quantification approach, peptides are labeled with stable isotopes that have a defined shift in their mass so that their observed mass in the MS^1 or MS^2 spectrum is shifted with regards to the non-labeled peptide. Two of the more common ones are called stable isotope labeling with amino acids in cell culture (SILAC) and isobaric tag for relative and absolute quantitation (iTRAQ).

iTRAQ is an isobaric labeling method to determine the amount of proteins in multiple samples in a single experiment and can be used to compare four or eight samples simultaneously. The iTRAQ reagents consist of a reporter group, reactive group, and a balancer group (Fig. 5). The reactive group of an iTRAQ reagent is used to label the peptides in multiple samples by binding covalently to the free amines on these peptides, which are generally present at the N-terminus and the lysine side chain. Several kinds of iTRAQ

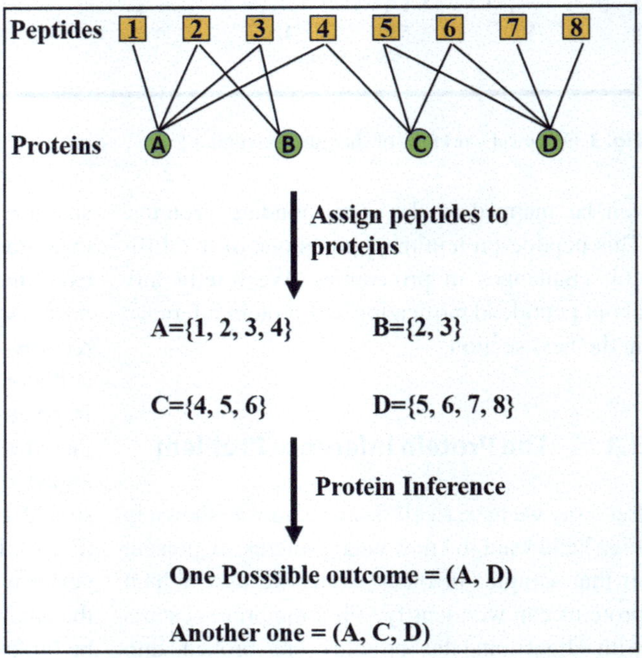

Fig. 4 Protein inference problem. Both A and D are identified by unique peptides, and the outcome (A, D) encompasses the entire peptide space. Therefore, the most widely accepted solution to protein inference in this hypothetical example would be (A, D). As C has evidence only from shared peptides, all inference algorithms will favor (A, D) over the theoretically also possible explanation (A, C, D)

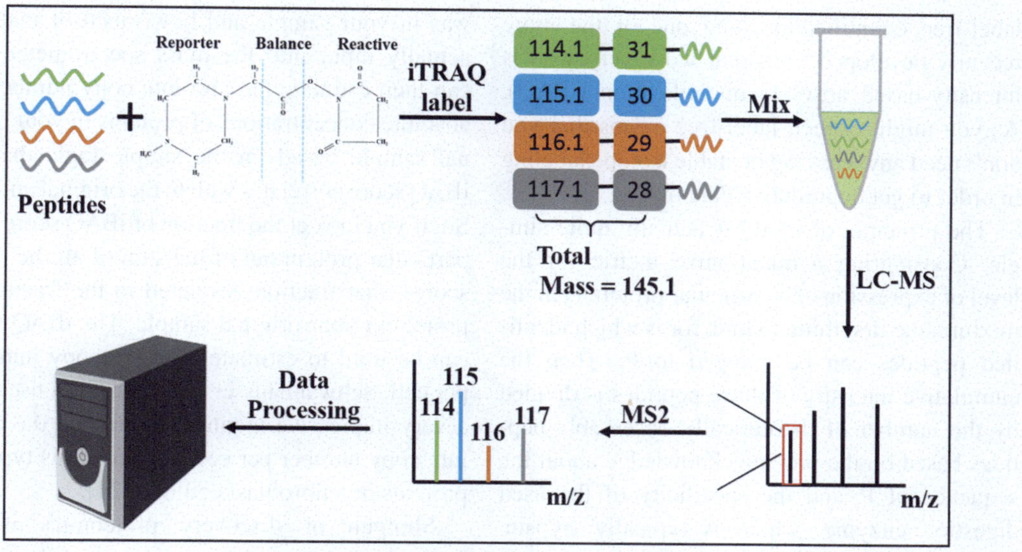

Fig 5 iTRAQ reagent-based shotgun proteomics using iTRAQ-4-plex as an example

reagents exist, but the cumulative mass of different groups in each of them is equal. Their chemistry has been optimized such a way that the labeled peptides always elute from the LC system at the same time and can be quantified using the so-called reporter ions (Fig. 5). A more detailed description on how to use iTRAQ reagents for protein complex and profiling studies can be found in this article [19].

SILAC is based on metabolically incorporating stable isotope-labeled amino acids into the entire proteome. In this technique cells are metabolically labeled with two amino acids (generally lysine and arginine), holding stable, nonradioactive isotopes. In SILAC, two populations of cells are grown in two different culture media, the light medium that contains amino acids with the natural isotopes and the heavy medium that contains stable isotope-labeled amino acids. After a sufficient number of cell divisions, all the proteins from the cells cultured in heavy medium contain amino acids in the heavy state analyzed with liquid chromatography with tandem mass spectrometry. The quantification of SILAC is based on determining the ratio of introduced isotope-labeled peptides to unlabeled peptides. The signal intensities from light and heavy samples allow for a quantitative comparison of their relative abundance in the mixture.

When studying protein-protein interactions using SILAC, protein complexes are immunoprecipitated from the mixture of SILAC-labeled cell lysates. Combined with SILAC, specifically interacting proteins can be efficiently distinguished from nonspecific background proteins. The abundance of specific interaction partners purified from the bait sample is significantly higher than the one from the control sample, resulting in quantified ratios much higher than 1. In contrast, the abundance of nonspecific background proteins should be comparable from both the bait and the control sample, such that their ratio is close to 1. SILAC-based quantitative proteomics can be used to identify the specifically interacting proteins in investigating exogenous PPIs, endogenous PPIs, or inducible PPIs [20].

3.3 Label-Free Quantification

All quantification methods described earlier involves some sort of labeling, which results in a defined and measurable mass shift so that you can know, based on the mass, which sample you are dealing with. But there is also the so-called label-free quantification, and one of the more recently developed ones that work very well is intensity-based absolute quantification (iBAQ). As you might expect, label-free means that you don't need any mass tag or stable isotope labeling in order to get quantitative data out.

The principle of iBAQ is actually quite simple. Considering a quantitative metric for the level of expression of a particular protein P_i in the mixture, the first thing to look for is which identified peptides can be mapped to P_i. Then the cumulative intensity of those peptides is divided by the number of theoretically observable peptides based on the previous knowledge about the sequence of P_i and the specificity of the used digestion enzyme, which is typically trypsin. This is to address the problem of bigger proteins just tend to generate more peptides because they're bigger. The idea is actually quite similar to the mRNA sequencing data analysis, where sum of the intensities of all the fragments of a transcript is divided by the transcript length to address the problem of longer transcripts just tend to generate more fragments because they are longer. This corrected number is called iBAQ score of P_i and can be calculated for any protein of interest. Schwanhäusser et al. showed that the iBAQ score correlates actually very well with the original amount of protein that was entered into the mass spectrum [21]. The iBAQ score of a protein can be mathematically defined as

$$\mathrm{iBAQ} = \frac{\sum_{j=1}^{n} I_j}{N} \tag{1}$$

where $I_j \rightarrow$ is the intensity value of the jth peptide

N is the total of theoretically observable peptides
n is the total number of observed peptides

This calculation is provided as an option in MS data processing software programs like MaxQuant [22].

Another very nice feature of iBAQ is that if you analyze the proteome very deeply, i.e., you quantify the levels of almost every protein in the mixture and you keep track of how much protein was in your sample and how much of that you actually input into the mass spectrometer, you can then estimate the absolute copy numbers or absolute concentrations of proteins in your original sample based on the simple logic that the iBAQ score correlates well to the original amount. So, if you look at the fraction of iBAQ score for a particular protein out of the sum of all the iBAQ scores, that fraction is related to the fraction of protein in your original sample. The iBAQ score can be used to estimate absolute copy numbers per cell. Schwanhäusser and colleagues used it in a very impressive manner to estimate the absolute copy number per cell for numerous types of proteins in a fibroblast cell line [21].

Shotgun or discovery proteomics is the method of choice when you want to identify as many proteins as possible in a sample. However, in some instances the goal is to consistently identify and precisely quantify the same set of proteins in different conditions. Then targeted proteomics might be the ideal technique, and here we briefly talk about the most popular targeted proteomics approach over the last few years: selected reaction monitoring (SRM) (Fig. 6).

SRMs are mostly measured on triple-quadrupole mass spectrometers, in which the first quadrupole Q1 works as a mass filter to select one specific peptide. The second quadrupole Q2 works as collision chamber to fragment the selected peptide. The third quadrupole Q3 again works as mass filter, but it filters the specific fragment ions of the selected peptide, which hit the detector. Finally, a SRM measurement records the pairs (precursor, fragmentation-ion) over time to compute a chromatographic trace, also known as SRM trace.

Using this SRM trace, the software programs can identify and quantify individual proteins in a highly reproducible manner, which is frequently not the case in shotgun approaches. The basic idea of SRM is depicted in Fig. 6. You will find this tutorial [23] discussing the application of SRM for quantitative proteomics very interesting.

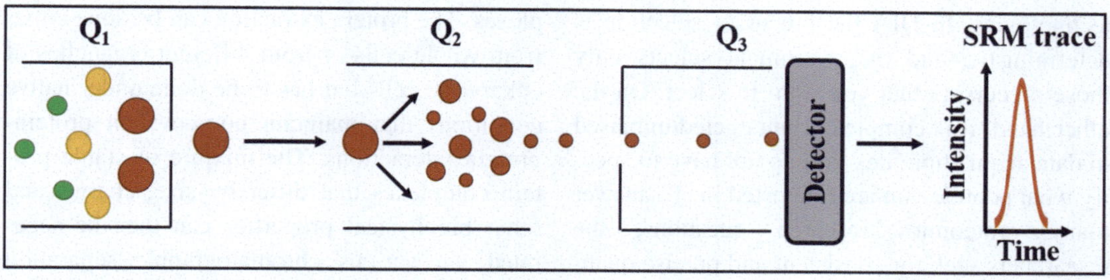

Fig. 6 SRM workflow; performed on a triple-quadrupole mass spectrometer

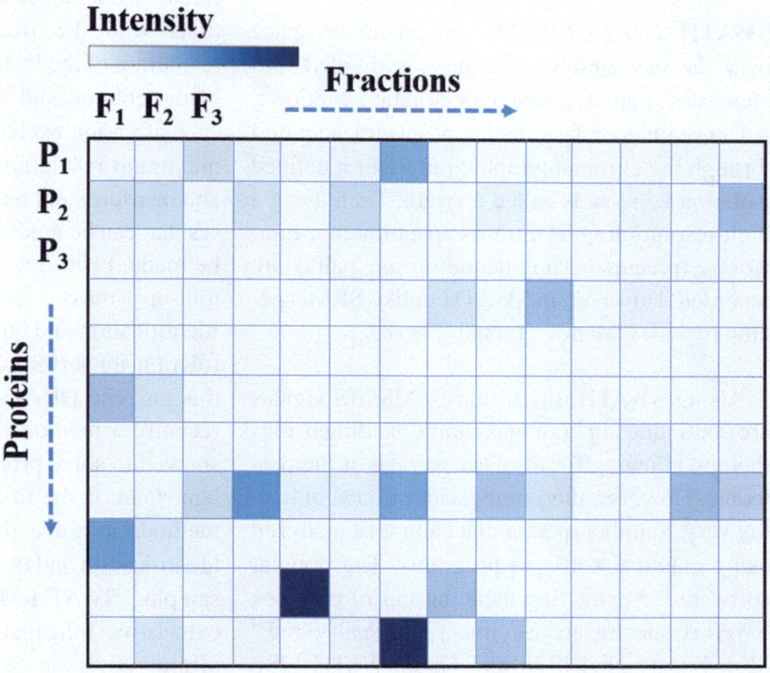

Fig. 7 Quantitative peptide/protein matrix; columns are fractions or the conditions and the rows are typically proteins/peptides

So why do we need something other than the DDA or SRM? What if we want to fill a matrix of samples or a matrix of peptides or proteins (Fig. 7) in a more complete way with high-quality quantitative intensity values?

In DDA, the instrument stochastically selects the *n* most highly abundant peptides for MS2 during the MS1 scan. Hence you get snapshots of MS2 spectra that correspond to the given point in time in the MS1 space. The key point here is that the precursor space is sampled discontinuously in both mass and the retention time (t_r) dimensions, and, because the on-the-fly heuristic is to select only the few most abundant peptides, the quantification is biased toward high-abundance species. Due to this DDA-based quantification results in many missing intensities corresponding to proteins/peptides.

In SRM, the duty of an MS instrument is to sample the precursor space in a deterministic way based on the peptides of interest. In this case the precursor space is sampled discontinuously in mass but continuously in the t_r dimensions. However, the instrument monitors only a relatively low number of precursors per run, although accurate and consistent quantification is obtained.

So, can we get SRM like accurate and consistent quantification at the same time covering entire precursor space? The answer is yes, if we do sequential window acquisition of all theoretical mass spectra (SWATH-MS), which is one of the latest data-independent acquisition (DIA)

technologies. In DIA the precursor selection is deterministic, and the instrument selects only those precursors that you want to select. On the other hand, it is completely untargeted/unbiased in data acquisition, i.e., you do not have to specify what peptide you are interested in. To answer many proteomics research questions, the researchers seek for consistent and precise quantification of all peptides in multiple samples, and SWATH aims to achieve exactly that.

SWATH During SWATH the precursor space over the measurable *m/z* range is divided into chunks of small *m/z* precursor isolation windows. An ensemble of fragment ion spectra acquired through the chromatographic range for a defined isolation window is called a swath. Then using a high-resolution Q-TOF mass spectrometer, transitions (precursor, fragmentation-ion pairs) are recorded. However, in SWATH unlike SRM, specific peptides are not targeted (Fig. 8).

Using SWATH, the resulting MS/MS signals are continuous in both mass and time dimensions thereby allowing for deeper coverage of the proteome. However, this comes with the cost of having very complex spectra that cannot be analyzed using classical software programs. The popular software for extracting meaning out of complex SWATH spectra are Skyline [25], PeakView™ [26], Spectronaut [27], and OpenSWATH [28]. Among them Skyline and OpenSWATH are free to use. Skyline provides a GUI and can also be used in a command line. OpenSWATH in contrast, has only a command line version making it more cumbersome to use. Also, in order to use OpenSWATH third-party software packages are needed to be installed as it uses multiple tools from other sources for processing DIA datasets.

However, the state-of-the-art methods for discovering soluble protein complexes relies on chromatographic fractionation of biological extracts coupled to precision mass spectrometry analysis. Therefore, we need the protein sample from which we want to discover protein complexes. The protein extraction can be done either from whole cells or from different organelles of eukaryotic cells but has to be done under native conditions that maintain non-covalent protein-protein interactions. The mixture of stable protein complexes that differ by size, charge, and other biophysical properties can then be separated via various chromatographic separation methods. A SEC separation, for example, will result in a separation by size. Complexes of different size will then be collected in different fractions with the fraction size determining the resolution of the later analysis. Similar to the size of the column and the length of the gradient in case of an ion exchange, a compromise between maximum resolution (i.e., smaller fraction size) and measurement time (number of LC-MS analyses that can be made in a reasonable time) have to be made. Each fraction is then prepared for bottom-up mass spectrometry and analyzed. Identification and quantification proteins in every fraction are performed by mass spectrometry. For this purpose DDA-MS has been used [6–8], but recently a method [29] based on DIA has also showed to solve protein complex detection problem. One of the major goals of the proteomics methods to solve this problem is to maximize identification and quantification of proteins in the sample. SWATH-MS will provide the most exhaustive information in this regard, since it affords coverage of the entire precursor space, which results in very few missing values in protein intensities. Due to the stochastic peak picking in DDA experiments (see paragraph above), missing values for protein present in the sample are a major problem in DDA-based quantitative proteomics.

To set up a SWATH data analysis pipeline, first a spectral library needs to be generated. A spectral library contains information such as non-redundant peptide transitions, t_r, etc., which any DIA data processing software uses to correlate with corresponding information from the peptides in the sample, allowing identification and quantification of proteins in the sample.

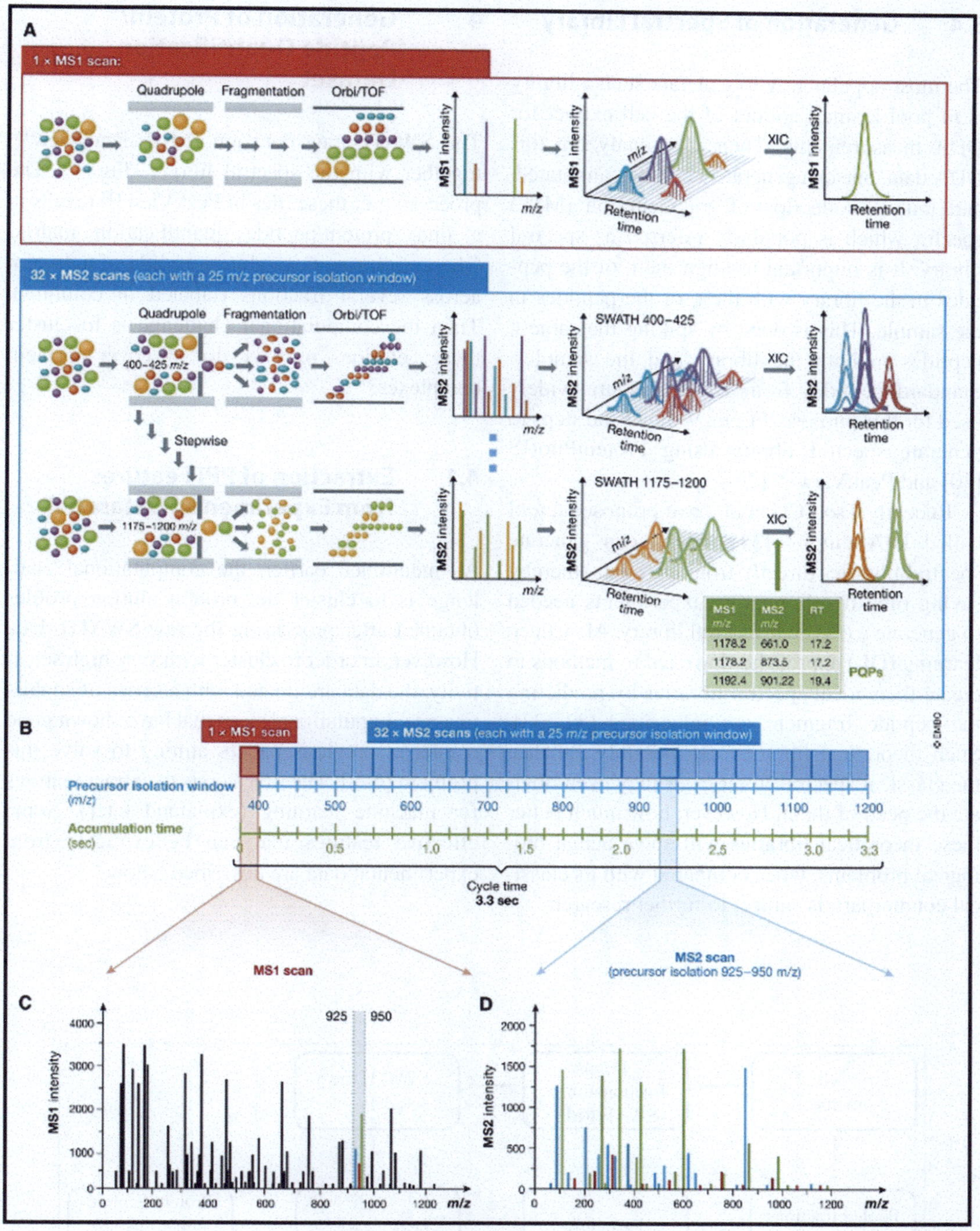

Fig. 8 Concept of SWATH-based MS. (**a**) SWATH-MS measurements are performed using a quadrupole as first mass analyzer and a TOF or Orbitrap as second mass analyzer. (**b**) SWATH data acquisition Scheme. (**c**) The MS1 full scan detects all peptide precursors eluting at a given time point. For example, in the mass range from 925 to 950 m/z, three co-eluting peptide species are detected (green, red, and blue). (**d**) The corresponding MS2 scan with a precursor isolation window of 925–950 m/z represents a mixed MS2 spectrum with fragments of all three peptide species [24]

3.4 Generation of Spectral Library

The most popular way to generate such a library is to pool a small aliquot of the cell extract for DDA measurements. Then after analyzing this DDA data, one can generate a curated, annotated, and unique collection of fragment ion (MS2) spectra which is popularly referred as spectral library. It is important to align the t_r of the peptides in the library with the t_r of the peptides in the sample. This is done by spiking the same t_r peptides in both the library and the samples. Standard peptides from Biognosys are widely used for this purpose. Figure 9 shows the steps to generate spectral library using ProteinPilot™ [30] and PeakView™ [26].

Recently Tsou CC et al. have proposed a tool called DIA-Umpire [31], which can generate spectral libraries directly from DIA data, thereby saving time by eliminating experiments needed to generate a classical spectral library. Also, deep learning (DL) approaches have led to methods to create theoretical spectral libraries by predicting the peptide fragment ion intensities [32, 33]. Such theoretical libraries can possibly enhance the classical spectral libraries or may even obviate the need of them. However, how much better these theoretical libraries will solve actual biological problems, when compared with its classical counterpart, is subject to further research.

4 Generation of Protein/Peptide Quantification Dataset

The SWATH raw data are fed to PeakView™ together with the spectral library (Fig. 9). The processing of these files in PeakView™ results in a final protein/peptide quantification matrix, where each row is an elution profile of a protein across several fractions (labeled in columns). Then the computational challenge is to cluster these elution profiles to discover protein complexes.

4.1 Extraction of PPI Features from Experimental Dataset(s)

As mentioned earlier, the computational challenge is to cluster the protein elution profiles obtained after processing the raw SWATH data. However, in order to cluster with very high sensitivity, the data are treated with a series of sophisticated computational steps that have shown good results in previous papers aiming to solve this problem [6–8]. The first step is to extract features for machine learning (explained later). Some effective features that can be extracted from experimental data are described below.

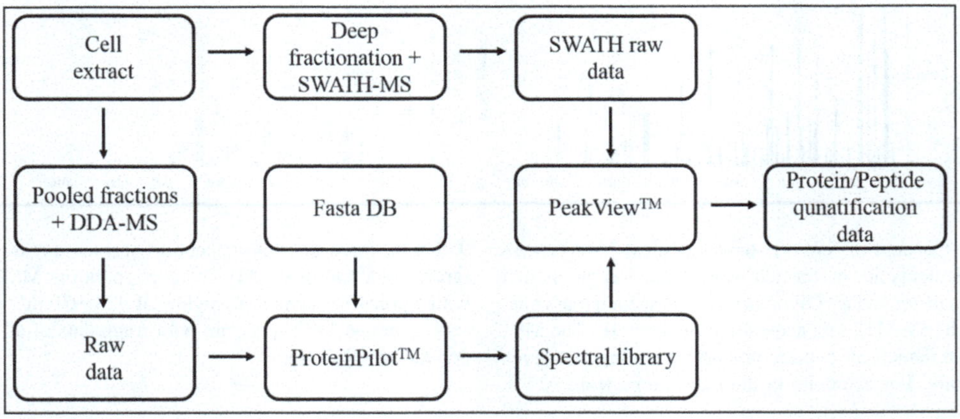

Fig. 9 Steps to produce spectral libraries for processing DIA-SWATH data

Pearson Correlation For each pair of protein elution profiles, a Pearson correlation can be computed, which serves as a feature for the PPIs. An elution profile contains mass spectrometry-quantified intensities of a protein in every fraction. For the elution profiles of two proteins *x* and *y*, the Pearson correlation (*r*) is mathematically defined as

$$r = \frac{\Sigma(x-\overline{x})(y-\overline{y})}{\sqrt{\Sigma(x-\overline{x})^2 \Sigma(y-\overline{y})^2}} \quad (2)$$

where $\overline{x}$ and $\overline{y}$ are the means of *x* and *y* respectively.

Weighted Cross Correlation (wcc) The Pearson correlation defined above does not consider shifts in the elution profiles. Using wcc we can compare an intensity value of a protein within a single fraction with the values in the chromatographic neighborhood of another protein. By weighing different fractions in the neighborhood, we can take into account the relative shifts in the elution profiles. Havugimana et al. [8] has ranked features used in their exploratory analysis of protein complex in human, and wcc was one of their top ranking feature. The similarity based on wcc is defined mathematically as [34].

$$S_{xy} = \frac{\int W(\Delta) C_{xy}(\Delta) d\Delta}{\sqrt{\int W(\Delta) C_{xx}(\Delta) d\Delta \int W(\Delta) C_{yy}(\Delta) d\Delta}} \quad (3)$$

where Δ implies relative shift between the elution profiles

C_{xx} and C_{yy} are the auto correlation functions
W is the weighting function.

Figure 10 demonstrates wcc using two sample vectors using weighing equal to one [35].

Co-apex Score determines how well apexes of a pair of protein elution profiles align in replicates. Hence the co-apex score for a PPI is a ratio between numbers of replicates in which elution profiles of the pair of proteins align the total number of replicate experiments.

These methods are widely used for extracting features from experimental datasets [7, 8].

4.2 Extraction of PPI Features from Literature(s)

Incorporation of genomics and proteomics evidences from literature can also improve the protein complex discovery [7, 8]. These features are usually stored in a database as evidence codes for PPIs. Some of the widely used databases are STRING [36] or HumanNet [37]. The latest one (v1) contains probabilistic gene network of 16,243 validated protein encoding genes of human covering 476,399 interactions. HumanNet is built under a modified Bayesian integration of 21 types of "omics" data, where each data type is weighed based on goodness of functional interaction between genes in Human. Every interaction in HumanNet is weighed using a log likelihood score which measures the odd of a functional interaction being true or false.

STRING has imported many PPIs from other databases in addition to the PPIs predicted by them. The PPIs in STRING forms a Bayesian network, which means each link has a probability associated with it. The latest STRING version (v11) contains PPIs from 5090 organisms with a total number of 3,091,648,416 edges. However, if you consider high scoring (>0.7) interactions only, then that number boils down to 152,484,793.

Some examples of literature extracted features for PPIs are presented below.

Conserved Neighborhood For the genes encoding two proteins, if their neighborhood is conserved across several genomes, then there might exist a possibility of functional interaction between the proteins [38].

Gene Fusion It has been determined using the Rosetta Stone method that sometimes two or more interacting proteins in a genome are fused into a single protein in another genome [39].

$$x = 0101$$
$$y = 1100$$

0 1 0 1	0 1 0 1	0 1 0 1	0 1 0 1	0 1 0 1	0 1 0 1	0 1 0 1
1 1 0 0	1 1 0 0	1 1 0 0	1 1 0 0	1 1 0 0	1 1 0 0	1 1 0 0
$O_{xy}^{i} = 0$	$O_{xy}^{i} = 0$	$O_{xy}^{i} = 0$	$O_{xy}^{i} = +1$	$O_{xy}^{i} = +1$	$O_{xy}^{i} = +1$	$O_{xy}^{i} = +1$

$$S_{xy} = \sum O_{xy}^{i} = 4$$

Fig. 10 Example showing similarity between two binary vectors using wcc. O_{xy}^{i} are the intermediate correlations

Coevolution The concept here is interacting proteins have a tendency to coevolve. In previous studies PPIs in human and bacteria were predicted using this feature [40, 41].

mRNA Co-expression It is known from the previous works that there is a positive correlation of mRNA expression patterns with interacting proteins, and often these patterns are conserved during the course of evolution [42–44].

Protein Domain Co-occurrence It has been shown that two proteins can interact with each other via some of the common/co-occurred domains [45].

Text Mining identifies facts and relationships between entities from the text. In this context entities mean proteins. It uses several methods to process text, and one of the most important of them is natural language processing (NLP). Published literature is a goldmine for PPI data. STRING uses PubMed as a resource to mine for PPIs. One of the major difficulties in extracting PPIs from literature is the enormous variety in the possible form of gene names.[1] For example, both Su(var)205 and HP1a should recognize the same named entity. Text mining in STRING is not as confident feature for a PPI as other features. Readers might find this introduction to biomedical NLP quite exciting [46].

[1]For text mining purposes, gene and protein names are almost always considered to be completely equivalent

4.3 Noise Modeling and Missing Value Imputation

The quantitative protein/peptide matrix yielded by DDA-based MS contains many zeros and/or low spectral counts. Although these zeros/low values often correlate very well with each other, they poorly contribute to the ML predictions [8]. Havugimana et al. proposed to artificially add noise into the data matrix to deal with this problem [8]. The missing value imputation has been tackled by researchers in many different ways. Cox et al. proposed to draw random values from a truncated normal distribution located close to the lower tail of the original data distribution [47]. Karpievitch et al. proposed random selection based on censoring probability from ANOVA model parameters [48]. Webb-Robertson et al. have reviewed several imputation methods and readers are encouraged to read this article [49].

5 Application of Machine Learning to Predict PPI

In order to discover complexes, we first need to predict every binary interaction that makes up the complex, followed by training a machine learning classifier. Machine learning (ML) is a branch of artificial intelligence that can detect patterns in the data when there are no precise rules to define those patterns. ML is well suited to solve the PPI prediction problem because there are no precisely defined rules that one can use to classify between a true and false PPI. Building a ML classifier is a multistage process (Fig. 11):

1. Decide the most suited type of training to solve the problem. Two kinds of training are possible: supervised and unsupervised. As the supervised approach is most popular to solve the current problem, we will restrict our discussion to it. In supervised mode we need to provide the ML algorithm with a labeled dataset. We call a set of features for PPIs a dataset. So, in a dataset each row implies a PPI, and each column is a feature of that PPI. Then we divide this dataset into three parts (not necessarily equal): training set, test set, and validation set. Then we label each PPI in the training and validation sets as either interacting/noninteracting or most often as positive/negative or 0/1 classes. Let's label a true interaction by 1 and a false interaction by 0. For the test set we provide no labels.
2. Train one ML algorithm or ensemble of algorithms using the training set. Training an ML algorithm involves tuning several parameters used by algorithm(s), a thorough discussion of which is beyond the scope of this chapter. However, for our PPI classification task, the major aim of training is to minimize classification error/loss. A kind of error that many ML algorithm try to minimize is sum squared error (SSE) defined as

$$SSE = \Sigma\left(y_i - \hat{y}_i\right)^2 \tag{4}$$

where $y_i \rightarrow$ actual label of a PPI and $\hat{y}_i \rightarrow$ predicted label for that PPI. In each training epoch, we check the SSE, and we keep training until the SSE has fallen below a desired threshold after a predetermined number of epochs. If the SSE is not able to reach the desired threshold even after training for many epochs, then we say that the model is underfit.

3. Once the SSE is adequately small, we then switch to validation of the algorithm(s). Here we use the validation dataset to check the SSE. If the gap between the training error and the validation error is too big, we say the model is overfit.
4. Overfitting and underfitting are two prime challenges that must be handled during training of any ML algorithm. For that reason, a cycle of training-validation-training is often performed. Only when the training error is small and the gap between the training error and the validation error is also small we can be confident about the training. Once we are sufficiently confident about the training, then only we should go for testing.

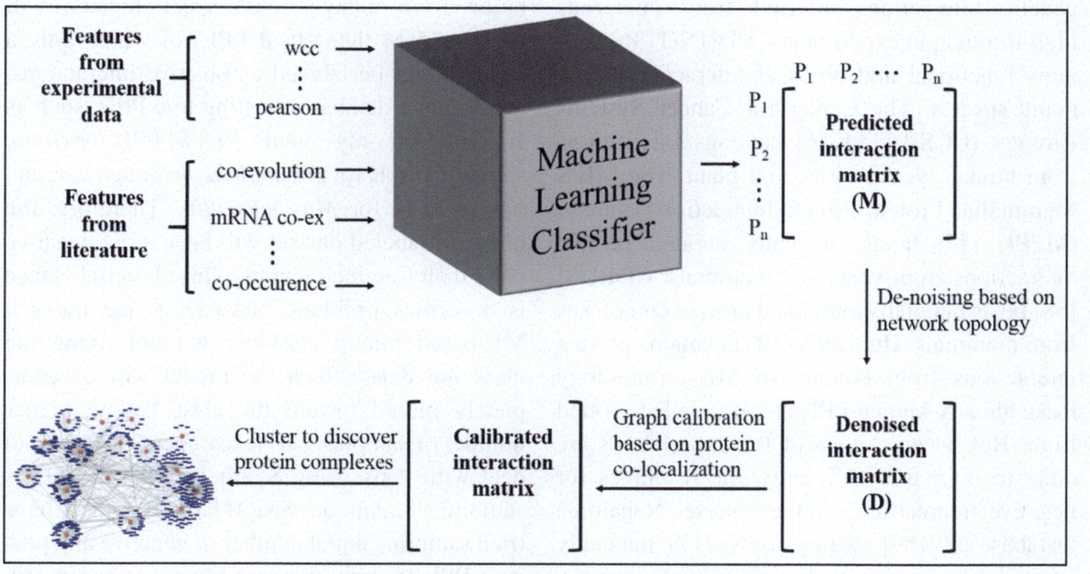

Fig. 11 Computational pipeline for putative complex discovery

5. Feed into the classifier a set of unlabeled PPIs (test dataset), and classifier spits out a probabilistic score for each PPI during testing. For a PPI if the score is zero or close to zero, then we consider that PPI as false interaction, but if the score is close to one or one, then we consider that as true PPI. In past many researchers have used ML to predict PPIs in various organisms [50–52]. ML-based PPI prediction also found its usage in the studies that were particularly interested in protein complex discovery [6–8].

Ground Truth In the previous section, we discussed about training a ML algorithm in a supervised manner, and we saw that labeling samples is key for the learning process. However, we did not discuss about how the samples (PPIs in our context) are labeled. Soon you will realize that this is the most difficult part of ML-based PPI prediction. Here the key questions are from where do we get true or positive PPIs? and how do you label a PPI as negative? Latter is much harder to answer than the former as you will see shortly.

There are many databases that host true PPIs. For example, BioGrid [53] hosts physical and genetic interactions inferred from numerous high-throughput experiments. STRING [36] contains functional and physical interactions from many species. The Center for Cancer Systems Biology (CCSB) [54–56] catalogs interactions from human, yeast, virus, and plant. The MIPS Mammalian Protein-Protein Interaction Database (MPPI) [57] holds manually curated protein interactions from yeast and mammals. CORUM [58] hosts manually annotated protein complexes from mammals. HumanNet [37] contains protein interactions from human. In ML terminology these already known PPIs are also called ground truth. But because most of the experiments are done to infer true PPIs only, the resources for negative interaction is rather sparse. Negatome Database 2.0 [59] catalogs only 2171 manually curated protein pairs that are less likely to interact. Therefore, people had to come up with heuristics to generate the negative interactions. Havugimana et al. defined a negative interaction pair by pairing two proteins annotated to be in two different complexes [8]. This idea is depicted in Fig. 12. However, as you might have already figured out, the validity of this heuristic depends on the already identified interactions, and absence of an interaction does not mean negative interaction. In another work, Crozier et al. [6] has defined negative pairs by choosing a pair of proteins randomly from the set of proteins present in their ground truth dataset. There is a possibility to introduce false-negative interactions in this heuristic too. As some of the randomly sampled PPIs can in fact be true interactions.

Imbalanced Dataset Problem When the number of samples in one class outnumbers the number of samples in other, then we have an imbalanced dataset. Let's say from a sample of a particular species, a proteomics experiment identifies and quantifies n proteins. Then all possible pairs of PPIs could be generated by those n proteins are

$$n_{C_2} = \frac{n!}{2!(n-2)!} \tag{5}$$

Plugging $n = 3000$, which is a realistic expectation from today's mass spectrometers, will yield ~4.5 M theoretical PPIs, of which only a handful will be labeled as positive interactions. Most public databases hosting true PPIs, such as BioGrid [53] only contain ~0.5 M PPIs for *Homo sapiens*, ~76 K for *Drosophila melanogaster*, and only ~30 K for *Mus musculus*. Therefore, the obtained labeled dataset will have more negative pairs than true interactions. This class imbalance is a serious problem, because if one trains a ML-based binary classifier without fixing the class imbalance, then the model will be completely biased toward the class having higher number of samples. ML researchers have tried to deal with it by using Synthetic Minority Oversampling Technique (SMOTE) [60]. Others have tried sampling equal number of negative and positive PPIs to train an ensemble of classifiers [6]. Nonetheless, it is important to note that class

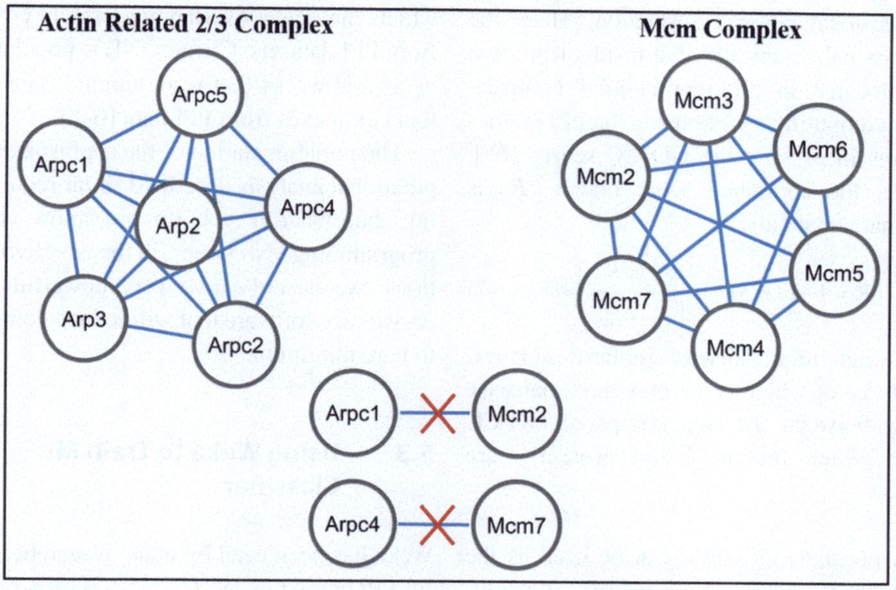

Fig. 12 A heuristic approach toward negative PPI generation

imbalance is inherent in the PPI prediction and it is crucial to deal with it.

Feature Selection It is important to understand the significance of features in ML classification. Using irrelevant features might lead to longer training times and may overfit the model [61]. Researchers have used innovative ways to improve classification through feature selection. Havugimana et al. [8] have used a greedy stepwise feature selection algorithm before training random forest classifier for PPI interaction prediction. There are many other ways to select features, and a discussion on that is beyond our scope, but you will find detailed information about it in this reference [62].

Classification Algorithm Random forest has been used by many researchers dealing with the PPI classification task [6, 8]. Some people have also used support vector machines [7, 63]. Recently, with the rising popularity of deep learning, many researchers have adopted deep neural network based architectures to solve this task [64–66]. The benefit of these deep architectures is that the feature selection is done automatically by the classifiers. However, these classifiers need a large number of data for training in order to yield good results. Our suggestion is to begin with a simple classifier and then iteratively add complexity to the classification module until the desired performance is achieved.

5.1 Denoising a Predicted PPI Matrix

Noise in the predicted interaction matrix can be further reduced by removing the edges lacking support from the network topology. Havugimana et al. [8] used a multistep diffusion procedure to measure connectivity, which can be defined mathematically as

$$C = e^{\lambda * M} - \lambda * M \quad (6)$$

where $M \rightarrow$ predicted PPI matrix

$\lambda \rightarrow$ Inverse of the largest eigenvalue of M.

Edges with connectivity less than a threshold τ are deleted from $\boldsymbol{M}$ to denoise the PPI interaction matrix. Let's denote this denoised matrix as $\boldsymbol{M}^{\dagger}$, which can then be calibrated based on the protein co-localization information. Here the idea is to penalize any PPI that results from two proteins located in different cellular compartments by computing a score using the PPI prediction scores in $M^{\dagger}$ and the GO-CC scores [67]. Therefore, the combined score matrix ($\boldsymbol{R}$) is defined mathematically as

$$\boldsymbol{R} = 1 - \left(1 - \boldsymbol{M}^{\dagger}\right) * \left(1 - \frac{S}{S_{\max}}\right) \quad (7)$$

where $S \rightarrow$ maximum pairwise similarities matrix, each cell (S_{ij}) of which is the maximum pairwise similarity between the two groups of GO-CC terms to which protein i and protein j are annotated.

$S_{\max} \rightarrow$ A normalizing factor can be used as the maximum value among all the semantic similarity scores.

In order to get a thorough understanding of the scoring presented in Eq. (7) readers are encouraged to read this article [68].

5.2 Methods to Cluster the Denoised PPI Matrix (*R*)

In order to identify the densely connected areas in $\boldsymbol{R}$, we need to cluster the PPIs contained in it. There are mainly two classes of clustering methods: hard clustering and soft clustering. In hard clustering a data point is never assigned to multiple clusters. K-means is a popular hard clustering algorithm. In soft clustering every data point receives a membership probability to belong to multiple clusters. In our case soft clustering methods are more suitable because a protein can participate in several different complexes. In other words, protein complex might overlap. Shuye et al. created a comprehensive CYC2008 catalogue, which contains 408 hand-curated heteromeric protein complexes in *S. cerevisiae* [69]. 207 out of 1628 proteins in CYC2008 participate in multiple complexes [70]. Therefore, we need a clustering algorithm that can allow a protein to participate in multiple complexes. Nepusz et al. has addressed these issues in ClusterONE [70], which can detect overlapping protein complexes from PPI datasets. ClusterONE is popularly used in several works that were aiming to detect protein complexes from PPI data [6–8].

The implementation of the sophisticated computational analysis described so far requires coding, but what if you do not know computer programming? No issues, in the next two subsections, we describe two very powerful and yet easy to use software that will reduce your worries to bare minimum.

5.3 Using Weka to Train ML Classifier

Weka has been used by many researchers in past for PPI prediction [8, 71, 72]. It is an acronym for "Waikato Environment for Knowledge Analysis." Weka provides a free workspace as a collection of feature extraction, data preprocessing, and ML algorithms programmed in JAVA [73]. It is available for download at the Weka download page. The biggest benefit of using Weka is that users can right away try many different ML algorithms on their datasets. In fact, no coding is required at all as it provides several GUIs (Fig. 13). First of them is called *Explorer*, where you can explore major Weka packages such as filters, classifiers, associations, and feature selection. It also provides an option to visualize your dataset and facilitates the visualization of prediction of classification and output of clustering in two dimensions only.

The other GUI *KnowledgeFlow* facilitates machine learning on very large datasets that cannot be fit into the computer's physical memory by allowing incremental data loading and processing. However, in this mode only ML algorithms that support incremental learning may be used.

Another GUI is *Experimenter*, which helps to figure out what combinations of parameter values and algorithms works best for the current problem. In *Experimenter* you can automatically compare several learning algorithms with different parameter settings on a collection of datasets. Then, by looking at the performance statistics and results from significance tests, you will be in

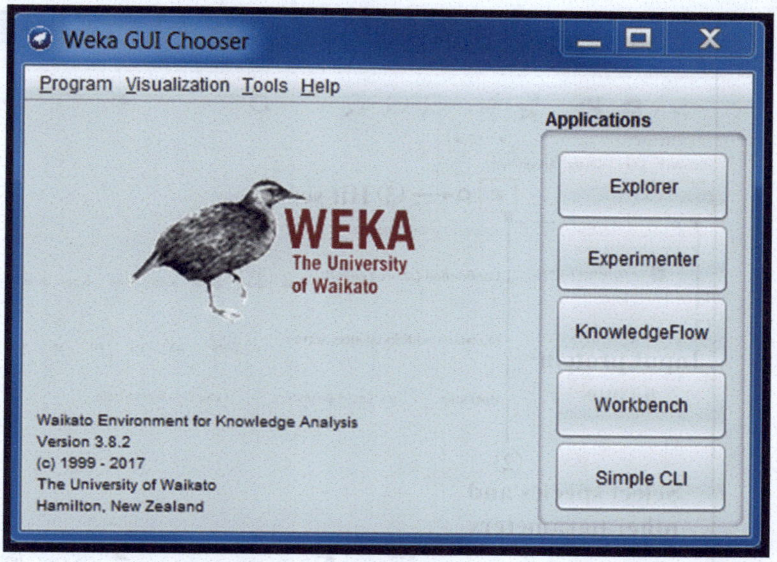

Fig. 13 Graphical User Interfaces (GUIs) provided in Weka

a better position to take a decision on what will be the best configuration for the given datasets.

Workbench GUI integrates the previous three GUIs into a unified and highly configurable GUI.

Weka also provides another GUI called *Simple CLI* for running Weka classification, filtering, feature selection, etc., on command line. This option is for those users who want to control memory usage because it allows a lower-level way of accessing Weka. Weka also allows to add your own classifiers to the software environment. For example, Havugimana et al. [8] used third-party implementation of random forest algorithm, as they found Weka implementation was too slow for their datasets. In order to get familiar with Weka, readers are encouraged to go through the tutorials provided by the Weka developers.

5.4 Network Analysis Using Cytoscape

Cytoscape is an open-source software platform for visualizing and analyzing data in two-dimensional matrix form [74]. It was originally designed for knowledge discovery from datasets generated by biological experiments, but over the years people added plugins to solve problems from other domains like social network analysis, semantic web, etc. The latest version of Cytoscape can be downloaded from the download page. With more than 200 plugins Cytoscape provides a great workbench for network data analysis. It is also possible to integrate PPIs from own measurements performed in the user's laboratory with the PPIs from several databases. For example, using stringApp users can import PPI from STRING DB and augment with their PPI network in Cytoscape as shown in Fig. 14.

Cytoscape also has an app for ClusterONE algorithm discussed earlier. We demonstrate how to use it in Cytoscape in Fig. 15. Readers are encouraged to read the officially released tutorials to get started with latest Cytoscape 3.

6 Conclusion

Protein complexes mediate virtually all biological functions within a cell. A sincere endeavor toward protein complex discovery is essential to unfold the mysteries of this complex cellular machinery. In this chapter we provided a platform to comprehend the conceptual and technical underpinnings of protein complex discovery. We tried to provide a brief introduction to almost a decade of research toward solving the protein complex prediction problem. We hope that our effort will inspire the readers to commence their future research into this very exciting and flour-

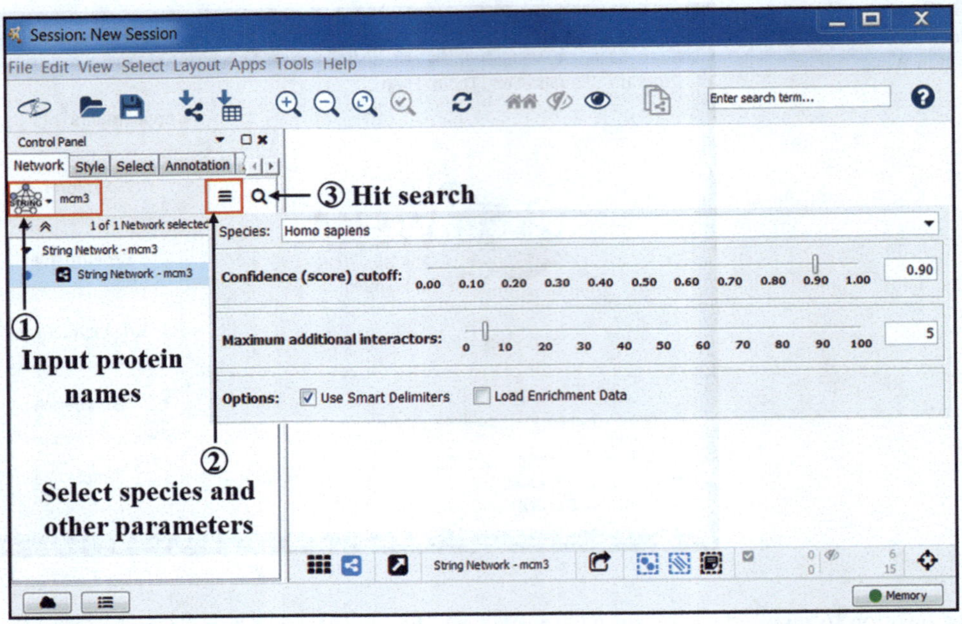

(A)

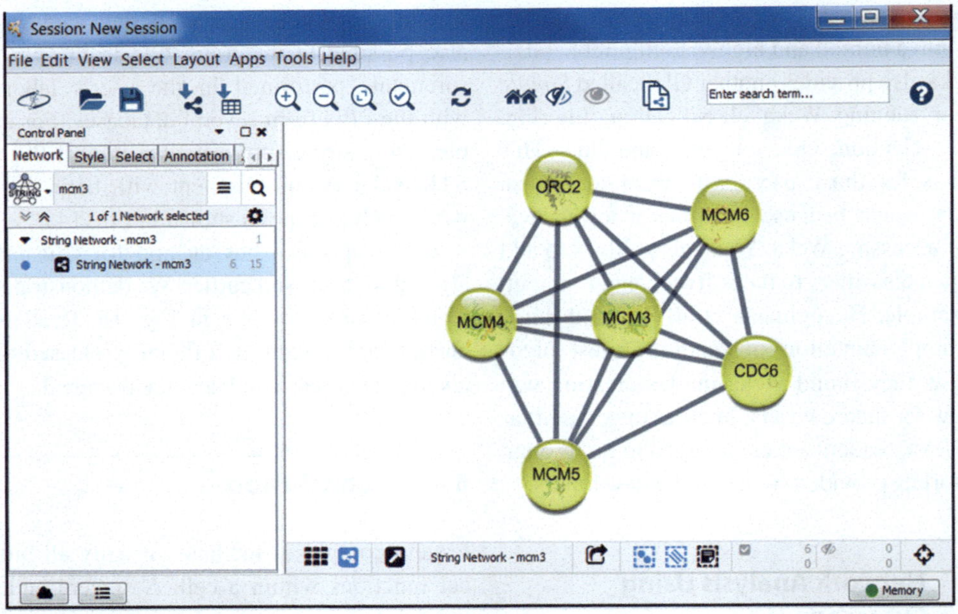

(B)

Fig. 14 stringApp can be used to import PPIs from STRING into the Cytoscape environment. (**a**) From a Cytoscape session, search interaction partners of a protein (mcm3 in this example) in the STRING database. (**b**) The extracted MCM complex from STRING database becomes available in the current Cytoscape session via stringApp

ishing research domain. We believe as the proteomics technologies are improving, more powerful mass spectrometers are emerging, and adoption of sophisticated computational methods are growing, in the future we will not only see tremendous improvement in the sensitivity and

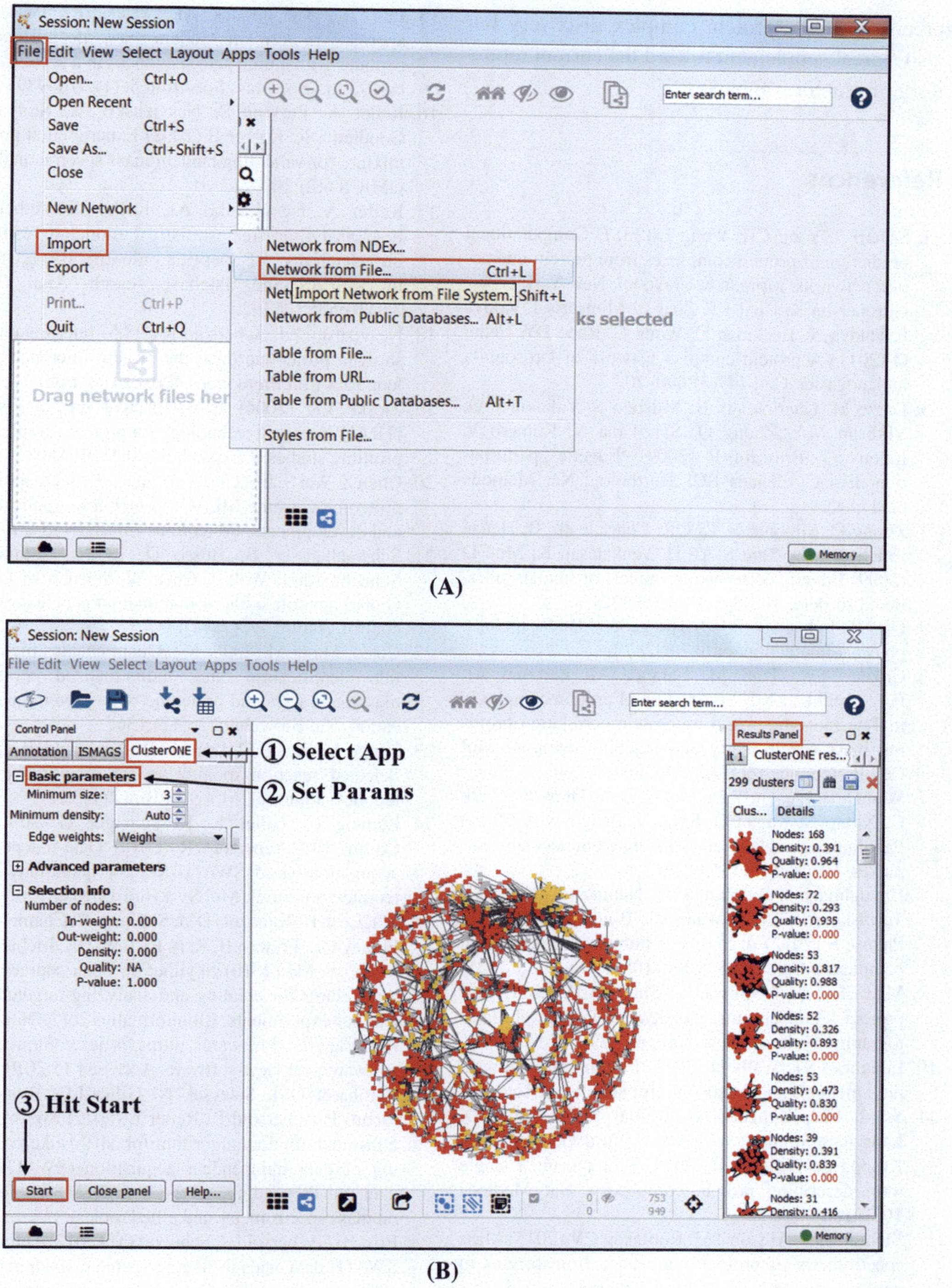

Fig. 15 Cytoscape ClusterONE app. (**a**) Loading a network file. (**b**) Running ClusterONE

precision of the protein complex discovery but also a great supplement toward the current repository of protein complexes.

References

1. Srihari S, Yong CH, Wong L (2017) Computational prediction of protein complexes from protein interaction networks. Morgan & Claypool, New York
2. Guruharsha K, Rual J-F, Zhai B, Mintseris J, Vaidya P, Vaidya N, Beekman C, Wong C, Rhee DY, Cenaj O (2011) A protein complex network of Drosophila melanogaster. Cell 147(3):690–703
3. Dreze M, Charloteaux B, Milstein S, Vidalain P-O, Yildirim MA, Zhong Q, Svrzikapa N, Romero V, Laloux G, Brasseur R (2009) 'Edgetic' perturbation of a C. elegans BCL2 ortholog. Nat Methods 6(11):843
4. Zhong Q, Simonis N, Li QR, Charloteaux B, Heuze F, Klitgord N, Tam S, Yu H, Venkatesan K, Mou D (2009) Edgetic perturbation models of human inherited disorders. Mol Syst Biol 5(1):321
5. Diss G, Lehner B (2018) The genetic landscape of a physical interaction. elife 7:e32472
6. Crozier TW, Tinti M, Larance M, Lamond AI, Ferguson MA (2017) Prediction of protein complexes in Trypanosoma brucei by protein correlation profiling mass spectrometry and machine learning. Mol Cell Proteomics 16(12):2254–2267
7. Wan C, Borgeson B, Phanse S, Tu F, Drew K, Clark G, Xiong X, Kagan O, Kwan J, Bezginov A (2015) Panorama of ancient metazoan macromolecular complexes. Nature 525(7569):339
8. Havugimana PC, Hart GT, Nepusz T, Yang H, Turinsky AL, Li Z, Wang PI, Boutz DR, Fong V, Phanse S (2012) A census of human soluble protein complexes. Cell 150(5):1068–1081
9. Madadlou A, O'Sullivan S, Sheehan D (2011) Fast protein liquid chromatography. In: Protein chromatography. Springer, New York, pp 439–447
10. Loughran ST, Walls D (2011) Protein chromatography: methods and protocols. Humana Press, Totowa
11. Smith LM, Kelleher NL, Linial M, Goodlett D, Langridge-Smith P, Goo YA, Safford G, Bonilla L, Kruppa G, Zubarev R (2013) Proteoform: a single term describing protein complexity. Nat Methods 10(3):186
12. Mehmood S, Allison TM, Robinson CV (2015) Mass spectrometry of protein complexes: from origins to applications. Annu Rev Phys Chem 66:453–474
13. Eidhammer I, Barsnes H, Eide GE, Martens L (2013) Computational and statistical methods for protein quantification by mass spectrometry. Wiley Online Library, West Sussex
14. Hupé P. https://commons.wikimedia.org/wiki/File:Mass_spectrometry_protocol.svg. vol 407 × 390. Wikipedia
15. Eng JK, McCormack AL, Yates JR (1994) An approach to correlate tandem mass spectral data of peptides with amino acid sequences in a protein database. J Am Soc Mass Spectrom 5(11):976–989
16. Keller A, Purvine S, Nesvizhskii AI, Stolyar S, Goodlett DR, Kolker E (2002) Experimental protein mixture for validating tandem mass spectral analysis. OMICS 6(2):207–212
17. Keller A, Nesvizhskii AI, Kolker E, Aebersold R (2002) Empirical statistical model to estimate the accuracy of peptide identifications made by MS/MS and database search. Anal Chem 74(20):5383–5392
18. Nesvizhskii AI, Aebersold R (2005) Interpretation of shotgun proteomic data: the protein inference problem. Mol Cell Proteomics 4(10):1419–1440
19. Zieske LR (2006) A perspective on the use of iTRAQ™ reagent technology for protein complex and profiling studies. J Exp Bot 57(7):1501–1508
20. Chen X, Wei S, Ji Y, Guo X, Yang F (2015) Quantitative proteomics using SILAC: principles, applications, and developments. Proteomics 15(18):3175–3192
21. Schwanhäusser B, Busse D, Li N, Dittmar G, Schuchhardt J, Wolf J, Chen W, Selbach M (2011) Global quantification of mammalian gene expression control. Nature 473(7347):337
22. Cox J, Mann M (2008) MaxQuant enables high peptide identification rates, individualized ppb-range mass accuracies and proteome-wide protein quantification. Nat Biotechnol 26(12):1367
23. Lange V, Picotti P, Domon B, Aebersold R (2008) Selected reaction monitoring for quantitative proteomics: a tutorial. Mol Syst Biol 4(1):222
24. Ludwig C, Gillet L, Rosenberger G, Amon S, Collins BC, Aebersold R (2018) Data-independent acquisition-based SWATH-MS for quantitative proteomics: a tutorial. Mol Syst Biol 14(8):e8126
25. MacLean B, Tomazela DM, Shulman N, Chambers M, Finney GL, Frewen B, Kern R, Tabb DL, Liebler DC, MacCoss MJ (2010) Skyline: an open source document editor for creating and analyzing targeted proteomics experiments. Bioinformatics 26(7):966–968
26. PeakView™ Software. https://sciex.com/products/software/peakview-software. Accessed 15 2019
27. Bernhardt OM, Selevsek N, Gillet LC, Rinner O, Picotti P, Aebersold R, Reiter L (2012) Spectronaut: a fast and efficient algorithm for MRM-like processing of data independent acquisition (SWATH-MS) data. In: Proceedings of the 60th ASMS conference on mass spectrometry and allied topics
28. Röst HL, Aebersold R, Schubert OT (2017) Automated SWATH data analysis using targeted extraction of ion chromatograms. In: Proteomics. Springer, Cham, pp 289–307
29. Heusel M, Bludau I, Rosenberger G, Hafen R, Frank M, Banaei-Esfahani A, van Drogen A, Collins BC, Gstaiger M, Aebersold R (2019) Complex-centric proteome profiling by SEC-SWATH-MS. Mol Syst Biol 15(1):e8438

30. Sean L, Seymour CLH ProteinPilot™ software overview. https://sciex.com/x59785. Accessed Sept 15 2019
31. Tsou C-C, Avtonomov D, Larsen B, Tucholska M, Choi H, Gingras A-C, Nesvizhskii AI (2015) DIA-Umpire: comprehensive computational framework for data-independent acquisition proteomics. Nat Methods 12(3):258
32. Gessulat S, Schmidt T, Zolg DP, Samaras P, Schnatbaum K, Zerweck J, Knaute T, Rechenberger J, Delanghe B, Huhmer A (2019) Prosit: proteome-wide prediction of peptide tandem mass spectra by deep learning. Nat Methods 16(6):509
33. Tiwary S, Levy R, Gutenbrunner P, Soto FS, Palaniappan KK, Deming L, Berndl M, Brant A, Cimermancic P, Cox J (2019) High-quality MS/MS spectrum prediction for data-dependent and data-independent acquisition data analysis. Nat Methods 16:519
34. de Gelder R, Wehrens R, Hageman JA (2001) A generalized expression for the similarity of spectra: application to powder diffraction pattern classification. J Comput Chem 22(3):273–289
35. Bodis L (2007) Quantification of spectral similarity: towards automatic spectra verification. PhD thesis, ETH Zurich
36. Szklarczyk D, Gable AL, Lyon D, Junge A, Wyder S, Huerta-Cepas J, Simonovic M, Doncheva NT, Morris JH, Bork P, Jensen LJ, Mering Christian V (2018) STRING v11: protein–protein association networks with increased coverage, supporting functional discovery in genome-wide experimental datasets. Nucleic Acids Res 47(D1):D607–D613. https://doi.org/10.1093/nar/gky1131
37. Lee I, Blom UM, Wang PI, Shim JE, Marcotte EM (2011) Prioritizing candidate disease genes by network-based boosting of genome-wide association data. Genome Res 21(7):1109–1121
38. Dandekar T, Snel B, Huynen M, Bork P (1998) Conservation of gene order: a fingerprint of proteins that physically interact. Trends Biochem Sci 23(9):324–328
39. Marcotte EM, Pellegrini M, Ng H-L, Rice DW, Yeates TO, Eisenberg D (1999) Detecting protein function and protein-protein interactions from genome sequences. Science 285(5428):751–753
40. Tillier ER, Charlebois RL (2009) The human protein coevolution network. Genome Res 19(10):1861–1871
41. Pazos F, Valencia A (2001) Similarity of phylogenetic trees as indicator of protein–protein interaction. Protein Eng 14(9):609–614
42. Eisen MB, Spellman PT, Brown PO, Botstein D (1998) Cluster analysis and display of genome-wide expression patterns. Proc Natl Acad Sci 95(25):14863–14868
43. Marcotte EM, Pellegrini M, Thompson MJ, Yeates TO, Eisenberg D (1999) A combined algorithm for genome-wide prediction of protein function. Nature 402(6757):83
44. Lee I, Date SV, Adai AT, Marcotte EM (2004) A probabilistic functional network of yeast genes. Science 306(5701):1555–1558
45. Wang Z, Zhang X-C, Le MH, Xu D, Stacey G, Cheng J (2011) A protein domain co-occurrence network approach for predicting protein function and inferring species phylogeny. PLoS One 6(3):e17906
46. Cohen KB, Demner-Fushman D (2014) Biomedical natural language processing, vol 11. John Benjamins Publishing Company, Amsterdam
47. Deeb SJ, D'Souza RC, Cox J, Schmidt-Supprian M, Mann M (2012) Super-SILAC allows classification of diffuse large B-cell lymphoma subtypes by their protein expression profiles. Mol Cell Proteomics 11(5):77–89
48. Karpievitch YV, Dabney AR, Smith RD (2012) Normalization and missing value imputation for label-free LC-MS analysis. BMC Bioinf 13(16):S5
49. Webb-Robertson B-JM, Wiberg HK, Matzke MM, Brown JN, Wang J, McDermott JE, Smith RD, Rodland KD, Metz TO, Pounds JG (2015) Review, evaluation, and discussion of the challenges of missing value imputation for mass spectrometry-based label-free global proteomics. J Proteome Res 14(5):1993–2001
50. Bock JR, Gough DA (2001) Predicting protein–protein interactions from primary structure. Bioinformatics 17(5):455–460
51. Shen J, Zhang J, Luo X, Zhu W, Yu K, Chen K, Li Y, Jiang H (2007) Predicting protein–protein interactions based only on sequences information. Proc Natl Acad Sci 104(11):4337–4341
52. Bradford JR, Westhead DR (2004) Improved prediction of protein–protein binding sites using a support vector machines approach. Bioinformatics 21(8):1487–1494
53. Oughtred R, Stark C, Breitkreutz B-J, Rust J, Boucher L, Chang C, Kolas N, O'Donnell L, Leung G, McAdam R (2018) The BioGRID interaction database: 2019 update. Nucleic Acids Res 47(D1):D529–D541
54. Yu H, Tardivo L, Tam S, Weiner E, Gebreab F, Fan C, Svrzikapa N, Hirozane-Kishikawa T, Rietman E, Yang X (2011) Next-generation sequencing to generate interactome datasets. Nat Methods 8(6):478
55. Yu H, Braun P, Yıldırım MA, Lemmens I, Venkatesan K, Sahalie J, Hirozane-Kishikawa T, Gebreab F, Li N, Simonis N (2008) High-quality binary protein interaction map of the yeast interactome network. Science 322(5898):104–110
56. Rolland T, Taşan M, Charloteaux B, Pevzner SJ, Zhong Q, Sahni N, Yi S, Lemmens I, Fontanillo C, Mosca R (2014) A proteome-scale map of the human interactome network. Cell 159(5):1212–1226
57. Pagel P, Kovac S, Oesterheld M, Brauner B, Dunger-Kaltenbach I, Frishman G, Montrone C, Mark P, Stümpflen V, Mewes H-W, Ruepp A, Frishman D (2004) The MIPS mammalian protein–protein interaction database. Bioinformatics 21(6):832–834. https://doi.org/10.1093/bioinformatics/bti115

58. Giurgiu M, Reinhard J, Brauner B, Dunger-Kaltenbach I, Fobo G, Frishman G, Montrone C, Ruepp A (2018) CORUM: the comprehensive resource of mammalian protein complexes – 2019. Nucleic Acids Res 47(D1):D559–D563
59. Blohm P, Frishman G, Smialowski P, Goebels F, Wachinger B, Ruepp A, Frishman D (2013) Negatome 2.0: a database of non-interacting proteins derived by literature mining, manual annotation and protein structure analysis. Nucleic Acids Res 42(D1):D396–D400
60. Chawla NV, Bowyer KW, Hall LO, Kegelmeyer WP (2002) SMOTE: synthetic minority over-sampling technique. J Artif Intell Res 16:321–357
61. Bermingham ML, Pong-Wong R, Spiliopoulou A, Hayward C, Rudan I, Campbell H, Wright AF, Wilson JF, Agakov F, Navarro P (2015) Application of high-dimensional feature selection: evaluation for genomic prediction in man. Sci Rep 5:10312
62. Guyon I, Elisseeff A (2003) An introduction to variable and feature selection. J Mach Learn Res 3(Mar):1157–1182
63. Guo Y, Yu L, Wen Z, Li M (2008) Using support vector machine combined with auto covariance to predict protein–protein interactions from protein sequences. Nucleic Acids Res 36(9):3025–3030
64. Sun T, Zhou B, Lai L, Pei J (2017) Sequence-based prediction of protein–protein interaction using a deep-learning algorithm. BMC bioinf 18(1):277
65. Tian K, Shao M, Wang Y, Guan J, Zhou S (2016) Boosting compound-protein interaction prediction by deep learning. Methods 110:64–72
66. Hashemifar S, Neyshabur B, Khan AA, Xu J (2018) Predicting protein–protein interactions through sequence-based deep learning. Bioinformatics 34(17):i802–i810
67. Consortium GO (2004) The Gene Ontology (GO) database and informatics resource. Nucleic Acids Res 32(suppl_1):D258–D261. https://doi.org/10.1093/nar/gkh036
68. Yang H, Nepusz T, Paccanaro A (2012) Improving GO semantic similarity measures by exploring the ontology beneath the terms and modelling uncertainty. Bioinformatics 28(10):1383–1389
69. Pu S, Wong J, Turner B, Cho E, Wodak SJ (2008) Up-to-date catalogues of yeast protein complexes. Nucleic Acids Res 37(3):825–831
70. Nepusz T, Yu H, Paccanaro A (2012) Detecting overlapping protein complexes in protein-protein interaction networks. Nat Methods 9(5):471
71. Qi Y, Klein-Seetharaman J, Bar-Joseph Z (2007) A mixture of feature experts approach for protein-protein interaction prediction. In: BMC bioinformatics, vol 10. BioMed Central, p S6
72. Zahiri J, Yaghoubi O, Mohammad-Noori M, Ebrahimpour R, Masoudi-Nejad A (2013) PPIevo: protein–protein interaction prediction from PSSM based evolutionary information. Genomics 102(4):237–242
73. Holmes G, Donkin A, Witten IH (1994) Weka: a machine learning workbench. University of Waikato, Hamilton
74. Shannon P, Markiel A, Ozier O, Baliga NS, Wang JT, Ramage D, Amin N, Schwikowski B, Ideker T (2003) Cytoscape: a software environment for integrated models of biomolecular interaction networks. Genome Res 13(11):2498–2504

Capillary Electrophoresis-Based N-Glycosylation Analysis in the Biomedical and Biopharmaceutical Fields

Renata Kun, Eszter Jóna, and Andras Guttman

Abstract

Glycomics has a growing interest in the biopharmaceutical industry and biomedical research requiring new high-performance and high-sensitivity bioanalytical tools. Analysis of N-glycosylation is very important during the development of protein therapeutics and it also plays a key role in biomarker discovery. The most frequently used glycoanalytical methods are capillary electrophoresis, liquid chromatography, and mass spectrometry. In this chapter, the capillary electrophoresis-based N-linked carbohydrate analysis methods are conferred with emphasis on its use in the biopharmaceutical and biomedical fields.

Keywords

Capillary electrophoresis · N-glycans · Biopharmaceuticals · Biomarkers

R. Kun
Horváth Csaba Memorial Laboratory of Bioseparation Sciences, Doctoral School of Molecular Medicine, Research Center for Molecular Medicine, Faculty of Medicine, University of Debrecen, Debrecen, Hungary

E. Jóna
Research Institute of Biomolecular and Chemical Engineering, University of Pannonia, Veszprem, Hungary

A. Guttman (✉)
Horváth Csaba Memorial Laboratory of Bioseparation Sciences, Doctoral School of Molecular Medicine, Research Center for Molecular Medicine, Faculty of Medicine, University of Debrecen, Debrecen, Hungary

Research Institute of Biomolecular and Chemical Engineering, University of Pannonia, Veszprem, Hungary

Abbreviations

AFP	alpha-fetoprotein
APTS	8-aminopyrene-1,3,6-trisulfonate
BHK	baby hamster kidney
CA	carbohydrate antigen
CEA	carcinoembryonic antigen
CE-LIF	capillary electrophoresis with laser-induced fluorescence detection
CHO	Chinese hamster ovary
HILIC	hydrophilic interaction liquid chromatography
HPLC	high-performance liquid chromatography
mAB	monoclonal antibody
MS	mass spectrometry
NS0	murine myeloma cell line
PNGase F	peptide-N-glycosidase F
PSA	prostate-specific antigen

A. V. Colnaghi Simionato (ed.), *Separation Techniques Applied to Omics Sciences*, Advances in Experimental Medicine and Biology 1336, https://doi.org/10.1007/978-3-030-77252-9_7

1 Introduction

Glycosylation of proteins is one of the most important post-translation modifications, which impacts their function and their lifespan, also taking part in important biochemical and physiological processes as well as in cell-cell interactions. The buildup of the glycan structure may therefore modify the cell functions and can serve as indicators of various diseases. In the course of N-glycosylation, the carbohydrate structures bind to the polypeptide chains of the proteins being synthesized (co-translational) and then modified post-translationally [1, 2]. The glycan structures are usually made up of several glycoforms, which significantly increase the structural diversity of glycans. This so-called microheterogeneity depends on the expression, concentration, and kinetic features of glycosyltransferases and glycosidases. The glycoforms might have various binding sites within a protein, which is referred to as glycosylation macroheterogeneity (site specificity) [3–6].

The structural diversity of recombinant glycoproteins is very important for the pharmaceutical industry to avoid unwanted side effects and allergic reactions. During the development of these new-generation medicines, the producing microorganism must be chosen carefully to avoid immunogenic effects caused by nonhuman glycan epitopes. The most immunogenic nonhuman sugar residues are alpha-1,3-galactosylation and N-glycolylneuraminic acid (Neu5Gc) [7–9].

2 The Biochemical Background of Glycosylation

Asparagine (N)-linked glycans have three main structural subtypes. If it contains only mannose in addition to the core structure (Fig. 1, left panel), it is called "high mannose" type (Fig. 1, right panel). Hybrid glycans consist of both mannose and other sugar units in addition to the core. Complex-type glycans have other (non-mannose) sugar units added to the core structure (Fig. 1, right panel) [3, 10].

Another major type of protein glycosylation (not discussed in this chapter) is O-glycosylation via Thr or Ser residues. O-Glycans are synthetized in the Golgi apparatus and they predominantly appear on the surface of cells synthetizing mucins and on epithelial cell surfaces rich in serine and threonine [2, 11–15].

3 Glycobiomarker Discovery

Modifications in N-glycan structures significantly influence the half-life of proteins, their maturity state, cellular adhesion characteristics, migration, tumor invasion, and the formation of metastases. Various serological assays are available to identify organ-specific tumor glycobiomarkers, and they provide information on the prognosis of the disease [8, 16–18]. Biomarker assays recognize the glycan structures on the surface of the cells, for example, carbohydrate antigen (CA) 19-9 (pancreas/colorectal/gastro-carbohydrate antigen), CA 72-4 (colorectal/gastric), CA 125 (ovary), CA 15-3 (breast), AFP-L3 (hepatic cells), and PSA (prostate-specific antigen). Carcinoembryonic antigen (CEA) is a general and diagnostically widespread tumor marker, rich in N-glycan structures. The serum level of these biomarkers can be specific for determination of tumor genesis; therefore, the mapping of new, more specific glycobiomarkers is desirable in the future (Fig. 2) [16–25].

4 Glycosylated Biopharmaceuticals

The asparagine-linked carbohydrate moieties also have high significance in the pharmaceutical industry, because most of the new-generation biotherapeutics are glycosylated protein-based medicines, manufactured by recombinant techniques. In addition to monoclonal antibody (mAb)-type drugs, a number of hormones, coagulation factors, and lecithin-type compounds are continuously entering the market these days. Modifications of their linked glycan structures

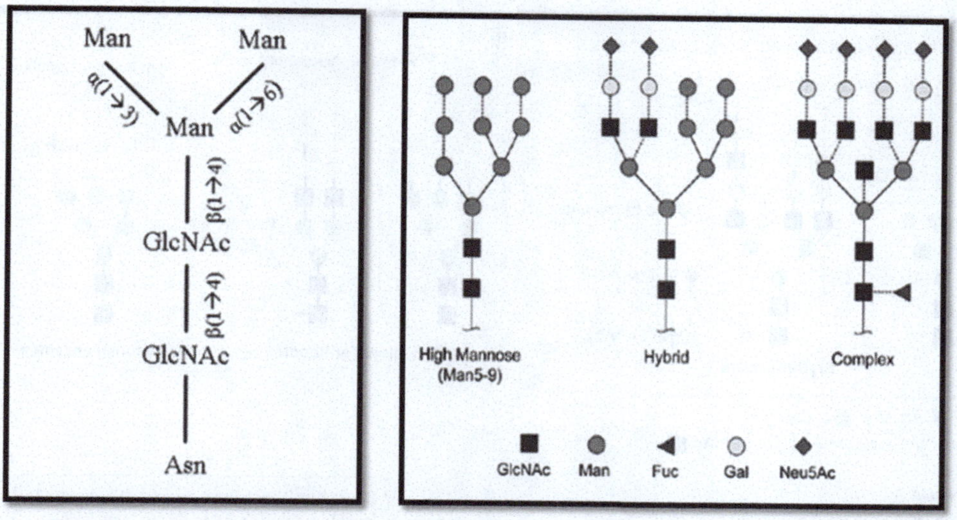

Fig. 1 The trimannosyl core structure of N-linked glycans (left panel) and the main N-glycan structure subtypes (right panel). Symbols: ■ N-Acetylglucosamine; ● Mannose; ◄ Fucose; ○ Galactose; ◆ N-acetylneuraminic acid. With permissions from [2, 10]

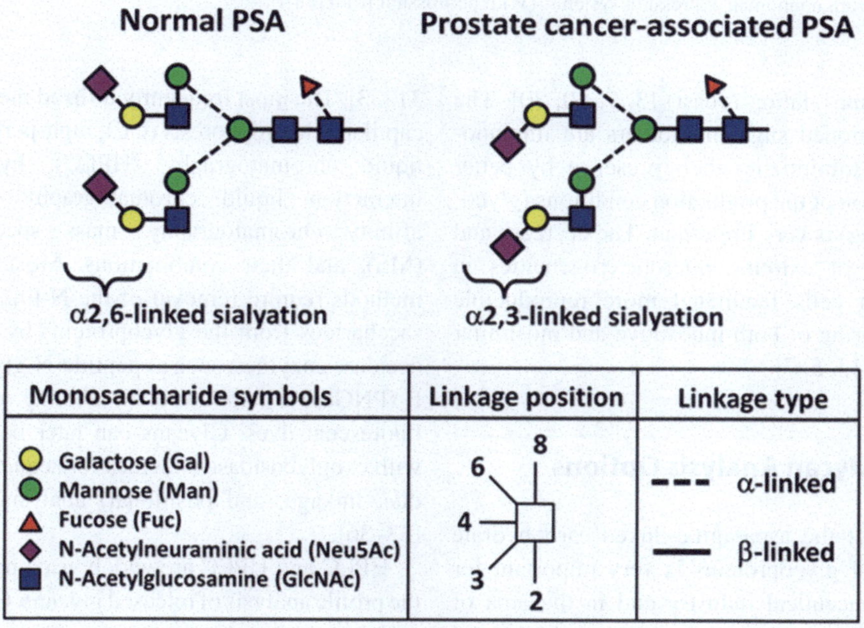

Fig. 2 Differences between normal and prostate cancer patients in the N-glycosylation of prostate-specific antigen (PSA). The terminal sialic acids are α2,3 linked on the aberrant PSA. (With permission from [21])

highly influence their efficiency, stability, safety, and half-life (Fig. 3) [6, 7, 26–29].

The glycosylation decoration of biological medicines greatly depends on the type of the producing microorganism/mammalian cell lines and the production environment. Nonhuman cell lines and microorganisms may synthesize immunogenic glycan residues such as N-glycolylneuraminic acid (Neu5Gc) (CHO cell line), alpha1,3-galactose epitope (BHK, NS0 cells), core alpha1,3-fucose (insects), beta1,2-xylose/core and alpha1,3-fucose (plants), and

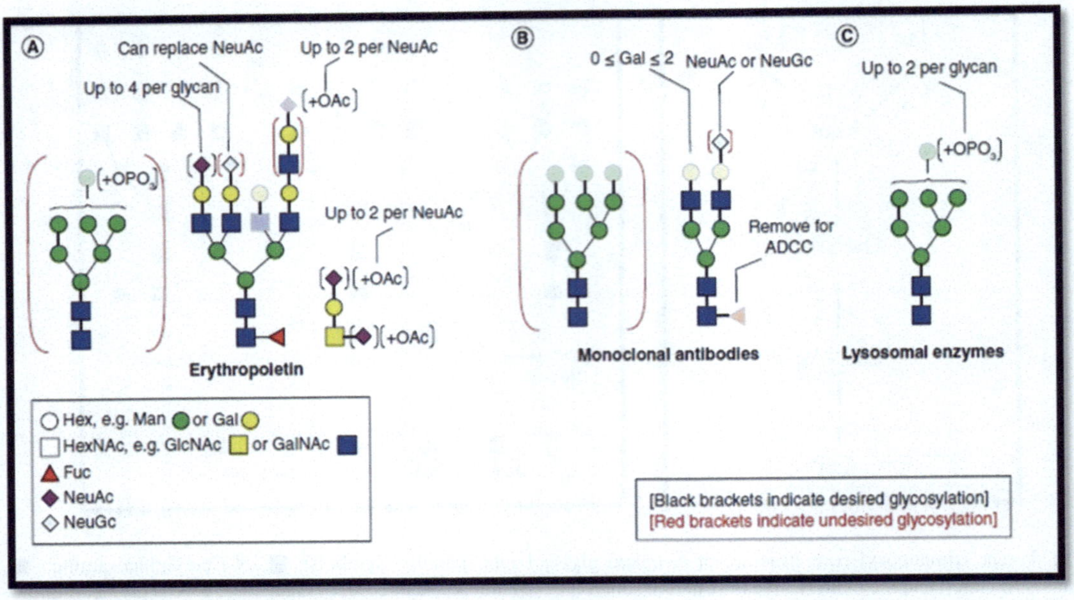

Fig. 3 Glycosylation of biopharmaceuticals. The red brackets indicate undesired/immunogenic sugar epitopes on glycopeptides from nonhuman expression systems. (With permission from [28])

hyper-mannosylation (yeast) [3, 7, 10, 30]. The abovementioned sugar monomers are immunogenic, so minimizing their presence by better optimization of the production conditions (glycoengineering) is very important. The decrease and avoidance of extreme microheterogeneities in production cells facilitated more reproducible manufacturing of both innovative and biosimilar medicines [3, 5, 7].

5 Glycan Analysis Options

Analysis of the asparagine-linked carbohydrate moieties of glycoproteins is very important for the pharmaceutical industry and in the area of biomarker research requiring high sensitivity and high-resolution separation and detection methods. These bioanalytical techniques should provide detailed N-glycan profile information, including data about linkages and positional isomers. High-sensitivity glycoanalytical tools are readily available today on a wide scale to map protein glycosylations, help to identify smaller structural dissimilarities in the carbohydrate structures, and discover new glycan epitopes [6, 31–33]. The most frequently utilized methods are capillary electrophoresis (CE), high-performance liquid chromatography (HPLC), hydrophilic interaction liquid chromatography (HILIC), affinity chromatography, mass spectrometry (MS), and their combinations. Most of these methods require removal of the N-linked oligosaccharides from the glycoproteins by endoglycosidase enzymes, such as peptide-N-glycosidase F (PNGase F), followed by labeling, e.g., with fluorescent dyes. Glycans can later be digested with exoglycosidase enzymes for sequence (residue, linkage, and positional) analysis (Fig. 4) [33–36].

HPLC and UPLC are widely used methods in the profile analysis of released glycans. Combined with mass spectrometry, it provides some structural information about the glycans [37, 38]. An excellent method for the N-glycosylation analysis of intact glycoproteins is MS combined with RP-HPLC, providing detailed information about the heterogeneity of carbohydrate binding sites, most frequently combined with ESI-TOF analyzer. MALDI is another efficient way for glycan analysis, but its accuracy is not as good as with HPLC. However, it is but appropriate for O-glycan

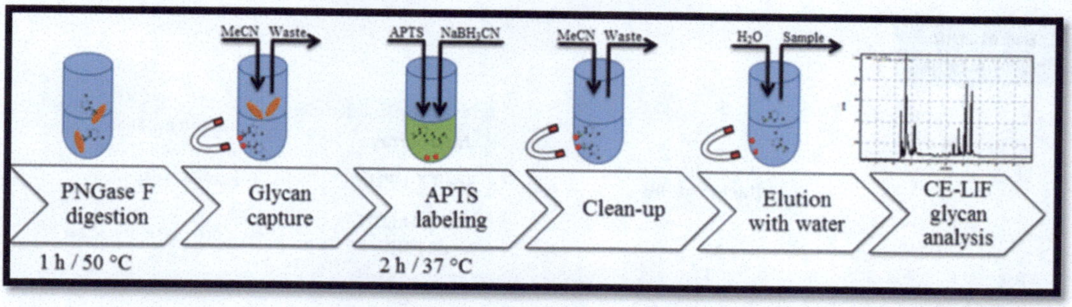

Fig. 4 Glycosylation analysis workflow including endoglycosidase (PNGase F) digestion, capturing the released glycans by magnetic beads, fluorophore labeling of the carbohydrates on the beads, cleanup, and CE-LIF analysis after elution. (With permission from [33])

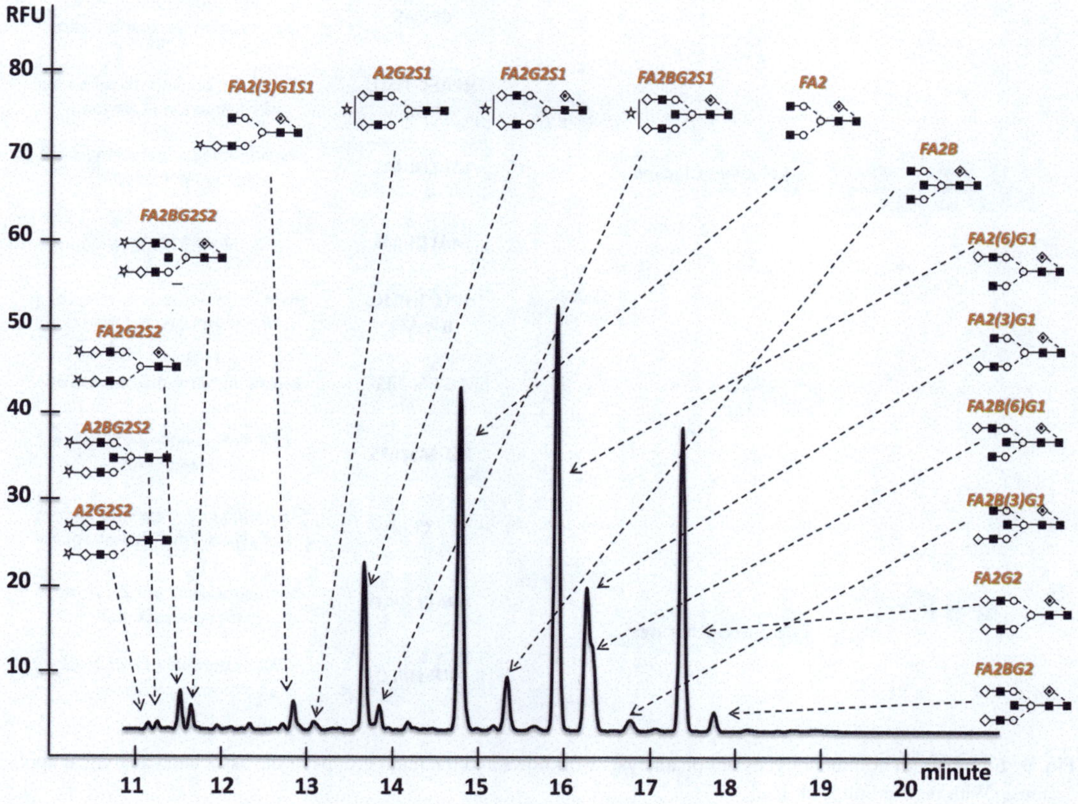

Fig. 5 Capillary electrophoresis analysis of endoglycosidase-released and 8-aminopyrene-1,3,6-trisulfontate-labeled human IgG N-glycans using laser-induced fluorescence detection

analysis/characterization [39–43]. The advent of HILIC chromatography in carbohydrate analysis was an important step forward, as this method is capable of efficiently separating N-glycans [34, 44].

Capillary electrophoresis (CE) has proven to be one of the most excellent separation methods for the analysis of complex N-glycan structures (Fig. 5). Coupled with laser-induced fluorescence detection (LIF), it is possible to reach very high sensitivity and resolution. In the case of carbohydrate sequencing, serial exoglycosidase digestion is required, similar to that of LC. The fluorescent labeling of glycans for CE-LIF is performed

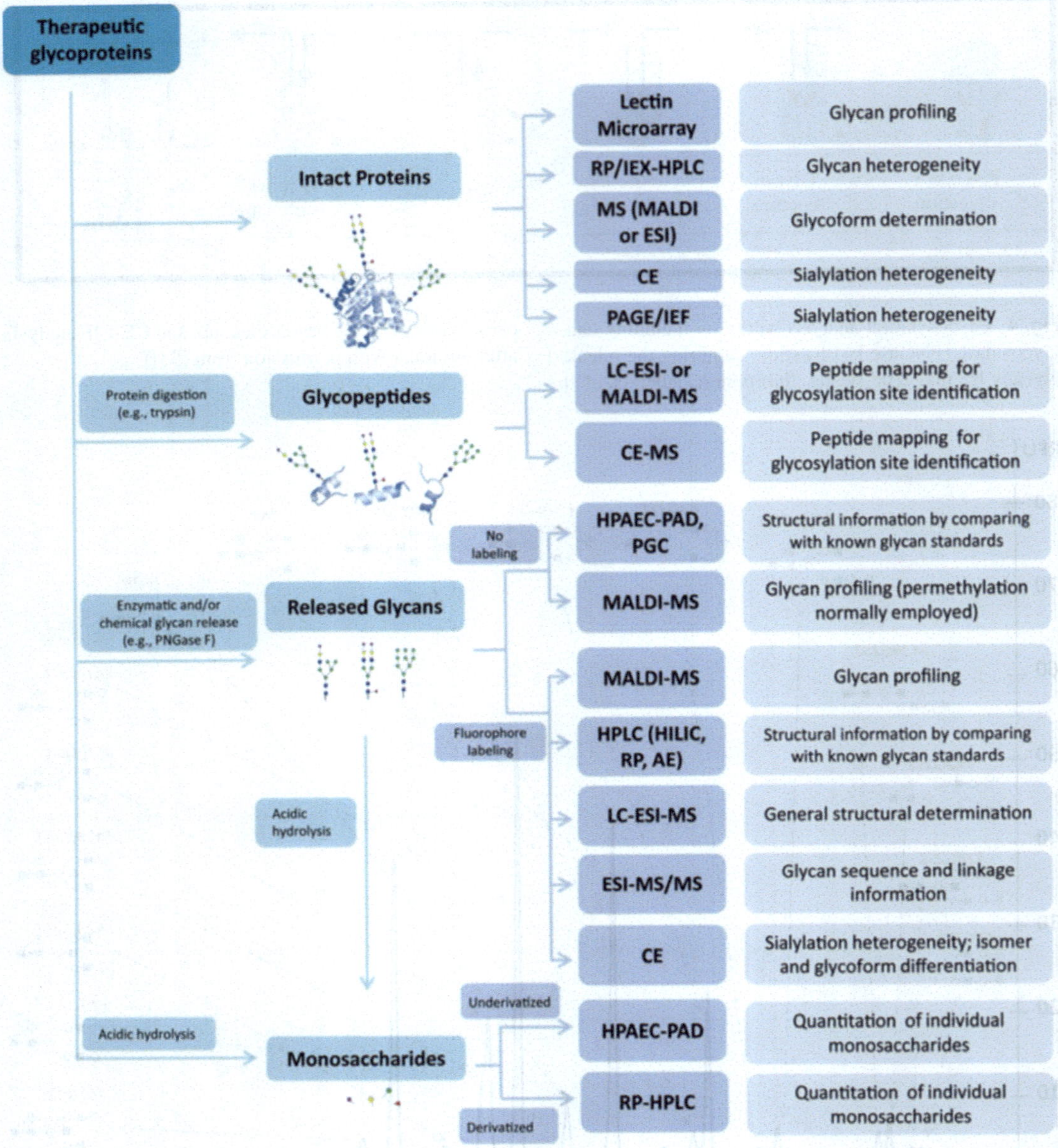

Fig. 6 Summary of common glycosylation analysis workflows and the main groups of the associated analytical applications. (With permission from [60])

using APTS (8-aminopyrene-1,3,6-trisulfonate). Various subsets of CE such as micellar electrokinetic chromatography or isoelectric focusing increase the efficiency and resolution of the separation at the glycopeptide and/or glycoprotein level with lower sample requirement than that of MS or HPLC [45–48].

These commonly used analytical methods can give valuable structural information about the main carbohydrate groups on glycopeptides, intact proteins in addition to released glycans and monosaccharides. Intact protein glycan mapping is possible with lectin microarrays and CE or MS [43, 46, 49, 50]. MS coupled with ESI or MALDI can determine various glycoforms. CE with MS can reveal sialylation heterogeneity of intact proteins and can give detailed site identification of glycopeptides as well [41, 51, 52]. Lectin microarrays can detect glycan-lectin interactions of intact proteins and give useful information about glycoconjugates, but cannot provide structural information [53]. MALDI-MS offers detailed structural information about intact glycoproteins including linkage and branches and can provide information about glycosylation site specificity, but with lower mass accuracy [54]. Charge-based electrophoresis such as capillary isoelectrofocusing (cIEF) is useful for quality control testing for sialylated species [55]. HPLC with ESI-MS or with MALDI-MS is widely used to gain information about protein glycosylation sites and occupancy also suitable for rapid glycopeptide profiling [56, 57]. RPLC-MS could provide complete sequence analysis of glycoproteins but needs higher sample amounts. On the other hand, CE in combination with MS offers high-resolution analysis of glycoproteins providing useful information about the entire glycan structures while only needing very low sample amounts [58, 59].

6 Summary

The growing interest in glycomics in the pharmaceutical industry and in biomedical research demanded the development of high-resolution and high-sensitivity glycoanalytical techniques. N-Glycosylation analysis is very important for the development of monoclonal antibody-based and new modality medicines, showing promising results in the field of biomarker research. Due to the complex structures of N-linked carbohydrates, their analysis is challenging that requires new-generation, high-resolution/high-sensitivity methods such as CE-LIF and various liquid chromatography-based methods preferably connected to mass spectrometry. Indeed, the additional mass spectrometry data provides deeper structural information, but mostly requires coupling to liquid phase separation methods as described in Fig. 6.

Acknowledgments The authors gratefully acknowledge the grants from the National Research, Development and Innovation Office (K116263, NN127062, 2018-2.1.17-TÉT-KR-2018-00010). This work was also sponsored by the BIONANO_GINOP-2.3.2-15-2016-00017 project of the Hungarian Government. This is contribution #161 from the Horváth Csaba Memorial Laboratory of Bioseparation Sciences.

References

1. Shental-Bechor D, Levy Y (2009) Folding of glycoproteins: toward understanding the biophysics of the glycosylation code. Curr Opin Struct Biol 19(5):524–533
2. Varki A et al (2015) Symbol nomenclature for graphical representations of Glycans. Glycobiology 25(12):1323–1324
3. Brooks SA (2006) Protein glycosylation in diverse cell systems: implications for modification and analysis of recombinant proteins. Expert Rev Proteomics 3(3):345–359
4. Kornfeld R, Kornfeld S (1985) Assembly of asparagine-linked oligosaccharides. Annu Rev Biochem 54:631–664
5. Marino K et al (2010) A systematic approach to protein glycosylation analysis: a path through the maze. Nat Chem Biol 6(10):713–723
6. Szekrenyes A et al (2015) Sample preparation for N-glycosylation analysis of therapeutic monoclonal antibodies by electrophoresis. Methods Mol Biol 1274:183–195
7. Dicker M, Strasser R (2015) Using glyco-engineering to produce therapeutic proteins. Expert Opin Biol Ther 15(10):1501–1516
8. Dube DH, Bertozzi CR (2005) Glycans in cancer and inflammation. Potential for therapeutics and diagnostics. Nat Rev Drug Discov 4(6):477–488
9. Shrimal S, Gilmore R (2013) Glycosylation of closely spaced acceptor sites in human glycoproteins. J Cell Sci 126(23):5513–5523
10. Brooks SA (2009) Strategies for analysis of the glycosylation of proteins: current status and future perspectives. Mol Biotechnol 43(1):76–88
11. Brockhausen I (1999) Pathways of O-glycan biosynthesis in cancer cells. Biochim Biophys Acta 1473(1):67–95
12. Hanisch FG (2001) O-glycosylation of the mucin type. Biol Chem 382(2):143–149
13. Ito Y et al (2005) Structural approaches to the study of oligosaccharides in glycoprotein quality control. Curr Opin Struct Biol 15(5):481–489
14. Spiro RG (1973) Glycoproteins. Adv Protein Chem 27:349–467
15. Spiro RG (2002) Protein glycosylation: nature, distribution, enzymatic formation, and disease implications of glycopeptide bonds. Glycobiology 12(4):43r–56r
16. Drake RR (2015) Glycosylation and cancer: moving glycomics to the forefront. Adv Cancer Res 126:1–10

17. Pearce OMT (2018) Cancer glycan epitopes: biosynthesis, structure and function. Glycobiology 28(9):670–696
18. Reis CA et al (2010) Alterations in glycosylation as biomarkers for cancer detection. J Clin Pathol 63(4):322–329
19. Gilgunn S et al (2013) Aberrant PSA glycosylation-a sweet predictor of prostate cancer. Nat Rev Urol 10(2):99–107
20. Goldstein MJ, Mitchell EP (2005) Carcinoembryonic antigen in the staging and follow-up of patients with colorectal cancer. Cancer Investig 23(4):338–351
21. Hatakeyama S et al (2017) Recent progress and perspectives on prostate cancer biomarkers. Int J Clin Oncol 22(2):214–221
22. Varki A, Cummings R, Esko J, et al., editors. (1999) Essentials of Glycobiology. Cold Spring Harbor (NY): Cold Spring Harbor Laboratory Press. Chapter 35, Glycosylation Changes in Cancer.
23. Li M, Song L, Qin X (2010) Glycan changes: cancer metastasis and anti-cancer vaccines. J Biosci 35(4):665–673
24. Locker GY et al (2006) ASCO 2006 update of recommendations for the use of tumor markers in gastrointestinal cancer. J Clin Oncol 24(33):5313–5327
25. Pinho SS, Reis CA (2015) Glycosylation in cancer: mechanisms and clinical implications. Nat Rev Cancer 15(9):540–555
26. Brinks V et al (2011) Quality of original and biosimilar Epoetin products. Pharm Res 28(2):386–393
27. Lingg N et al (2012) The sweet tooth of biopharmaceuticals: importance of recombinant protein glycosylation analysis. Biotechnol J 7(12):1462–1472
28. Oh MJ et al (2016) Analytical detection and characterization of biopharmaceutical glycosylation by MS. Bioanalysis 8(7):711–727
29. Wacker C et al (2011) Glycosylation profiles of therapeutic antibody pharmaceuticals. Eur J Pharm Biopharm 79(3):503–507
30. Hoja-Lukowicz D et al (2013) L1CAM from human melanoma carries a novel type of N-glycan with Galbeta1-4Galbeta1- motif. Involvement of N-linked glycans in migratory and invasive behaviour of melanoma cells. Glycoconj J 30(3):205–225
31. Bielik AM, Zaia J (2010) Historical overview of glycoanalysis. Methods Mol Biol 600:9–30
32. Dimitrov DS, Marks JD (2009) Therapeutic antibodies: current state and future trends--is a paradigm change coming soon? Methods Mol Biol 525:1–27, xiii
33. Szigeti M et al (2016) Fully automated sample preparation for ultrafast N-glycosylation analysis of antibody therapeutics. J Lab Autom 21(2):281–286
34. Melmer M et al (2010) HILIC analysis of fluorescence-labeled N-glycans from recombinant biopharmaceuticals. Anal Bioanal Chem 398(2):905–914
35. Reusch D et al (2015) Comparison of methods for the analysis of therapeutic immunoglobulin G Fc-glycosylation profiles-part 2: mass spectrometric methods. MAbs 7(4):732–742
36. Vreeker GC, Wuhrer M (2017) Reversed-phase separation methods for glycan analysis. Anal Bioanal Chem 409(2):359–378
37. Veillon L et al (2017) Characterization of isomeric glycan structures by LC-MS/MS. Electrophoresis 38(17):2100–2114
38. Zhou S et al (2017) Direct comparison of derivatization strategies for LC-MS/MS analysis of N-glycans. Analyst 142(23):4446–4455
39. Dell A, Morris HR (2001) Glycoprotein structure determination by mass spectrometry. Science 291(5512):2351–2356
40. Geyer H, Geyer R (2006) Strategies for analysis of glycoprotein glycosylation. Biochim Biophys Acta 1764(12):1853–1869
41. Grünwald-Gruber C et al (2017) Determination of true ratios of different N-glycan structures in electrospray ionization mass spectrometry. Anal Bioanal Chem 409(10):2519–2530
42. Higel F et al (2013) Reversed-phase liquid-chromatographic mass spectrometric N-glycan analysis of biopharmaceuticals. Anal Bioanal Chem 405(8):2481–2493
43. Zhang Z, Pan H, Chen X (2009) Mass spectrometry for structural characterization of therapeutic antibodies. Mass Spectrom Rev 28(1):147–176
44. Ruhaak LR et al (2008) Hydrophilic interaction chromatography-based high-throughput sample preparation method for N-glycan analysis from total human plasma glycoproteins. Anal Chem 80(15):6119–6126
45. Guttman A (1996) High-resolution carbohydrate profiling by capillary gel electrophoresis. Nature 380(6573):461–462
46. Guttman A (2013) Capillary electrophoresis in the N-glycosylation analysis of biopharmaceuticals. Trac Trends Anal Chem 48:132–143
47. Kamoda S, Ishikawa R, Kakehi K (2006) Capillary electrophoresis with laser-induced fluorescence detection for detailed studies on N-linked oligosaccharide profile of therapeutic recombinant monoclonal antibodies. J Chromatogr A 1133(1–2):332–339
48. Ma S, Nashabeh W (1999) Carbohydrate analysis of a chimeric recombinant monoclonal antibody by capillary electrophoresis with laser-induced fluorescence detection. Anal Chem 71(22):5185–5192
49. North SJ et al (2009) Mass spectrometry in the analysis of N-linked and O-linked glycans. Curr Opin Struct Biol 19(5):498–506
50. Zhang L, Luo S, Zhang B (2016) The use of lectin microarray for assessing glycosylation of therapeutic proteins. MAbs 8(3):524–535
51. Gennaro LA, Salas-Solano O, Ma S (2006) Capillary electrophoresis-mass spectrometry as a characterization tool for therapeutic proteins. Anal Biochem 355(2):249–258
52. Nishikaze T (2017) Sensitive and structure-informative N-glycosylation analysis by MALDI-MS; ionization, fragmentation, and derivatization. Mass Spectrom (Tokyo, Jpn) 6(1):A0060–A0060

53. Hu S, Wong DT (2009) Lectin microarray. Proteomics Clin Appl 3(2):148–154
54. Mechref Y, Novotny MV, Krishnan C (2003) Structural characterization of oligosaccharides using MALDI-TOF/TOF tandem mass spectrometry. Anal Chem 75(18):4895–4903
55. Markely LR et al (2016) A high-throughput capillary isoelectric focusing immunoassay for fingerprinting protein sialylation. Biotechnol Prog 32(1):235–241
56. Gillmeister MP et al (2009) An HPLC-MALDI MS method for N-glycan analyses using smaller size samples: application to monitor glycan modulation by medium conditions. Glycoconj J 26(9):1135–1149
57. Han L, Costello CE (2013) Mass spectrometry of glycans. Biochem Biokhimiia 78(7):710–720
58. Pioch M, Bunz SC, Neususs C (2012) Capillary electrophoresis/mass spectrometry relevant to pharmaceutical and biotechnological applications. Electrophoresis 33(11):1517–1530
59. Zhou S, Hu Y, Mechref Y (2016) High-temperature LC-MS/MS of permethylated glycans derived from glycoproteins. Electrophoresis 37(11):1506–1513
60. Zhang L, Luo S, Zhang B (2016) Glycan analysis of therapeutic glycoproteins. MAbs 8(2):205–215

Practical Considerations in Method Development for Gas Chromatography-Based Metabolomic Profiling

Andre Cunha Paiva, Amilton Moreira de Oliveira, Juliana Crucello, Roselaine Facanali, and Leandro Wang Hantao

Abstract

This chapter discusses the fundamentals of gas chromatography (GC) to improve method development for metabolic profiling of complex biological samples. The selection of column geometry and phase ratio impacts analyte mass transfer, which must be carefully optimized for fast analysis. Stationary phase selection is critical to obtain baseline resolution of critical pairs, but such selection must consider important aspects of metabolomic protocols, such as derivatization and dependence of analyte identification on existing databases. Sample preparation methods are also addressed depending on the sample matrix, including liquid-liquid extraction and solid-phase microextraction.

Keywords

Comprehensive two-dimensional gas chromatography · Liquid-liquid extraction · Mass spectrometry · Metabolic profiling · Metabolomics · Sample preparation · Untargeted analysis

1 Introduction to Modern Gas Chromatography

Gas chromatography (GC) is undoubtedly an important technique with over 40 years in development for the analysis of volatile and semi-volatile organic compounds [1]. Today, GC has become an important solution in metabolomic investigations due to its unparalleled chromatographic efficiency and improved peak capacity.

The core of the chromatographic separation is the differential migration attained by the different solute bands as they migrate downstream in the GC column [2]. Hence, chromatographic resolution arises from this differential migration, which is dependent on the distribution constant of the analyte between gas phase and stationary phase (K_D), while minimizing band broadening during such process [3].

For most metabolomic applications, high-resolution separations are attained using wall-coated open tubular capillary (WCOT) columns with liquid phases. Accordingly, analytes partition between the gas phase and liquid stationary phase, wherein the net retention mechanism is a combination of vapor pressure and solubility [4, 5].

A. C. Paiva · A. M. de Oliveira · J. Crucello · R. Facanali · L. W. Hantao (✉)
Institute of Chemistry, University of Campinas (UNICAMP), Campinas, SP, Brazil
e-mail: wang@unicamp.br

A. V. Colnaghi Simionato (ed.), *Separation Techniques Applied to Omics Sciences*, Advances in Experimental Medicine and Biology 1336, https://doi.org/10.1007/978-3-030-77252-9_8

2 Rate Theory

Solid phases are not employed in GC-based methods for metabolic profiling because adsorption is the dominant retention mechanism [2]. Although strong intermolecular interactions are established with analytes in such solid phases, the slow mass transfer causes excessive band broadening and readily diminishes the peak capacity of the GC method. Liquid polymeric phases, on the other hand, exhibit absorption (i.e., partition) as the dominant retention mechanism [4, 5]. So, faster mass transfer is attained in the chromatographic process generating sharp peaks and, thereby, improving peak capacity (n_c).

For temperature-programmed separations, the n_c parameter is much more informative for chromatographic method development, as it accounts for the usable retention space and average peak width of the analytes. Conversely, plate count (N) is better suited to evaluate column quality and durability. The theoretical peak capacity of a chromatographic method may be estimated by $\sqrt{N}$ [2], while the experimental n_c may be determined by dividing the usable retention window by the average peak width at base (w_b) [6].

A consistent knowledge of the mass transfer process in open capillary columns is an important tool in method development to generate fast, but powerful, separations for routine metabolomic investigations. The rate theory explains the band broadening in time and it has been successfully applied to elution chromatography. More specifically, the Golay equation correlates the minimum plate height ($H_{\min}$) of a model peak with the average linear velocity of carrier gas ($\bar{\mu}$), as shown in Eq. 1.

$$H = \frac{2D_G}{\bar{\mu}} + \frac{2kd_f^2\bar{\mu}}{3(1+k)^2 D_S} + \frac{\left(1+6k+11k^2\right)r_C^2\bar{\mu}}{24(1+k)^2 D_G} \quad (1)$$

wherein D_G and D_S are the molecular diffusion coefficients of the solute in the gas phase and stationary phase, respectively, k is the retention factor, d_f is the film thickness of the stationary phase, and r_C is the radius of the capillary column. In accordance to the extended van Deemter equation, the longitudinal diffusion (B term) is inversely proportional to $\bar{\mu}$, while the combined resistance to mass transfer (C term) is directly proportional to $\bar{\mu}$. A Golay plot of plate height versus linear velocity of carrier gas is illustrated in Fig. 1 [7].

In practice, analysis time may be effectively reduced without deteriorating chromatographic resolution by using fast and ultrafast GC columns, embodied by short columns with lengths from 5 to 20 meters and inner diameters from 0.10 to 0.18 mm internal diameter (ID). It is noteworthy that linear velocities below the optimum value are discouraged, as the average diffusion coefficient of an analyte in gas phase is 1000-fold larger than that of the condensed (liquid) phase [2]. Hence, in such conditions, band broadening by longitudinal diffusion of the solute (B term) will be severely aggravated. As a rule of thumb, linear velocity values from 30 to 45 cm s^{-1} are a good starting point in any GC method development [2].

When dealing with standard methods, GC method translation is essential to develop and implement fast gas chromatography. Simple and easy translation may be attained by using software (freeware) available online, e.g., Restek's EZ Method Translator (https://www.restek.com/chromatogram/ezgc/). The benefits of using freeware to optimize GC methods with practical examples are described elsewhere [8], while a few key theoretical and practical concepts are explained by Klee and Blumberg [9]. The impact of fast GC is illustrated by the analysis of a complex mixture using a gold-standard method and a translated method employing a fast 5 m × 0.05 mm ID capillary column. In this example, a significant improvement in sample throughput was attained by reducing analysis time by a factor of 8.3 [7]. Although the aforementioned example does not entail a metabolomic-based investigation, it clearly shows that there are many opportunities for method development and optimization in routine metabolic profiling using GC-based protocols. For example, an exploratory GC-based method that employs a GC oven programing from 60 °C

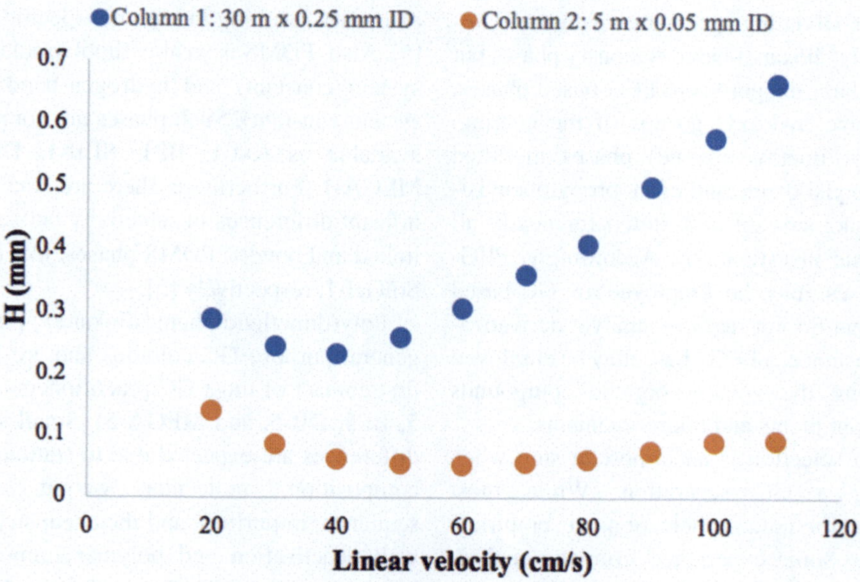

Fig. 1 Golay curve determined experimentally using *n*-tetradecane (n-C_{14}) and isothermal analysis at 130 °C. The plate height was calculated using ten values of average velocities of carrier gas ($\bar{\mu}$). A general reduction of resistance to mass transfer was observed for short wall-coated open tubular (WCOT) columns with smaller inner diameters. (Data retrieved from Mondello et al. [7])

to 300 °C at 3 °C min^{-1} and a 30 m × 0.25 mm ID (0.25 μm film thickness) column requires an analysis time of 80 min. This method translates into a sample throughput of approximately 18 samples per day. If careful method translation is performed, an equivalent method may be attained using a GC oven programing from 60 °C to 300 °C at 10 °C min^{-1} and a 10 m × 0.10 mm ID (0.10 μm film thickness) column GC. The analysis time is then reduced to 24 min, enabling the user to analyze 60 samples per day – a 3.3-fold improvement in laboratory productivity.

3 Column Chemistry

The unique physicochemical properties of GC stationary phases must include high viscosity, wide liquid range, low volatility, and high thermal stability [10, 11]. Viscosity and thermal stability directly limit the maximum allowable operating temperature (MAOT) of a GC column.

Stationary phases possess a tendency to flow and pool at high temperatures, so a highly viscous phase is required to maintain the homogeneity of the coating throughout its use [10]. In other words, GC phases must exhibit a low viscosity drop with increasing temperatures, in order to ensure separation efficiency and lifetime of the column. Also, organic compounds tend to decompose at elevated temperatures with exposure to humidity and oxygen. Accordingly, GC phases also exhibit such behavior and it may change the selectivity of the column during its lifetime. Hence, the available chemistry of GC stationary phases is limited to poly(ethyleneglycol) (PEG), modified poly(siloxanes), and some ionic liquids [10–12].

In GC-based metabolomic experiments, a derivatization step is necessary to deactivate reactive hydrogens from polyols (e.g., sugars), amino acids, and organic acids producing higher vapor pressure analogues, increasing thermal stability, and/or improving peak shapes. Perfluoro acid anhydride acylation reagents, such as trifluoroacetic anhydride (TFAA), generate acid byproducts, which may alter the composition of stationary phase [13, 14]. For this reason, it is recommended to evaporate the sample extract to dryness to remove excess reagent [13, 14].

Conversely, silylating agents are much less damaging to poly(siloxane)-based stationary phases but are particularly dangerous to PEG-based phases, due to active hydroxyl groups of the coating. Thus, poly(dimethylsiloxane) phases modified with diphenylsiloxane and cyanopropylphenylsiloxane monomers are best suited for nearly all metabolomic investigations. Accordingly, PEG-based phases may be employed in GC-based methods that do not required analyte derivatization. For instance, a PEG phase may be employed for profiling the volatile organic compounds (VOCs) from plants and microorganisms.

Column selection is an important step when developing a GC separation. While most approaches for column selection are empirical and require some experience from the analyst, polarity scales may be used to assist the user in finding the best stationary phase. The Abraham's solvation parameter model [15] describes and estimates the strength of individual solvation interactions between probe molecules and the stationary phase. The contributions of specific intermolecular interactions may be estimated by Eq. 2.

$$\log k = c + eE + sS + aA + bB + lL \quad (2)$$

The parameters e, s, b, a, and l are the system constants, which measures the ability of the stationary phase to interact with the analytes by electron lone pair interactions (e), dipole-type interaction (s), dispersive interaction/cavity formation (l), hydrogen-bond basicity (a), and hydrogen-bond acidity (b). The parameters E, S, A, B, and L are the solute descriptors [15]. Although seemingly complex, the solvation parameter model readily reveals the solvation characteristics of WCOT columns and allows the identification of stationary phases with equivalent selectivity. Thus, despite the overwhelming quantity of commercially available GC columns, the solvation parameter model reveals that most columns possess redundant solvation capabilities, thereby facilitating method translation and column selection [5].

Poly(dimethylsiloxane) (PDMS) stationary phases are characterized by low cohesion with favorable cavity formation/dispersion (l system constant), resulting in high nonspecific retention [5]. Also, PDMS is weakly dipolar/polarizable (s system constant) and hydrogen-bond basic (a system constant). Such phases are commercially available as Rxi-1, BP1, SLB-1, DB-1, and MEGA-1. Furthermore, there are small but significant differences in selectivity between cross-linked and bonded PDMS phases, like DB-1 and SolGel-1, respectively [5].

Poly(dimethyldiphenylsiloxane) phases are general-purpose GC columns and are likely the first contact of most GC practitioners (e.g., Rxi-5, BP5, DB-5, and MEGA-5). Small selectivity differences are expected due to stationary phase composition, molecular weight, monomer sequence, impurities, and the chemistry used for wall deactivation and polymer immobilization [5]. A general trend for increasing the diphenylsiloxane monomer content in such phases from 5% to 65% is the capacity of the stationary phase to engage in strong dipole-type interactions (s system constant), while smaller increases in electron lone pair interactions (e system constant) and hydrogen-bond basicity (a system constant) are also detected (Fig. 2) [5]. Poly(dimethyldiphenylsiloxane) phases with 50% diphenyl monomer incorporation are excellent alternatives for high-temperature separations when dipole-type interactions are needed to resolve the critical pair of analytes. Such columns typically exhibit MAOT values of 350 °C.

The incorporation of cyanopropylphenylsiloxane and cyanopropylmethylsiloxane monomers into poly(dimethylsiloxane) phases produces a larger increase in the phase's ability to engage in strong hydrogen-bond basic interactions and dipole-type interactions, compared to poly(dimethyldiphenylsiloxane)-based sorbents [5]. Thus, such phases are best suited for the chromatographic resolution of positional isomers such as trimethylsilyl (TMS) derivatives of monosaccharides [16] and methyl esters of saturated and polyunsaturated fatty acids (FAME) [17]. Poly(dimethylcyanopropylphenylsiloxane) phases are excellent alternatives for high-temperature separations when dipole-type interactions (s system constant) and hydrogen-bond basicity (a system constant) are needed to

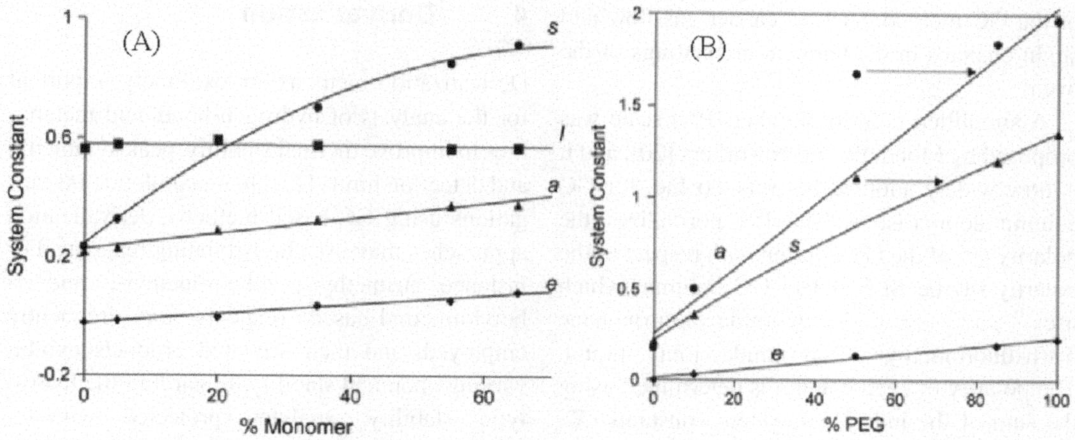

Fig. 2 Characterization of the solvation properties of poly(dimethyldiphenyl)siloxane (**a**) and mixed poli(dimethylsiloxane)/poly(ethyleneglycol) (**b**) stationary phases with varying monomer incorporation. The system constants measure the ability of the stationary phase to interact with the analytes by electron lone pair interactions (*e*), dipole-type interaction (*s*), dispersive interaction/cavity formation (*l*), hydrogen-bond basicity (*a*), and hydrogen-bond acidity (*b*). (Reprinted from Poole and Poole [5]. Copyright 2008, with permission from Elsevier)

resolve overlapped peaks, but exhibit lower MAOT values (e.g., 260–320 °C) compared to poly(dimethyldiphenylsiloxane) phases with 50% diphenyl monomer content. Similarly, poly(ethyleneglycol) (PEG) phases are hydrogen-bond basic and are nearly as dipolar/polarizable as poly(dimethylcyanopropylphenylsiloxane) phases but generally exhibit lower MAOT values (e.g., 240 °C to 280 °C), with the exception of MEGA-WAX HT that possesses a MAOT value of 300 °C. In Fig. 2, the impact of the PEG monomer content to the solvation characteristics of mixed PEG/PDMS phases is also shown. The interested reader is directed elsewhere [5] for the solvation characteristics of additional GC columns.

The Rohrschneider–McReynolds [18, 19] system is one of the oldest and widely used models for the characterization of GC stationary phases, as shown in Eq. 2 [10]. This model assumes that the intermolecular forces established between solute probe and stationary phase are additive and can be estimated from the contribution of individual probe molecules. Therefore, benzene (*a*) measures the capacity of the phase to engage in dispersion interactions (*X*'), butanol (*b*) measures hydrogen-bond basicity of the phase (*Y*'), 2-pentanone (*c*) measures dipolar-type interactions (*Z*'), nitropropane (*d*) measures the Lewis basicity of the phase (*U*'), and pyridine (*e*) estimates the hydrogen-bond acidity of the stationary phase (*S*') (Eq. 3).

$$\Delta I = aX' + bY' + cZ' + dU' + eS' \qquad (3)$$

The Kovats retention indexes of the five probe solutes are determined experimentally using isothermal analysis and *n*-alkane series. The relationship between the retention index (*I*) and retention times of analytes is shown in Eq. 4. The differences between the Kovats indexes (Δ*I*) calculated on the stationary phase and a squalene phase are used to estimate the solvation properties of the stationary phase by using multiple linear regression analysis.

$$I = 100 \times \left(n + \log \frac{t_{R,sol} - t_{R,n}}{t_{r,N} - t_{R,n}} \right) \qquad (4)$$

where t_R is the retention time of the solute (*sol*) eluting between two adjacent *n*-alkanes (*n* and *N*), and *n* is the carbon number of the less retained alkane. The retention index is the most reliable approach to standardize the retention coordinates of an analyte, as it accounts for small deviations of the GC method, such as carrier gas flow and slight changes in the temperature settings of the oven.

A simplified polarity number (PN) scale was proposed by Mondello and coworkers [20], and it is now widely adopted by Supelco Inc. for GC column comparisons. The PN normalizes the polarity (P) of the GC column with respect to the polarity of the SLB-IL100 GC column, which uses a 1,9-di(3-vinylimidazolium)nonane bis(trifluoromethylsulfonyl)imide ionic liquid. The polarity of each column is determined using the sum of the individual system constants (*X*', *Y*', *X*', *U*', *S*'), as shown in Eq. 5. Figure 3 shows the application of the PN scale that may assist analysts in the initial steps of method development. However, the solvation parameter model is more adequate for accurate discussions involving molecular interactions and selectivity.

$$PN = 100'\left(P / P_{SLB\ \ IL100}\right) \qquad (5)$$

4 Derivatization

Derivatization methods are extremely important for the analysis of hydrogen-bond acid metabolites to improve thermal stability, peak symmetry, and detection limits [21]. In metabolomic investigations using GC-based methods, derivatization approaches may use alkylsilylating reagents. For instance, trimethyl-, alkyldimethyl-, and *t*-butyldimethyl-based reagents are frequently employed, and their silylated products exhibit varying chemical stability. Regarding the hydrolytic stability, analytes protected with *t*-butyldimethylsilyl possess the best hydrolytic stability (i.e., 1000- to 10,000-fold) compared to trimethylsilyl derivatives [21]. It is noteworthy that polyols such as carbohydrates are incompletely derivatized with *t*-butyldimethyl-based reagents, like *N*-methyl-*N*-*t*-butyldimethylsilyltri fluoroacetamide (MTBSTFA), because of the steric hindrance of such reagents [22].

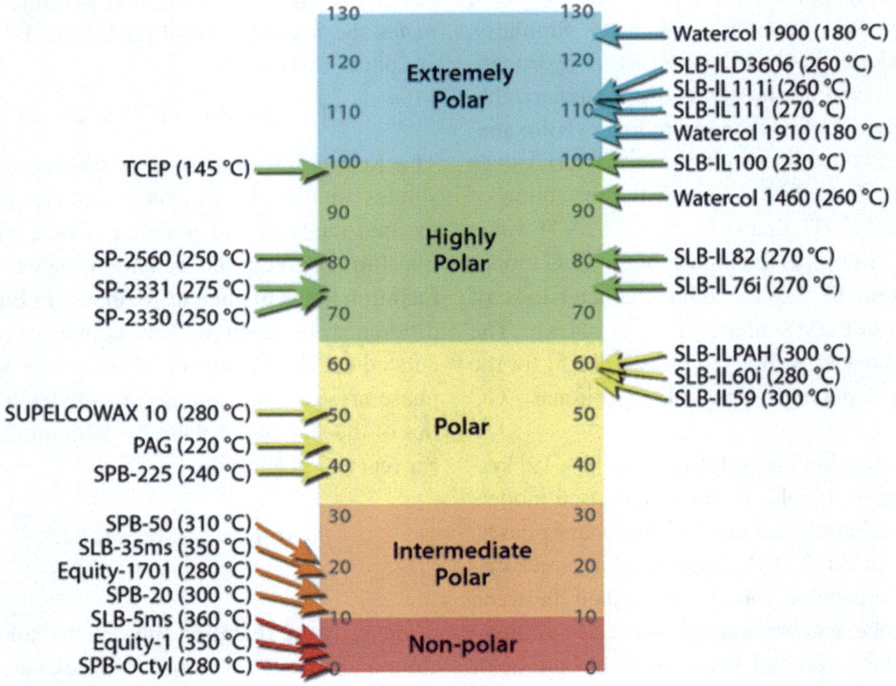

Fig. 3 Polarity number (PN) calculated from the Rohrschneider–McReynolds constants. As a general rule of thumb, high PN values are associated with stationary phases that may engage in strong dipole-type interactions and exhibit pronounced hydrogen-bond basicity. Polar phases are also usually very cohesive. (Reprinted from Supelco Inc., 2019 (https://www.sigmaaldrich.com/technical-documents/articles/analytical/gc-column-selection-guide.html))

In Poole's review [21], the interested reader may find an extensive list of reactions for alkylsilyl derivatives, including the silyl-acceptor reactivity order: alcohols (more reactive) > phenols > carboxylic acids > amines > amides (less reactive) [23]. For example, primary alcohols and amines are less reactive than their tertiary counterparts [21, 23].

Derivatization of biological extracts with silylating reagents must be performed with great caution. Some compounds, such as carbonyl-containing metabolites, form additional unexpected derivatives (i.e., by-products, artifacts) generating multiple peaks for the same analyte. This artifact formation is troublesome since it produces an inaccurate profile of metabolites and, thus, erroneous biological interpretations. For instance, many authors mistakenly attribute the formation of artifacts to metabolite degradation, while the actual reason for such phenomena is the use of an inadequate derivatization protocol. Consequently, metabolite quantitation is also severely jeopardized since multiple peaks are produced from the same metabolite.

Biological samples or extracts containing high quantities of inorganic compounds, such as cell cultures, are also very susceptible to the generation of silylated artifacts. For instance, traces of phosphate buffer saline (PBS) are notoriously known to generate the TMS products of phosphate. Little has published an extensive list of artifacts in trimethylsilyl derivatization reactions and has proposed ways to avoid them [24].

Carbonyl-containing metabolites are particularly challenging to derivatize because by-products are formed during silylation. Carbonyl moieties engage in keto-enol tautomerization by movement of the alpha hydrogen and reorganization of the bonding electrons. Hence, carbonyl-containing metabolites experience keto-enol equilibrium and produce the TMS-carbonyl ester and TMS-enol ether, as shown in Fig. 4. Analogously, ketones and aldehydes also produce TMS artifacts. An important route to eliminate carbonyl-related artifacts is by first protecting the carbonyl moiety, followed by silylating the active hydrogen-containing group. Protection of the carbonyl may be accomplished

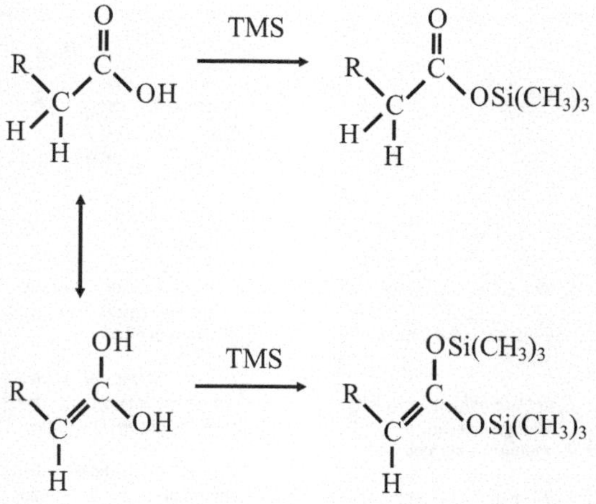

Fig. 4 Derivatization of organic acids with trimethylsilyl reagents. A common class of by-products are the enol ethers formed in solution due to the keto-enol tautomerization of carbonyl-containing metabolites. This artifact formation may be mitigated by protecting the carbonyl moiety with an oximation reaction

by using a methoximation (MeOX) reaction, which is particularly suited for metabolic investigations, since metabolites containing oxime functional groups are rare in natural products [25]. Next, the acidic hydrogens may be deactivated using trimethyl- or *t*-butyldimethyl-silylating reagents.

Qualitative analyses that rely on mass spectral and retention index databases (e.g., NIST, Wiley, FiehnLib [26], Binbase [27]) require some standardization of sample preparation. For instance, the mass spectral database compiled by Fiehn and coworkers requires MeOX/TMS derivatization [27]. An example of protocol for preparation of mammalian samples for GC-MS metabolome analysis is provided below [22]. The interested reader is directed to ref. 22 for the detailed method for blank and quality control GC runs. A general guideline for sample preparation of mammalian cells is provided in Fig. 5.

In the aforementioned protocol, analyte derivatization is completed upon injection of the sample solution into the heated GC injector. Thus, a clean and inert environment is required to ensure proper reaction of the silylating reagent and metabolites. If dirty liners and corroded base plates are used in the injection port, then partial derivatization is observed. For instance, serine

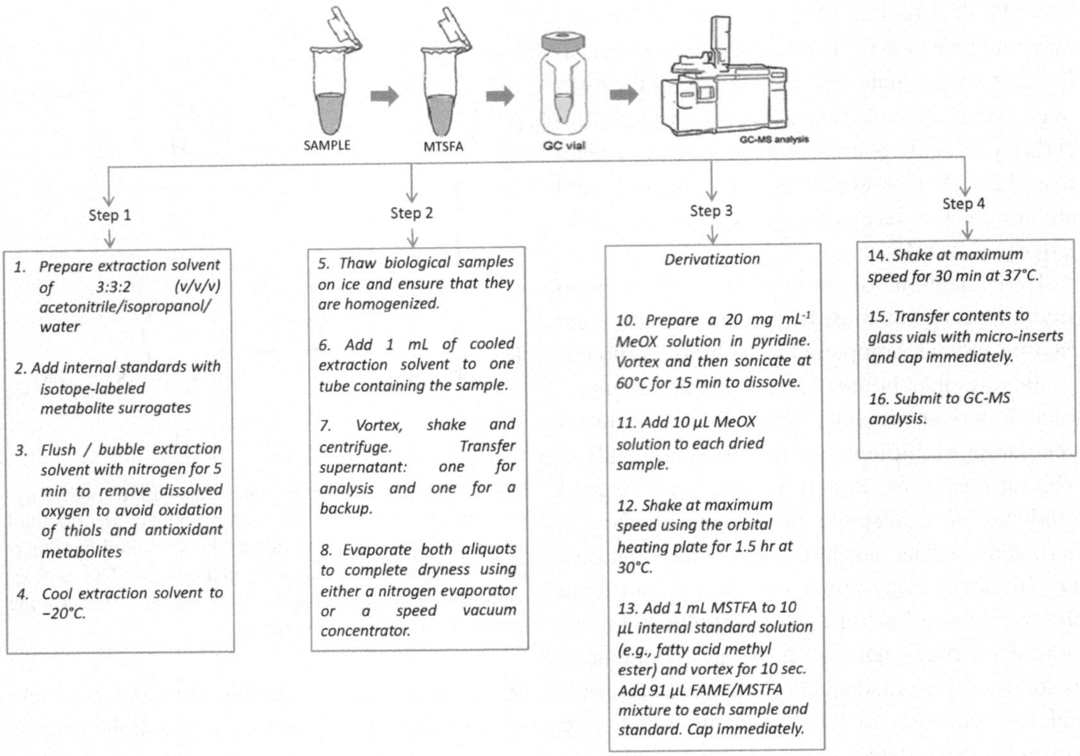

Fig. 5 Sample preparation protocol of mammalian samples for GC-MS metabolome analysis. A detailed discussion is available in Ref. [22] for the interested reader

can be used to monitor the quality of the deactivation reaction [22]. Complete derivatization is attained when only the *N,O,O-tris*-trimethylsilylated serine peak is detected. The presence of the *O,O-bis*-trimethylsilylated serine indicates partial derivatization.

Modification to sample preparation protocol may be necessary depending on the complexity of the biological sample and efficiency of the derivatization reaction. For instance, silylation reactions may also be executed using harsher conditions, as shown in Fig. 6 [24].

In addition, other methods for analyte derivatization are available depending on the extraction technique and complexity of the biological sample. For instance, liquid matrices free of proteins and lipids may be derivatized using in-port derivatization. Estrogens, namely, estrone, estradiol, estriol, and ethinyl estradiol, may be converted into their TMS derivatives using a programmable temperature vaporizing (PTV) injector, operating on large volume injection (LVI) mode [28]. The reagent solution comprises a mixture of BSTFA with 1% (v/v) of the catalyst, trimethylchlorosilane (TMCS). The derivatization solution may be prepared daily by mixing 150 µL of BSTFA+TMCS solution, 45 µL of pyridine, and 5 µL of internal standard into an amber vial. In-port derivatization is accomplished by sequentially injecting 25 µL of the derivatization solution and 25 µL of the sample [28] into the cold inlet. Next, the PTV inlet is heated to 100 °C and the vaporized solvent is flushed through the split vent. After 1 min, the injector is rapidly heated to 300 °C for analyte transference to the GC column for analysis. For such procedures, it is important to maintain a clean GC inlet with deactivated surfaces (e.g., deactivated glass wool, deactivated liners, and gold-coated or Siltek-deactivated base plates) to ensure complete reactions.

For solvent-free extraction techniques such as solid-phase microextraction (SPME) [23], on-fiber derivatization methods are available, which

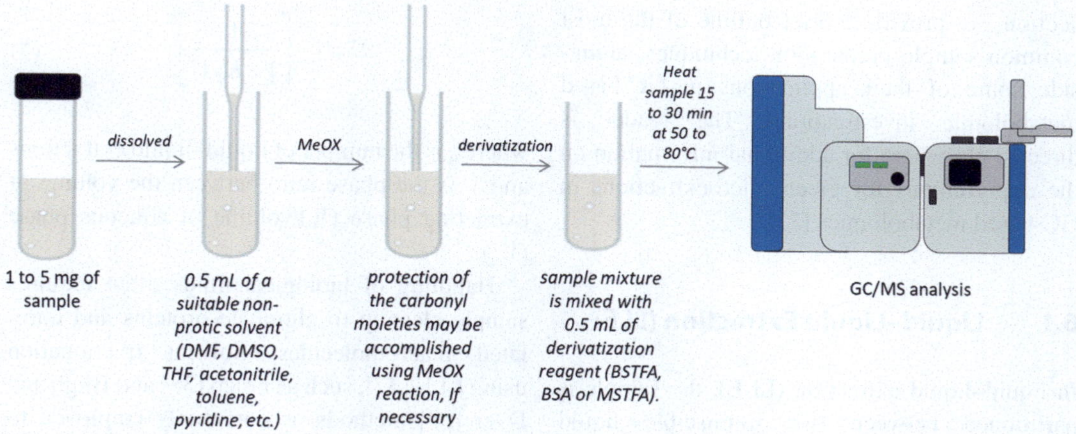

Fig. 6 General trimethylsilyl derivatization protocol for GC-MS analysis. A detailed discussion is available in Ref. [24] for the interested reader

has facilitated sample preparation in standard laboratories [29–31] and also for on-field analysis using reusable standard reagent generating devices [32].

5 Liquid Sample Introduction

Modern GC-MS has been the technique of choice for the analysis of volatile and semi-volatile organic compounds. Significant progress was accomplished in column technology and injection systems extending the application range of GC to molecular weights up to 1,500 Da [33]. Questionably, GC is often not explored for "thermostability" reasons. For instance, analytes that are considered thermolabile, and used as examples to illustrate degradation in elevated temperature high-performance liquid chromatography (HPLC), can often be analyzed by GC [33]. This example was demonstrated by GC-MS analysis of a hormone (ethinylestradiol), drugs (thalidomide, nifedipine, torcetrapib, and maraviroc), and a pesticide (methiocarb). The authors used a standard 30 m × 0.25 mm ID HP-5MS GC column and a cool-on-column injector – which can be emulated using a programmable temperature vaporization (PTV) injector. All solutes eluted with good peak shapes and without any signs of degradation.

Modern GC instrumentation has accelerated the access to highly versatile and efficient injection systems, such as the PTV, by improving module robustness and providing competitive prices. In the upcoming years, we expect to see a broader expansion of multivendor products, such as the multi-mode injection system OPTIC-4 from GL Sciences (Eindhoven, the Netherlands), which adds flexibility to method development by allowing direct introduction of solid samples.

6 Sample Preparation for GC Analysis

The chemical nature of primary and secondary metabolites produces an exceptionally broad range of properties, such as solvent solubility and vapor pressure [23]. For instance, it was estimated that the plant kingdom as a whole contained over 200,000 metabolites and phytochemicals, while it has registered more than 41,815 entries on the human metabolome database [34]. Consequently, isolation of all metabolites using a single technique is virtually impossible.

Sample preparation is a vital part of the analytical process, considering that sample collection and preparation require more than 80% of analysis time [35]. Hence, in the following section, we provide a brief outline of the most common sample preparation techniques, alongside some of their applications in GC-based metabolomic investigations. The reader is directed elsewhere for additional information on the opportunities for green microextractions in GC-based metabolomics [23].

6.1 Liquid–Liquid Extraction (LLE)

In liquid–liquid extraction (LLE), the sample is partitioned between two immiscible liquid phases. Such phases are selected to maximize the differences in solubility, thereby decreasing the effect of interfering compounds in GC analysis [36]. In general, one phase is aqueous and the second an organic solvent. Hence, hydrophilic metabolites will preferably partition into the aqueous phase, while hydrophobic metabolites will migrate to the organic phase. Such differential distribution is controlled by the distribution coefficient (K_D), as shown in Eq. 6.

$$K_D = C_o / C_{aq} \quad (6)$$

where C_o and C_{aq} are the concentrations of the analyte in the organic and aqueous phase, respectively.

Metabolite extraction may be enhanced by carefully selecting the organic phase. For ionizable analytes, the pH value of the aqueous phase may be adjusted to suppress metabolite ionization. Also, salts may be added to the aqueous phase to improve K_D values (salting out effect). Furthermore, a volatile solvent is required for analyte preconcentration prior to GC analysis.

Higher extraction efficiencies are attained by performing multiple LLE stages using smaller solvent volumes (e.g., 5 × 20 mL), instead of executing a single extraction with a large volume of extracting phase (e.g., 100 mL), as shown in Eq. 7 [37]. For instance, the fraction of extracted analyte (E) from a single extraction of 100 mL of urine with 100 mL of dichloromethane and a K_D value of 5 would be 0.83% or 83%. However, if five successive extractions were performed with 20 mL of organic phase, then the E value would increase to 0.97% or 97%.

$$E = 1 - \left[\frac{1}{1 + K_D V} \right]^z \quad (7)$$

where z is the number of liquid–liquid extractions and V is the phase ratio between the volume of extracting phase (V_o)/volume of aqueous phase (V_{aq}).

Handling of biological fluids often requires sample cleanup to eliminate proteins and unrelated macromolecules. Solvent fractionation using LLE [37], such as Folch [38] and Bligh and Dyer [39] methods, is commonly employed to separate the polar and nonpolar organic fractions [40, 41]. In summary, LLE has been successfully applied to isolate metabolites from urine [42], metabolic profiling of bacteria [43], viruses [44], and benthic amphipod (*Diporeia* spp.) [45].

6.2 Solid-Phase Extraction (SPE)

While LLE is suitable for low-complexity matrices, techniques that employ immobilized sorbents are best suited for sample preparation of complex mixtures, such as solid-phase extraction (SPE). In its simplest form, SPE comprises a plastic syringe packed with a sorbent phase that resembles the packing of a HPLC column. For instance, a 3–5 mL PP syringe may be packed with 0.1–5.0 g of solid phase. The sorbent phase comprises porous particles with 40 μm diameter coated with extracting phases. The nature of the coating dictates the selectivity of the SPE method. An excellent review on the principles of SPE is available in the literature [46].

The SPE procedure may include one or more of the following steps: sample loading onto SPE packing, sample cleanup, and solvent desorption of analytes. Sample cleanup and analyte desorption are accomplished by carefully selecting mixtures of organic solvents with varying selectivity. Depending on the combination of such steps, SPE can be used to remove interfering compounds, preconcentration of analytes, desalting, phase exchange, in-situ derivatization, and sample storage/transport [36]. A general guideline for

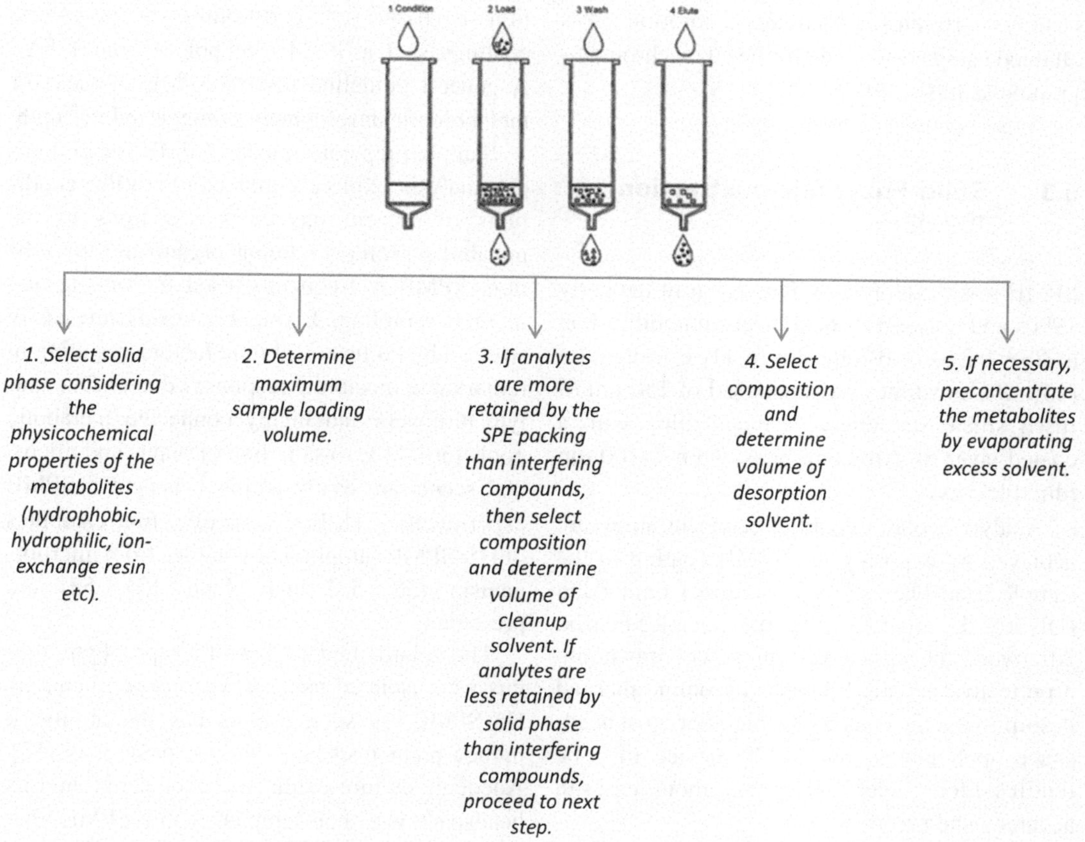

Fig. 7 General guideline for SPE method development

SPE method development for metabolomics is suggested in Fig. 7.

Supported liquid extraction using deactivated diatomaceous earth is highly effective for extraction of organic compounds from aqueous matrices compared with LLE [47, 48]. In this method, the aqueous sample is loaded onto the dry solid support and allowed to adsorb to the surface of diatomaceous earth. Next, an immiscible organic solvent is percolated through the SPE packing, extracting the organic metabolites. This procedure is extremely effective and fast compared with traditional LLE because of improved mass transfer, while bypassing the formation of emulsions during analyte extraction [36]. Another interesting SPE phase is Agilent's Enhanced Matrix Removal – Lipids (EMR–Lipids) that has enabled "one-pot" sample cleanup of serum samples by precipitating proteins and removing lipids in a single step [23].

A case study using SPE for sample cleanup is provided here. Hydroxylated fatty acids (OH-FAs) are formed in all branches of the arachidonic acid (AA) cascade from polyunsaturated fatty acids (PUFA). OH-FAs act as potent lipid mediators and serve as activity marker for pathways of the AA cascade, particularly the lipoxygenase branch [49]. Cell culture medium samples and plasma samples were mixed with acetonitrile and internal standard solution, followed by centrifugation at 20,000 × g at 4 °C for 5 min for protein removal. Next, the extracts were loaded onto a water-wettable macroporous copolymer consisting of a balance ratio of divinylbenzene-vinylpyrrolidone (Oasis HLB) SPE device (Waters Corporation). Matrix cleanup was obtained by washing the device with acidified aqueous solution to remove hydrogen-bond basic interferents. Analyte desorption was attained using acetonitrile followed by chromatographic analysis [49].

6.3 Solid-Phase Microextraction (SPME)

SPME was developed by Pawliszyn in the early 1990s and it uses a sorbent layer immobilized on a fiber-based substrate for analyte extraction [23]. The substrates are composed of 150 μm of fused silica, stableflex, or metal alloy with a coated layer of extracting phase with 7–100 μm film thickness.

Analyte isolation and preconcentration are achieved by exposing the SPME coating to the sample headspace (HS) or by direct immersion (DI) of the coating into the sample matrix. Afterward, the sorbent coating is withdrawn into a protective needle, followed by online thermal desorption in the heated GC inlet. For most headspace applications, the SPME device may be reutilized for at least 100 times without losses in accuracy and precision.

SPME has become one of the most popular solvent-free and equilibrium-based techniques for sampling and sample preparation in bioanalytical chemistry. The principle of SPME relies on the transport of analytes from the sample matrix to the extraction phase by convection through the bulk and diffusion through the boundary layer. The equilibration time and temperature are critical parameters in method development. As a general rule of thumb, higher temperatures increase mass transfer, allowing the multiphasic system to equilibrate faster, but often decrease the mass extracted at equilibrium. Ionic strength also plays an important role in SPME experiments, as it often leads to higher extraction efficiencies, i.e., salting-out effect. Furthermore, the distribution coefficients determine the affinity of analyte toward the sorbent phase, which largely depends on the molecular interactions that may be established during analyte uptake [23]. Porous solids, such as poly(divinylbenzene) (DVB) and Carboxen, exhibit a large internal surface area to promote the sorption of analytes by strong intermolecular interactions. Conversely, analyte sorption is based on partitioning in polymeric coatings, such as PDMS and poly(acrylate) (PA). A general guideline for HS-SPME method for metabolomic development is suggested in Fig. 8.

Sample preparation using SPME for biological analysis explores mild equilibration conditions to prevent unwanted alterations to the metabolic profile of living organisms. In addition, SPME is frequently used for on-site and in vivo sampling. Living organisms are easily stressed by biotic and abiotic factors, resulting in remarkable metabolic responses due to fast analyte turnovers and highly connected metabolic pools [50]. In vivo sampling of plants and microbial scents are easily attained using HS-SPME experiments [51]. In this section, four examples of HS-SPME sampling of volatiles from microorganism [52, 53] and plants [54, 55] are presented.

The volatile metabolites of a saprophytic fungus were isolated using in vitro experiments by HS-SPME, as such fungus has the ability to induce plant resistance against pathogens [52]. Adequate culture media was inoculated, and its headspace was then sampled with a SPME fiber and chromatographed every 24 h over 7 days. This information enabled the determination of the inoculation period, during which the concentration of volatile metabolites was maximized. Several volatile metabolites not previously described in the literature on biocontrol fungi were observed, as well as sesquiterpenes and aliphatic alcohols [52]. In a similar fashion, characterization of biocontrol agents may be a viable and low-cost alternative in the control of different plant pathogens [53]. Metabolites from some yeast strains can prevent infection, decreasing host tissue colonization and reducing pathogen survival and sporulation with varying degrees of efficiency. In vivo sampling of nine yeast strains by HS-SPME enabled the correlation between antiphytopathogenic activities with the volatile metabolite profiles of such microorganisms. Nine compounds were associated with the volatile metabolome of strains with higher antifungal action: ethyl propionate, 3-methyl-1-butanol, 2-methyl-1-butanol, ethyl butyrate, 3-methylbutyl acetate, 3-methylbutyl

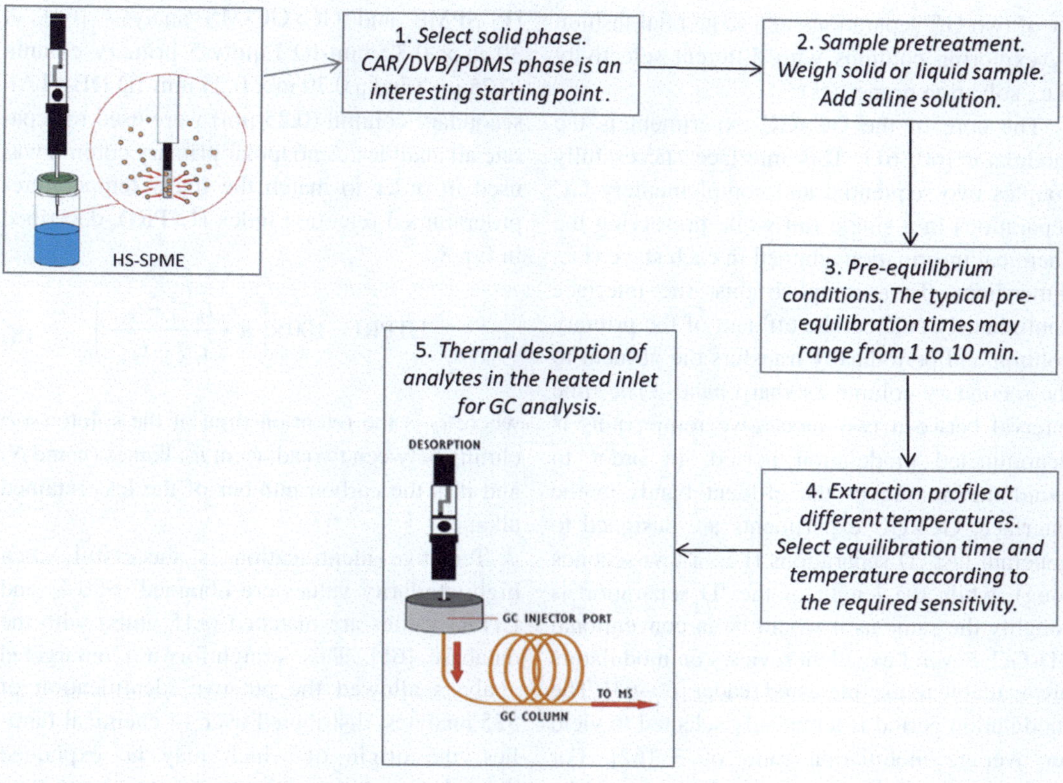

Fig. 8 General guideline for SPME method development

propionate, *n*-pentyl butyrate, phenylethyl alcohol, and 2-phenethyl acetate [53].

The volatile fraction of *Eucalyptus* leaves was isolated by HS-SPME and analyzed by a GC-based method for metabolic profiling [54]. By comparing the chemical profiles of samples from healthy and diseased plants, a correlation between the metabolic profile and the presence of the infection was determined. The method was checked to be independent of factors such as the age of the harvested plants, indicating that such approach could be used for disease diagnosis [54]. In a different report, the use of volatiles isolated from the leaves of plant clones by HS-SPME to determine which specimens were susceptible to rust disease was described [55]. The proposed method allowed the differentiation between susceptible and non-susceptible clones and determination of three potential resistance biomarkers.

Recent developments in SPME include the use of conventional and over-coated sorbent phases for in vivo sampling of metabolites by direct immersion solid-phase microextraction (DI-SPME) [23]. An additional layer of PDMS over a commercial PDMS/DVB coating prevents device fouling during direct sampling of complex matrices [56, 57], enabling +130 DI-SPME extraction cycles per fiber. In addition, the suitability of a PDMS/DVB/PDMS coating was further evaluated using a high-fat sample matrix, demonstrating the importance of coating rinsing and washing steps for improved sorbent lifetime [58].

7 Comprehensive Two-Dimensional Gas Chromatography

Multidimensional separations are powerful methods in which at least two independent separative steps are combined to improve analyte resolution [59]. Comprehensive two-dimensional gas chromatography (GC×GC) harnesses the peak capacity of two GC separations in a sequential fashion by exploring columns with different selectivity (i.e., solvation properties).

The core of the GC×GC experiment is the modulator [60, 61]. This interface successfully couples two sequential and complementary GC separations in a single run while preserving the chemical information attained in each stage (i.e., dimension). To accomplish this, the interface continuously collects the effluent of the primary column and periodically transfers the effluent to the secondary column as sharp bands. The time interval between two successive reinjections is denominated modulation period. In order to avoid the mixing of the effluent bands in the interface, GC×GC experiments are designed to generate fast ^{2}D separations (i.e., a few seconds long), while the length of the ^{1}D separation is roughly the same as it would be in conventional 1D-GC. Several excellent reviews on modulators are available to the interested reader [60–63]. The modulation period is tentatively selected to yield an average modulation ratio of 3 [62]. For instance, a modulation period of 4 s would be advised for a method wherein the peaks exhibit average $^{1}w_b$ values of 12 s.

Figure 9 shows a chromatogram of a mixture of terpenes and sesquiterpenes. The GC×GC chromatograms are typically visualized as contour plots, wherein each peak possesses two retention coordinates, $^{1}t_R$ and $^{2}t_R$. Peaks that would otherwise overlap in conventional GC analysis, i.e., elute in the same vertical line ($1t_R$), are now clearly resolved in the GC×GC chromatogram. Thus, GC×GC separations can offer unprecedented peak capacities without extending analysis time. As a consequence, reliable qualitative analysis is attained due to pure mass spectra. Also, accurate and precise quantitative analysis may be performed because of higher peak purity [23].

An interesting application of GC×GC is the exometabolome profiling of yeasts, which are important for general quality of products and can contribute to product differentiation. *Saccharomyces* spp. are widely used in the food and beverages industries. Hence, a metabolomic strategy for comprehensive mapping of cellular metabolites was accomplished by combining HS-SPME and GC×GC-MS analysis [64]. A 30 m × 0.32 mm-ID Equity-5 primary column (0.25 μm) and a 0.79 m × 0.25 mm-ID DB-FFAP secondary column (0.25 μm) were used to separate all analytes. A nonpolar primary column was used in order to match the linear temperature-programmed retention index (LTPRI), described in Eq. 8.

$$\mathrm{LTPRI} = 100 \times \left(n + \frac{t_{R,sol} - t_{R,n}}{t_{r,N} - t_{R,n}} \right) \tag{8}$$

where t_R is the retention time of the solute (*sol*) eluting between two adjacent *n*-alkanes (*n* and *N*) and *n* is the carbon number of the less retained alkane.

Tentative identification is successful when high similarity values are obtained (>80%) and LTPRI values are matched (±15 units) with the database [65]. This straightforward untargeted analysis allowed the putative identification of 525 analytes, distributed over 14 chemical families, the origin of which may be explained through the pathway network associated with yeast metabolism. While GC×GC alleviates chemical characterization of complex mixtures, processing of large metabolomic data remains unchallenged, and it is considered the bottleneck of most metabolomic investigations, regardless of the instrumental platform [66].

8 Coupling to Mass Spectrometry

The prime requisite for coupling gas chromatography-based separations with mass spectrometry is the minimum spectral acquisition rate. For most applications, a minimum acquisition of 7 to 10 data points across the chromatographic peak is necessary to ensure reliable peak reconstruction and reproducible peak areas [28, 29]. Hence, a minimum acquisition rate of 2 spectra s^{-1} may be used for conventional 1D-GC separations with average w_b values of 5 s. Conversely, for GC×GC separations, a minimum acquisition rate of 30 spectra s^{-1} is required considering an average $^{2}w_b$ of 0.25 s.

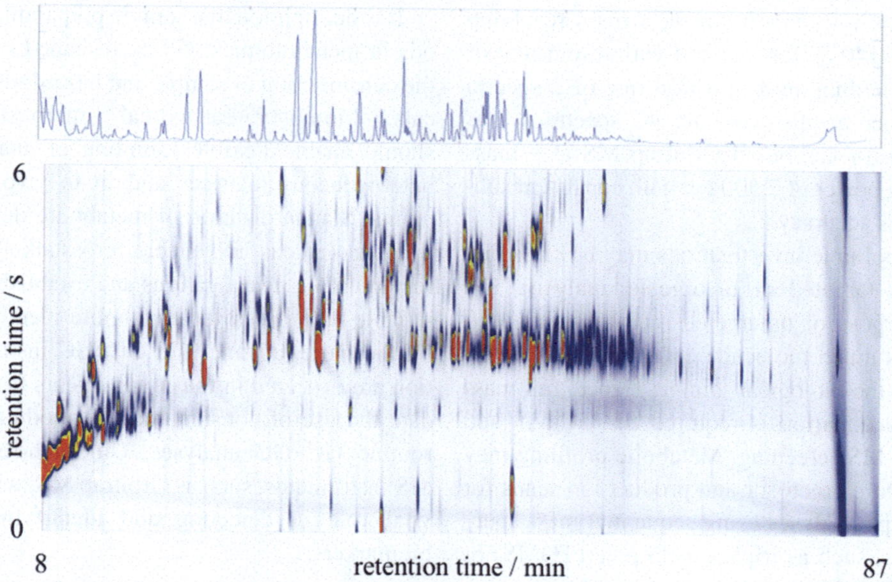

Fig. 9 Comparative analysis of the chromatographic profiles of a mixture of terpenes and sesquiterpenes, obtained using a conventional GC-MS (top) and GC×GC-MS (bottom) equipment. Columns: ^{1}D – 30 m × 0.25 mm-ID MEGA-5 (0.25 μm); ^{2}D – 5 m × 0.25 mm-ID MEGA-17 (0.20 μm). Modulation period: 5 s. Scanning range: *m/z* 40–400 at 31 spectra s^{-1}. Instrument: TRACE1310 GC and ISQ QD single quadrupole mass spectrometer (ThermoFisher Scientific) fitted with an INSIGHT flow modulator (SepSolve Analytical). Courtesy of Hantao, Institute of Chemistry, University of Campinas

The type of mass analyzer severely influences the quality of the mass spectrum and sensitivity of the overall GC-MS method. The most popular mass analyzer is the transmission quadrupole (QMS). Such scanning instrument is prone to spectral skewing if the acquisition rate is incompatible with the peak width of the analyte. In chromatography, there are transitory changes in solute concentration (e.g., Gaussian peak profile), and scanning instruments may exhibit mass spectral skewing, i.e., the spectra profile is heavily influenced by the chromatographic peak shape [23]. Hence, scanning mass analyzers must operate at faster acquisition rates to avoid spectral skewing [67]. For this reason, QMS with scanning speeds up to 20,000 u s^{-1} is currently available. Unfortunately, the QMS experiences some loss in sensitivity due to poor ion transmission if the maximum scanning speed of the mass analyzer is employed. So, a general recommendation is to avoid using fast scanning rates unless mandatory, such as the case of GC×GC-QMS instruments. Conversely, non-scanning instruments such as the time-of-flight (TOFMS) and Fourier transform (FTMS) mass spectrometers do not exhibit skewing [68].

TOF mass analyzers are available in two configurations, low mass resolution/low mass accuracy and high mass resolution/high mass accuracy, both suited for GC and GC×GC applications [69, 70]. The former is a benchtop instrument. These instruments are significantly more sensitive than the QMS in full spectra acquisition (i.e., untargeted analysis), while QMS offers competitive results when operated in selected ion monitoring (SIM) for targeted analysis. TOFMS is also available with high resolving power (up to 50,000) and mass accuracy (1–3 ppm) by extending the flight path of the mass analyzer [71]. Most TOF mass spectrometers can be operated at acquisition rates between 50 and 200 spectra s^{-1}.

The latest and most exciting introduction of ultrahigh-resolution and high-mass accuracy mass analyzer is the Orbitrap, which is a FTMS based on an electrostatic trap [72, 73]. This mass analyzer delivers sub-ppm accurate mass measurements [74], and the resolving power is highly dependent on the transient length and, thus, acquisition rate. For instance, a mass resolving power of 140,000 is attained with a transient of 64 ms, yielding an acquisition rate of 2 spectra s^{-1}. Higher acquisition rate, 42 spectra s^{-1}, is available for GC-based Orbitrap MS at a mass resolving power of 7,500 but with consistent sub-ppm mass accuracy.

Metabolomic investigations may be classified in either targeted or untargeted analysis. The main purpose of untargeted metabolomics is to gather as much molecular information as possible from the biological sample. Hence, full mass spectra acquisition is required for GC-MS and GC×GC-MS screening. Metabolic profiling may also explore precursor and product ion scans for group-type analyses using sequential mass spectrometers, such as triple quadrupole (TQMS) or hybrid QTOFMS. Also, analyte identification may be accomplished by using mass spectral similarity search with retention index filtering against databases. When using high-accuracy mass spectrometers, the accurate mass of the metabolite must be measured for assignment of reliable elemental formula.

Targeted metabolomics naturally benefits from the improved resolving power of GC and GC×GC, and additional mass selectivity is available due to QMS and TQMS coupling. For instance, exceptional selectivity may be attained by selected ion monitoring (SIM) with GC(×GC)-QMS or selected reaction monitoring (SRM) with GC(×GC)-TQMS [75]. Alternatively, improved selectivity may also be obtained with extracted ion chromatograms (EIC), especially from high mass accuracy data, such as the high-accuracy TOFMS and the Orbitrap.

9 Perspectives

The development of GC-based methods for metabolic profiling is one of the leading edges of modern analytical chemistry. The successful application of both targeted and untargeted analysis using GC-MS and GC×GC-MS will certainly benefit adjacent disciplines to metabolomics by providing critical insights into the biochemistry of complex living systems.

It is our opinion that sample preparation methods in metabolomics will be reshaped to reduce the consumption of sample and hazardous chemicals. Moreover, analytical microextractions should enable flexible sampling of analytes in heterogeneous matrices and in in vivo experiments. Standardization of metabolite derivatization is critical to enable crosstalk between metabolomic investigations and establishment of reliable data bases for metabolite identification. Furthermore, advances in GC×GC instrumentation are expected in the coming years to consolidate the use of consumable-free modulators for routine GC×GC analyses. Ultrahigh-resolution MS instruments, such as Orbitrap MS, will play a vital role in detecting and identifying novel biomarkers.

References

1. McNair HM, Miller JM (2009) Basic gas chromatography. Wiley, New York
2. Grob RL, Barry EF (2004) Modern practice of gas chromatography. Wiley, New York
3. Miller JM (2009) Chromatography: concepts and contrasts. Wiley, New York
4. Poole CF, Poole SK (2009) Foundations of retention in partition chromatography. J Chromatogr A 1216:1530–1550. https://doi.org/10.1016/j.chroma.2008.10.092
5. Poole CF, Poole SK (2008) Separation characteristics of wall-coated open-tubular columns for gas chromatography. J Chromatogr A 1184:254–280. https://doi.org/10.1016/j.chroma.2007.07.028
6. Gough DV, Song DH, Schöneich S et al (2019) Development of ultrafast separations using negative pulse partial modulation to enable new directions in gas chromatography. Anal Chem 91:7328–7335. https://doi.org/10.1021/acs.analchem.9b01085
7. Mondello L, Tranchida PQ, Casilli A et al (2004) Fast GC analysis with a 50 μm ID column: theory, practical aspects, and application to a highly complex sample. J Sep Sci 27:1149–1156. https://doi.org/10.1002/jssc.200401832
8. Majors RE, Lynam K (2011) Method Translation in Gas Chromatography. http://www.chromatographyonline.com/method-translation-gas-chromatography-0. Accessed 10 July 2019
9. Klee MS, Blumberg LM (2002) Theoretical and practical aspects of fast gas chromatography and method translation. J Chromatogr Sci 40:234–247. https://doi.org/10.1093/chromsci/40.5.234

10. Yao C, Anderson JL (2009) Retention characteristics of organic compounds on molten salt and ionic liquid-based gas chromatography stationary phases. J Chromatogr A 1216:1658–1712. https://doi.org/10.1016/j.chroma.2008.12.001
11. Jennings W (1980) Gas chromatography with glass capillary columns. Academic Press, New York
12. Ho TD, Zhang C, Hantao LW et al (2014) Ionic liquids in analytical chemistry: fundamentals, advances, and perspectives. Anal Chem 86:262–285. https://doi.org/10.1021/ac4035554
13. Sensue A (2012) Capillary GC Column Killers – Part 1. https://blog.restek.com/?p=5001. Accessed 10 July 2019
14. Sensue A (2012) Capillary GC Column Killers – Part 2. https://blog.restek.com/?p=5003. Accessed 10 July 2019
15. Abraham MH (1993) Scales of solute hydrogen-bonding: their construction and application to physicochemical and biochemical processes. Chem Soc Rev 22:73–83. https://doi.org/10.1039/CS9932200073
16. Zweckmair T, Schiehser S, Rosenau T et al (2017) Improved quantification of monosaccharides in complex lignocellulosic biomass matrices: a gas chromatography-mass spectrometry based approach. Carbohydr Res 446–447:7–12. https://doi.org/10.1016/j.carres.2017.04.011
17. Harynuk J, Wynne PM, Marriott PJ (2006) Evaluation of new stationary phases for the separation of fatty acid methyl esters. Chromatographia 63:S61–S66. https://doi.org/10.1365/s10337-006-0714-0
18. Rohrschneider L (1969) Eine method zur charakterisierung von gas-chromatographischen trennflüssigkeiten: II. Die berechnung von retentionsverhältnissen. J Chromatogr A 39:383–397. https://doi.org/10.1016/S0021-9673(01)98027-6
19. McReynolds WO (1970) Characterization of some liquid phases. J Chromatogr Sci 8:685–691. https://doi.org/10.1093/chromsci/8.12.685
20. Ragonese C, Sciarrone D, Tranchida PQ et al (2011) Evaluation of a medium-polarity ionic liquid stationary phase in the analysis of flavor and fragrance compounds. Anal Chem 83:7947–7954. https://doi.org/10.1021/ac202012u
21. Poole CF (2013) Alkylsilyl derivatives for gas chromatography. J Chromatogr A 1296:2–14. https://doi.org/10.1016/j.chroma.2013.01.097
22. Fiehn O (2016) Metabolomics by gas chromatography-mass spectrometry: the combination of targeted and untargeted profiling. Curr Protoc Mol Biol 114:30.4.1–30.4.32. https://doi.org/10.1002/0471142727.mb3004s114
23. Belinato JR, Dias FFG, Caliman JD et al (2018) Opportunities for green microextractions in comprehensive two-dimensional gas chromatography/mass spectrometry-based metabolomics – a review. Anal Chim Acta 1040:1–18. https://doi.org/10.1016/j.aca.2018.08.034
24. Little JL (1999) Artifacts in trimethylsilyl derivatization reactions and ways to avoid them. J Chromatogr A 844:1–22. https://doi.org/10.1016/S0021-9673(99)00267-8
25. Madala NE, Piater LA, Steenkamp PA et al (2014) Multivariate statistical models of metabolomic data reveals different metabolite distribution patterns in isonitrosoacetophenone-elicited *Nicotiana tabacum* and *Sorghum bicolor* cells. Springerplus 3:254–264. https://doi.org/10.1186/2193-1801-3-254
26. Kind T, Wohlgemuth G, Lee DY et al (2009) FiehnLib: mass spectral and retention index libraries for metabolomics based on quadrupole and time-of-flight gas chromatography/mass spectrometry. Anal Chem 81:10038–10048. https://doi.org/10.1021/ac9019522
27. Skogerson K, Wohlgemuth G, Barupal DK et al (2011) The volatile compound BinBase mass spectral database. BMC Bioinformatics 12:321–335. https://doi.org/10.1186/1471-2105-12-321
28. González A, Clavijo S, Cerdà V (2019) Estrogens determination exploiting a SIA-LOV system prior in-port derivatization-large volume injection-programmable temperature vaporization-gas chromatography. Talanta 194:852–858. https://doi.org/10.1016/j.talanta.2018.10.105
29. Parkinson DR, Warren JM, Pawliszyn J (2010) Analysis of ergosterol for the detection of mold in soils by automated on-fiber derivatization headspace extraction-SPME-GC/MS. Anal Chim Acta 661:181–187. https://doi.org/10.1016/j.aca.2009.12.027
30. Rodríguez I, Carpinteiro J, Quintana JB et al (2004) Solid-phase microextraction with on-fiber derivatization for the analysis of anti-inflammatory drugs in water samples. J Chromatogr A 1024:1–8. https://doi.org/10.1016/j.chroma.2003.10.049
31. Schmarr HG, Sang W, Ganß S et al (2008) Analysis of aldehydes via headspace SPME with on-fiber derivatization to their O-(2, 3, 4, 5, 6-pentafluorobenzyl)oxime derivatives and comprehensive 2D-GC-MS. J Sep Sci 31:3458–3465. https://doi.org/10.1002/jssc.200800294
32. Poole JJ, Grandy JJ, Gómez-Ríos GA et al (2016) Solid phase microextraction on-fiber derivatization using a stable, portable, and reusable pentafluorophenyl hydrazine standard gas generating vial. Anal Chem 88:6859–6866. https://doi.org/10.1021/acs.analchem.6b01449
33. Sandra P, David F, Pereira A et al (2010) Green chromatography (part 2): the role of GC and SFC. http://www.chromatographyonline.com/green-chromatography-part-2-role-gc-and-sfc . Accessed 11 July 2019
34. Gromski PS, Muhamadali H, Ellis DI et al (2015) A tutorial review: metabolomics and partial least squares-discriminant analysis – a marriage of convenience or a shotgun wedding. Anal Chim Acta 879:10–23. https://doi.org/10.1016/j.aca.2015.02.012
35. Gõmez-Ríos GA, Reyes-Garcés N, Pawliszyn J (2015) Evaluation of a multi-fiber exchange solid-phase microextraction system and its application to on-site sampling. J Sep Sci 38:3560–3567. https://doi.org/10.1002/jssc.201500158

36. Majors RE, Wilmington DE (2013) Sample preparation fundamentals for chromatography. https://www.agilent.com/cs/library/primers/Public/5991-3326EN_SPHB.pdf. Accessed 11 July 2019
37. Iverson SJ, Lang SLCC, Cooper MH (2001) Comparison of the bligh and dyer and folch methods for total lipid determination in a broad range of marine tissue. Lipids 36:1283–1287. https://doi.org/10.1007/s11745-001-0843-0
38. Folch J, Lees M, Stanley GHS (1957) A simple method for the isolation and purification of total lipides from animal tissues. J Biol Chem 226:497–509. https://doi.org/10.1007/s10858-011-9570-9
39. Bligh EG, Dyer WJ (1959) A rapid method of total lipid extraction and purification. Can J Biochem Physiol 37:911–917. https://doi.org/10.1139/o59-099
40. Boyaci E, Rodríguez-Lafuente Á, Gorynski K et al (2015) Sample preparation with solid phase microextraction and exhaustive extraction approaches: comparison for challenging cases. Anal Chim Acta 873:14–30. https://doi.org/10.1016/j.aca.2014.12.051
41. Álvarez-Sánchez B, Priego-Capote F, Castro MDL (2010) Metabolomics analysis II. Preparation of biological samples prior to detection. TrAC – Trends Anal Chem 29:120–127. https://doi.org/10.1016/j.trac.2009.12.004
42. Vasquez NP, Crosnier de bellaistre-Bonose M, Lévêque N et al (2015) Advances in the metabolic profiling of acidic compounds in children's urines achieved by comprehensive two-dimensional gas chromatography. J Chromatogr B Analyt Technol Biomed Life Sci 1002:130–138. https://doi.org/10.1016/j.jchromb.2015.08.006
43. Loots DT, Swanepoel CC, Newton-Foot M et al (2016) A metabolomics investigation of the function of the ESX-1 gene cluster in mycobacteria. Microb Pathog 100:268–275. https://doi.org/10.1016/j.micpath.2016.10.008
44. Meissner-Roloff RJ, Koekemoer G, Warren RM et al (2012) A metabolomics investigation of a hyper- and hypo-virulent phenotype of Beijing lineage M. tuberculosis. Metabolomics 8:1194–1203. https://doi.org/10.1007/s11306-012-0424-6
45. Maity S, Jannasch A, Adamec J et al (2012) Starvation causes disturbance in amino acid and fatty acid metabolism in *Diporeia*. Comp Biochem Physiol Biochem Mol Biol 161:348–355. https://doi.org/10.1016/j.cbpb.2011.12.011
46. Hennion M (1999) Solid-phase extraction: method development, sorbents, and coupling with liquid chromatography. J Chromatogr A 856:3–54. https://doi.org/10.1016/S0021-9673(99)00832-8
47. Yoshikawa S, Nagano C, Kanda M et al (2017) Simultaneous determination of multi-class veterinary drugs in chicken processed foods and muscle using solid-supported liquid extraction clean-up. J Chromatogr B Anal Technol Biomed Life Sci 1057:15–23. https://doi.org/10.1016/j.jchromb.2017.04.041
48. Rositano J, Harpas P, Kostakis C et al (2016) Supported liquid extraction (SLE) for the analysis of methylamphetamine, methylenedioxymethylamphetamine and delta-9-tetrahydrocannabinol in oral fluid and blood of drivers. Forensic Sci Int 265:125–130. https://doi.org/10.1016/j.forsciint.2016.01.017
49. Ostermann AI, Willenberg I, Weylandt KH et al (2015) Development of an online-SPE-LC-MS/MS method for 26 hydroxylated polyunsaturated fatty acids as rapid targeted metabolomics approach for the LOX, CYP, and autoxidation pathways of the arachidonic acid cascade. Chromatographia 78:415–428. https://doi.org/10.1007/s10337-014-2768-8
50. Hashemi B, Zohrabi P, Kim KHH et al (2017) Recent advances in liquid-phase microextraction techniques for the analysis of environmental pollutants. TrAC Trends Anal Chem 97:83–95. https://doi.org/10.1016/j.trac.2017.08.014
51. Augusto F, Valente ALP (2002) Applications of solid-phase microextraction to chemical analysis of live biological samples. TrAC – Trends Anal Chem 21:428–438. https://doi.org/10.1016/S0165-9936(02)00602-7
52. Lima PF, Furlan MF, Ribeiro FAL et al (2015) In vivo determination of the volatile metabolites of saprotroph fungi by comprehensive two-dimensional gas chromatography. J Sep Sci 38:1924–1932. https://doi.org/10.1002/jssc.201401404
53. Belinato JR, Kupper KC, Augusto F (2018) *In vivo* investigation of the volatile metabolome of antiphytopathogenic yeast strain active against *Penicillium digitatum* using comprehensive two-dimensional gas chromatography and multivariate data analysis. Microchem J 141:362–368. https://doi.org/10.1016/j.microc.2018.05.047
54. Hantao LW, Aleme HG, Passador MM et al (2013) Determination of disease biomarkers in *Eucalyptus* by comprehensive two-dimensional gas chromatography and multivariate data analysis. J Chromatogr A 1279:86–91. https://doi.org/10.1016/j.chroma.2013.01.013
55. Hantao LW, Todelo BR, Ribeiro FAL et al (2013) Comprehensive two-dimensional gas chromatography combined to multivariate data analysis for detection of disease-resistant clones of *Eucalyptus*. Talanta 116:1079–1084. https://doi.org/10.1016/j.talanta.2013.08.033
56. Souza Silva ÉA, Pawliszyn J (2012) Optimization of fiber coating structure enables direct immersion solid phase microextraction and high-throughput determination of complex samples. Anal Chem 84:6933–6938. https://doi.org/10.1021/ac301305u
57. Risticevic S, Souza-Silva EA, DeEll JR et al (2016) Capturing plant metabolome with direct-immersion in vivo solid phase microextraction of plant tissues. Anal Chem 88:1266–1274. https://doi.org/10.1021/acs.analchem.5b03684
58. De Grazia S, Gionfriddo E, Pawliszyn J (2017) A new and efficient solid phase microextraction approach for analysis of high fat content food samples using a matrix-compatible coating. Talanta 167:754–760. https://doi.org/10.1016/j.talanta.2017.01.064

59. Seeley JV, Seeley SK (2013) Multidimensional gas chromatography: fundamental advances and new applications. Anal Chem 85:557–578. https://doi.org/10.1021/ac303195u
60. Edwards M, Mostafa A, Gorecki T (2011) Modulation in comprehensive two-dimensional gas chromatography: 20 years of innovation. Anal Bioanal Chem 401:2335–2349. https://doi.org/10.1007/s00216-011-5100-6
61. Tranchida PQ, Purcaro G, Dugo P et al (2011) Modulators for comprehensive two-dimensional gas chromatography. TrAC – Trends Anal Chem 30:1437–1461. https://doi.org/10.1016/j.trac.2011.06.010
62. Khummueng W, Harynuk J, Marriott PJ (2006) Modulation ratio in comprehensive two-dimensional gas chromatography. Anal Chem 78:4578–4587. https://doi.org/10.1021/ac052270b
63. Bahaghighat HD, Freye CE, Synovec RE (2019) Recent advances in modulator technology for comprehensive two dimensional gas chromatography. TrAC – Trends Anal Chem 113:379–391. https://doi.org/10.1016/j.trac.2018.04.016
64. Martins C, Brandão T, Almeida A et al (2017) Metabolomics strategy for the mapping of volatile exometabolome from *Saccharomyces* ssp. widely used in the food industry based on comprehensive two-dimensional gas chromatography. J Sep Sci 40:2228–2237. https://doi.org/10.1002/jssc.201601296
65. Zellner BA, Bicchi C, Dugo P et al (2008) Linear retention índices in gas chromatographic analysis: a review. Flavour Fragr J 23:297–314. https://doi.org/10.1002/ffj.1887
66. Bean HD, Hill JE, Dimandja JD (2015) Improving the quality of biomarker candidates in untargeted metabolomics via peak table-based alignment gas chromatography-mass spectrometry data. J Chromatogr A 1394:111–117. https://doi.org/10.1016/j.chroma.2015.03.001
67. Mondello L (2011) Comprehensive chromatography in combination with mass spectrometry. John Wiley & Sons, USA. https://doi.org/10.1002/9781118003466
68. Watson JT, Orrin OD, Sparkman D (2007) Introduction to mass spectrometry: instrumentation, applications and strategies for data interpretation. John Wiley & Sons, USA
69. Cordero C, Guglielmetti A, Bicchi C et al (2019) Comprehensive two-dimensional gas chromatography coupled with time off light mass spectrometry featuring tandem ionization: challenges and opportunities for accurate fingerprinting studies. J Chromatogr A 1597:132–141. https://doi.org/10.1016/j.chroma.2019.03.025
70. Franchina FA, Purcaro G, Burklund A et al (2019) Evaluation of different adsorbent materials for the untargeted and targeted bacterial VOC analysis using GCxGC-MS. Anal Chim Acta 1066:146–153. https://doi.org/10.1016/j.aca.2019.03.027
71. Tranchida PQ, Franchina FA, Dugo P et al (2016) Comprehensive two-dimensional gás chromatography-mass spectrometry: recent evolution and current trends. Mass Spectrom Rev 35:524–534. https://doi.org/10.1002/mas.21443
72. Olsen JV, Godoy LMF, Li G et al (2005) Parts per million mass accuracy on an Orbitrap mass spectrometer via lock mass injection into a C-trap. Mol Cell Proteomics 4:2010–2021. https://doi.org/10.1074/mcp.T500030-MCP200
73. Hu Q, Noll RJ, Li H et al (2005) The Orbitrap: a new mass spectrometer. J Mass Spectrom 40:430–443. https://doi.org/10.1002/jms.856
74. Weidt S, Haggarty J, Kean R et al (2016) A novel targeted/untargeted GC-Orbitrap metabolomics methodology applied to *Candida albicans* and *Staphylococcus aureus* biofilms. Metabolomics 12:189–199. https://doi.org/10.1007/s11306-016-1134-2
75. Zoccali M, Cappello S, Mondello L (2018) Multilevel characterization of marine microbial biodegradation potentiality by means of flow-modulated comprehensive two-dimensional gas chromatography combined with a triple quadrupole mass spectrometer. J Chromatogr A 1547:99–106. https://doi.org/10.1016/j.chroma.2018.03.013

Capillary Electrophoresis-Mass Spectrometry for Metabolomics: Possibilities and Perspectives

Nicolas Drouin and Rawi Ramautar

Abstract

Capillary electrophoresis-mass spectrometry (CE-MS) is a very useful analytical technique for the selective and highly efficient profiling of polar and charged metabolites in a wide range of biological samples. Compared to other analytical techniques, the use of CE-MS in metabolomics is relatively low as the approach is still regarded as technically challenging and not reproducible. In this chapter, the possibilities of CE-MS for metabolomics are highlighted with special emphasis on the use of recently developed interfacing designs. The utility of CE-MS for targeted and untargeted metabolomics studies is demonstrated by discussing representative and recent examples in the biomedical and clinical fields. The potential of CE-MS for large-scale and quantitative metabolomics studies is also addressed. Finally, some general conclusions and perspectives are given on this strong analytical separation technique for probing the polar metabolome.

N. Drouin · R. Ramautar (✉)
Division of Systems Biomedicine and Pharmacology, Leiden Academic Centre for Drug Research, Leiden University, Leiden, The Netherlands
e-mail: r.ramautar@lacdr.leidenuniv.nl

Keywords

Capillary electrophoresis · Mass spectrometry · Metabolomics · Interfacing designs · Applications

Abbreviations

APFO	Ammonium perfluorooctanoate
BGE	Background electrolyte
CEC	Capillary electrochromatography
CE-MS	Capillary electrophoresis-mass spectrometry
CGE	Capillary gel electrophoresis
cIEF	Capillary isoelectric focusing
CZE	Capillary zone electrophoresis
EOF	Electro-osmotic flow
HILIC	Hydrophilic interaction liquid chromatography
ITP	Isotachophoresis
MEKC	Micellar electrokinetic chromatography
MSI	Multi-segment injection
NACE	Non-aqueous CE
SDS	Sodium dodecyl sulphate
TOF-MS	*Time-of-flight* mass spectrometry
μ_{ep}	Electrophoretic mobility

A. V. Colnaghi Simionato (ed.), *Separation Techniques Applied to Omics Sciences*, Advances in Experimental Medicine and Biology 1336, https://doi.org/10.1007/978-3-030-77252-9_9

1 Introduction

The aim of using metabolomics is to obtain insight into a well-defined biological question or problem [1]. For this purpose, targeted and untargeted metabolomics studies can be used. In the first approach, the focus is on the (quantitative) analysis of a set of well-defined metabolites or metabolite classes using a tailor-made sample preparation strategy, while in the second approach the focus is on analysing a broad range of metabolite classes without a priori knowledge of their nature or identity. The biological question often dictates whether a untargeted or targeted approach needs to be considered. Both approaches can be employed in a single metabolomics study, where the first approach is generally used to find potential biomarkers, and the second approach is then used to verify the results obtained with the first approach, preferably employing standardized protocols as required for biomedical and clinical studies. Ultimately, the use of metabolomics should provide an answer to the proposed biological question.

At present, the Human Metabolome Database is comprised of more than 115,000 metabolite entries, of which a major part consists of lipids and exogenous compounds derived from nutrients and drugs [2]. In order to probe as many (endogenous) metabolites as possible in a given biological sample, a combination of analytical techniques with complementary separation mechanisms is needed. For example, for the characterization of the human serum metabolome, multiple analytical separation techniques have been used in order to capture a broad range of endogenous metabolites [3]. In this case, more than 4000 metabolites were detected at concentrations spanning more than nine orders of magnitude.

Currently, nuclear magnetic resonance (NMR) spectroscopy, liquid chromatography (LC) and gas chromatography (GC) coupled to mass spectrometry (MS) are generally used for metabolomics studies [4–6]. Concerning chromatographic-based separation techniques, notably the use of LC columns based on sub-2 μm porous particles and/or core-shell silica particles has gained increased interest in metabolomics studies, as they can provide relatively fast separations with a high peak capacity [7]. In general, reversed-phase LC columns are employed for metabolomics, which can be used for the analysis of a wide range of metabolite classes. For the analysis of polar and charged metabolites, mainly ion-pair reversed-phase LC-MS and hydrophilic interaction liquid chromatography (HILIC) are considered. However, the use of ion-pair agents in LC-MS may result in severe ion suppression and may contaminate the ion source and ion optics [8]. Moreover, ion-pair agents may contribute to column instability and increased re-equilibration time. In HILIC a polar stationary phase is used in combination with aqueous organic eluents for the analysis of polar metabolites. This approach has gained interest as a complementary chromatographic separation technique for the profiling of polar metabolites over the past few years [9].

CE-MS is an analytical technique not commonly used in metabolomics, while it has very strong features for the analysis of highly polar and charged metabolites. Lack of (practical) expertise with this method and the perception that CE-MS is a technically challenging and not reproducible approach as compared to chromatographic-based methods may explain its limited use in metabolomics. However, a few recent studies clearly demonstrate the value of CE-MS for metabolic profiling of large sample sets [10, 11]. For example, the group of Soga and co-workers introduced the first CE-MS approaches for metabolomics in 2003 [12] and more recently has assessed the long-term performance of CE-MS for metabolic profiling of more than 8000 human plasma samples from the Tsuruoka Metabolomics Cohort Study over a 52-month period [11]. The study provided an absolute quantification for 94 polar metabolites in plasma with a similar or better reproducibility when compared to other analytical platforms.

In CE, compounds are separated on the basis of their intrinsic electrophoretic mobility, i.e. charge-to-size ratio. As such, the separation mechanism of CE is fundamentally different from chromatographic-based separation techniques. Both CE and HILIC can be employed for the analysis of polar and charged metabolites; however, some crucial differences exist [13, 14]. For example, a relatively larger amount of the sample can be injected in HILIC-MS, which is an advantage when sample amount is not an issue. On the other hand, significantly higher separation efficiencies can be obtained by CE, and it is especially useful for the efficient analysis of low-abundance polar and charged metabolites in low sample amounts [15]. Moreover, analysis times are significantly shorter for CE because re-equilibration of the HILIC column is often a time-consuming process [5]. In addition, HILIC requires an important fraction of non-polar solvent in the sample, potentially leading to solvent incompatibility with the most polar compounds and reducing the polarity range of HILIC [16]. The clear complementary role of CE-MS in comparison to other analytical techniques for metabolomics studies has been recently demonstrated by various research groups [14, 17–21].

The development of new interfacing designs over the past few years resulted in a relatively increased interest of the CE-MS approach for metabolomics studies, but also in other fields. Moreover, progress has been made in improving the migration time repeatability and to further shorten the total analysis time. For a comprehensive overview of these developments in CE-MS for metabolomics, the reader is referred to the following reviews [22–28]. The aim of this chapter is to give an overview of the possibilities of CE-MS for metabolomics studies. Various CE separation modes employed for CE-MS-based metabolomics are discussed. Main aspects related to hyphenation of CE to MS are considered, and special attention is devoted to the use of new interfacing techniques including its implications for metabolomics studies. Subsequently, the utility of CE-MS for targeted and untargeted metabolomics is illustrated on the basis of the discussion of relevant biomedical and clinical examples. Finally, some general conclusions and perspectives are provided.

2 CE-MS Methodology

2.1 CE Systems for Metabolomics

CE has been used for the analysis of (endogenous) metabolites in various biological samples for more than a few decades now. Jellum et al. reported one of the first assays for the profiling of organic acids in human body fluids in order to screen for various metabolic diseases [29–31]. In one of the studies, about 50 metabolites could be observed in human urine within a short analysis time (<15 min) using minimal sample pretreatment [30]. Compound identification was performed by comparison of migration time and UV-visible diode-array spectra against known standards. The group of Barbas and co-workers also played an important role in the development of CE assays for the selective determination of organic acids in human urine, which were used for the diagnosis of inborn errors of metabolism [32].

Like in LC, CE can be used in a number of separation modes, such as capillary zone electrophoresis (CZE), referred to as "CE" in this chapter, micellar electrokinetic chromatography (MEKC), non-aqueous CE (NACE), capillary electrochromatography (CEC), capillary gel electrophoresis (CGE), capillary isotachophoresis (cITP) and capillary isoelectric focusing (cIEF). From the viewpoint of coupling CE to electrospray ionization (ESI) mass spectrometry, CE is used as the main separation mode as volatile buffers are required. For example, ammonium acetate, acetic acid and formic acid are frequently employed as background electrolyte (BGE) in CE-MS studies. The use of these BGE systems may not essentially provide the same level of separation performance as obtained with the typically employed phosphate- or borate-based BGEs in CE-UV mode. Though only a limited number of BGEs can be considered in CE-MS method development, the CE separation can be improved by adding organic modifiers to the BGE. For example, Mayboroda et al. improved the separation of leucine from isoleucine by using a BGE of 2 mol/L formic acid (pH 1.8) containing 20% (v/v) methanol [33].

In CE, compounds are separated according to differences in their intrinsic electrophoretic mobilities, which is dependent on the size (hydrodynamic radius) and charge of the ion, as well as the viscosity of the BGE. Under constant separation conditions, the electrophoretic separation of metabolites is solely based on differences in charge-to-size ratio. The use of high voltages leads to relatively fast separation times, as the migration time of compounds is inversely proportional to the electric field. Soga and coworkers developed the first CE-MS methods for untargeted metabolic profiling of biological samples [12, 34]. For the global analysis of cationic metabolites (basic compounds), a bare fused-silica capillary using 1 mol/L formic acid (pH 1.8) as BGE has been employed, while a cationic polymer-coated capillary using 50 mmol/L ammonium acetate (pH 8.5) as BGE has been employed for the global analysis of anionic metabolites (acidic compounds). In this latter method, neutral compounds are dragged to the detector with the electro-osmotic flow (EOF). However, as their charge is zero, they cannot be electrophoretically resolved. The use of both methods allowed the detection of more than 1600 molecular features in *Bacillus subtilis* extracts, of which 150 could be identified.

In MEKC, micelles are used as pseudo-stationary phases in the BGE allowing the concomitant separation of both charged and neutral analytes. However, the coupling of MEKC to MS is not straightforward due to the use of non-volatile micelles (often sodium dodecyl sulphate (SDS)). On the other hand, volatile surfactants may be considered to overcome these issues. For example, Moreno-González et al. developed a MEKC-MS method for the analysis of amino acids in human urine using ammonium perfluorooctanoate (APFO) as a volatile surfactant [35].

In CEC, separation of compounds is based on differences in electrophoretic mobility and the partition between a stationary and mobile phase. The mobile phase flows through the column by the EOF, generated by the silanol groups of the stationary phase. Neutral compounds move through the packed column by the EOF and are separated by partitioning between the stationary and mobile phase, whereas charged compounds move through the packed column with the additional contribution from the intrinsic electrophoretic mobility of the analytes. Given the separation mechanism of CEC, this CE mode is useful for the efficient separation of a wide range of compounds. Recently, Wu et al. reported the development of a pressurized CEC method hyphenated to MS for profiling metabolites in human urine [36]. A home-made sheathless interface was constructed for hyphenating CEC to MS in order to take advantage from the intrinsically low flow rate of pressurized CEC. The analytical performance of the CEC-MS method was evaluated with a metabolite test mixture and pooled human urine. Limits of detection (LODs) for test compounds ranged from 18 to 88 ng/mL. Multivariate data analysis of urine metabolite profiles distinguished lung cancer patients from controls. For metabolite identification, the selectivity provided by CEC was critical in order to distinguish fragment ions of glutamine conjugates from coeluting metabolites. Three glutamine conjugates, i.e. phenylacetylglutamine, acylglutamine C8:1 and acylglutamine C6:1, were identified among 16 distinct metabolites in this study. The utility of CEC-MS for metabolomics studies has hardly been explored so far. A potential reason for that may be related to the design of consistent CEC columns.

The other CE separation modes, i.e. CGE and cIEF, are generally used for protein analysis. Still, cIEF may have potential for metabolomics studies as it is very suited for the high-resolution separation of amphoteric compounds according to their isoelectric point (pI). In cIEF, a mixture of ampholytes and sample fills the capillary. A basic catholyte, generally sodium hydroxide, is placed at the cathode, and an acidic anolyte, e.g. phosphoric acid, is placed at the anode. Due to the ampholyte properties, a pH gradient is established in the capillary under the influence of an electric field. Amphoteric compounds are focused until the region where the pH and the pI of the amphoteric compounds are equal; thus, the net charge of the compound is zero, and, as a result, the compound will not migrate. Electrophoretic or hydrodynamic mobilization is then used to move the individual components to the detector [37].

When employing CE systems based on bare fused-silica capillaries, variability of migration times due to adsorption of sample matrix components to the inner capillary wall may become an issue, especially when the purpose is to analyse large sample cohorts by CE-MS. Obviously, migration time variability should be very low for comparative metabolic profiling studies. A way to minimize this is by using coated capillaries. For an overview of the use of coated capillaries in CE-MS and more specifically for CE-MS-based metabolomics studies, we refer to the following papers [38, 39].

2.2 Interfacing Designs and Their Implications for Metabolomics

For metabolomics studies, MS is an indispensible analytical tool, especially from the viewpoint of the reliable identification of compounds present in a given biological sample. For untargeted metabolomics, *time-of-flight* MS (TOF-MS) instruments are generally employed for obtaining full-scan MS recordings with a high mass accuracy. Moreover, due to its high spectral acquisition rate, TOF-MS instruments are fully compatible with highly efficient CE separations (i.e. very sharp peaks are generated). For targeted metabolomics, TOF-MS, ion trap MS and triple quadrupole MS systems are generally employed in combination with CE [5, 40–43].

For the coupling of CE to MS, a special interface is required in order to perform the electrophoretic separation independent from ESI-MS, i.e. the electric fields used for both processes should not interfere with each other. Such an interfacing design for coupling CE to MS was first developed by Smith and co-workers, in which configuration a co-axial solvent (i.e. sheath-liquid) was delivered as a terminal electrolyte reservoir [44]. This interfacing design was further improved and commercially available since 1995 as a co-axial sheath-liquid interface. In this design, the CE capillary is inserted into a larger diameter tube (Fig. 1). The sheath-liquid, to which the CE terminating voltage is applied, is provided via an outside tube and merges with the CE effluent at the capillary outlet. A gas flow is often applied via a third co-axial capillary to facilitate spray formation in the ESI source. In principle, the flow rate and composition of the sheath-liquid dictate the ESI process in this particular configuration, and, as such, the selection and optimization of the sheath-liquid is very critical. In general, the sheath-liquid is composed of a mixture of water and organic modifier, including a small percentage of a volatile acid or base (e.g. formic acid or ammonium acetate), and provided at flow rates typically between 2 and 10 μL/min.

Until now, the co-axial sheath-liquid interface (or conventional interface) has been most often used for coupling CE to MS in a wide range of application fields [45–48]. Although this interfacing design can be used in a rather robust way [49], the CE effluent is significantly diluted, thereby often resulting in low- to mid-μmol/L detection limits for metabolites when using standard injection volumes [50, 51]. In-capillary preconcentration techniques, such as transient isotachophoresis or dynamic pH junction, can be used to further improve the detection sensitivity of sheath-liquid CE-MS for metabolomics studies [52, 53]. Another way to enhance detection sensitivity is by adding modifiers to the sheath-liquid [54]. For example, enhanced charging or supercharging of compounds in CE-MS has been explored for the analysis of intact proteins by the addition of various supercharging agents, such as 3-nitrobenzyl alcohol and sulfolane, to the sheath-liquid in order to modulate the charge-state distribution of proteins [55]. The potential of using this strategy in CE-MS-based metabolomics studies has not been explored so far.

Over the last decade, modified versions of the conventional co-axial sheath-liquid interface have been developed in which the sheath-liquid is provided at a significantly lower flow rate [56]. In some cases, such designs have been designated as a liquid junction interface; however, in our opinion, they can be considered as low-flow or miniaturized sheath-liquid interfaces. Maxwell et al. developed such a low-flow interface by using a flow-through micro-vial interface, which was

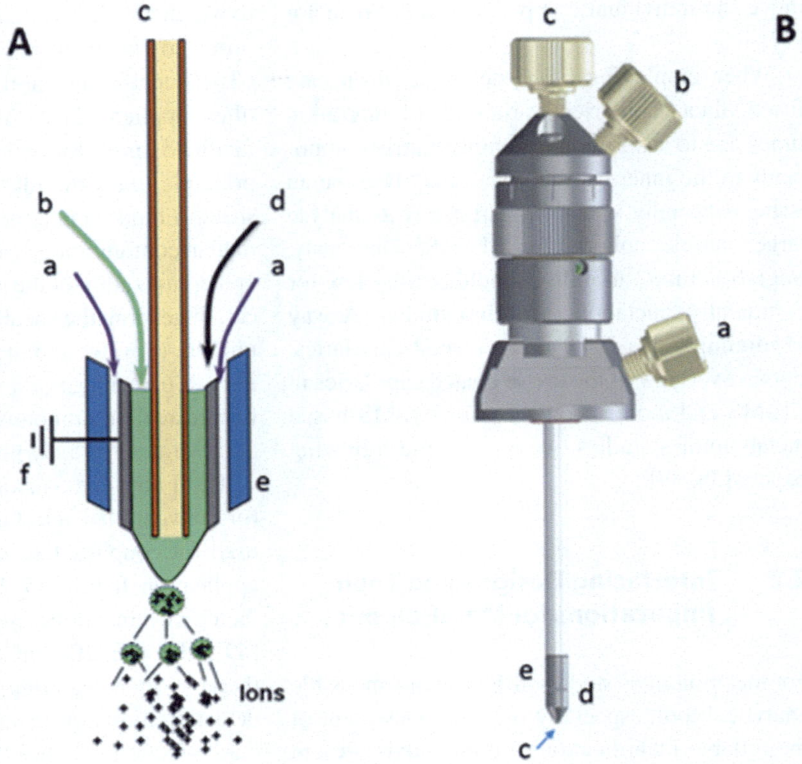

Fig. 1 Graphic representation of an Agilent co-axial sheath-liquid CE-MS interface: (a) nebulizing gas, (b) sheath-liquid, (c) CE capillary with BGE, (d) stainless steel spray needle 0.4 mm i.d., 0.5 mm o.d., (e) outer tube and (f) ground connection. (**b**) Engineering sketch of the co-axial sheath-liquid CE-MS interface. (Reproduced from reference [56] with permission)

created by inserting the separation capillary into a tapered stainless steel hollow electrospray emitter [57]. The small volume between the capillary and inner walls of the needle electrode tip formed a flow-through micro-vial that acted as both the outlet vial and the terminal electrode (Fig. 2). The flow-through micro-vial also allowed the addition of a sheath-liquid solution at low flow rates to provide a stable flow to the needle tip, which increased the compatibility of the CE effluent with ESI while minimizing sample dilution. The bevel on the emitter tip resulted in increased spray stability and effectively moved the ionization site away from the outlet.

The analytical performance of CE-MS using the flow-through micro-vial interface for metabolomics was recently assessed by Lindenburg et al. [58]. Using a standard metabolite mixture, the flow-through micro-vial (500 nL/min flow rate) and the standard sheath-liquid CE-MS interfaces (4 μL/min flow rate) were compared (Fig. 3). The LODs obtained with the flow-through micro-vial interface ranged from 0.01 to 3 μmol/L, which was on average a fivefold improvement as compared to the LODs obtained by conventional CE-MS. The potential of CE-MS using the flow-through micro-vial interface utility for metabolic profiling of large sample sets has not been shown so far. A drawback of this interface is the low tolerance to high CE currents as formation of electrolysis gases may accumulate in the spray needle and, as a result, hindering electrical contact. A nice feature of this approach is that it allows the use of CE capillaries with different inner diameters.

Next to the sheath-liquid interfaces, sheathless interfaces can be employed for coupling CE to MS. In such designs the CE voltage is applied directly to the BGE at the capillary outlet [59]. For conductivity, the metal may be coated on the end of the tapered separation capillary. Alternatively, a metal-coated, full metal or conductive polymeric sprayer tip may be connected to the CE outlet [56]. A closed electrical circuit may also be created by inserting a metal microelectrode through the capillary wall into the

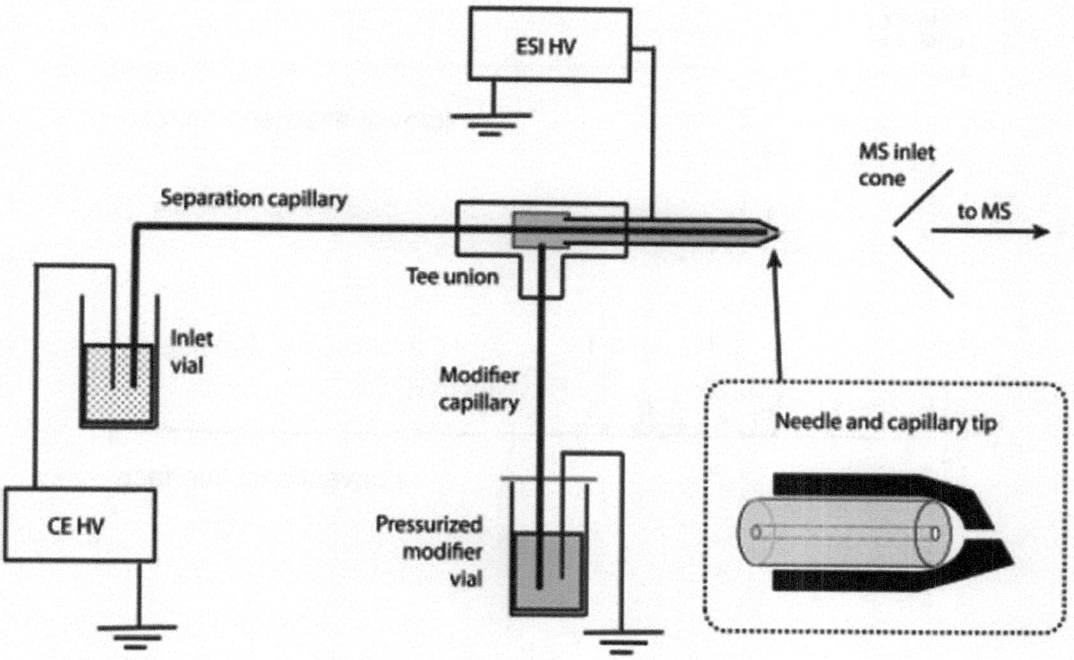

Fig. 2 Schematic illustration of the flow-through micro-vial interface apparatus, including a dissected view of the needle tip with inserted capillary (inset). (Reproduced from reference [57] with permission)

BGE, by direct introduction of a micro-electrode into the capillary [60, 61] or by a crack in the silica wall close to the capillary tip [62].

A CE-MS method using a sheathless porous tip interface, which was first developed by Moini [63], has been developed for the global profiling of cationic metabolites in human urine [53]. The porous tip interface was designed by removing the polyimide coating of the capillary outlet and etching the capillary wall with 49% solution of hydrofluoric acid to a thickness of about 5 μm (Fig. 4). The electrical connection to the capillary outlet was obtained by inserting the etched conductor into an ESI needle, which was filled with BGE. The sheathless porous tip interface was mainly useful for interfacing narrow capillaries (<30 μm inner diameter) and for low-flow-rate (<100 nL/min) nano-ESI-MS analyses. Using human urine, this approach allowed obtaining a highly information-rich metabolic profile as compared to CE-MS employing a sheath-liquid interface. Around 900 molecular features were observed with sheathless CE-MS, while only 300 were detected by sheath-liquid CE-MS. Overall, the improved detection sensitivity of sheathless CE-MS significantly improved the coverage of the urinary metabolome. However, with a single porous tip capillary emitter around 100 pre-treated biological samples can be analysed at this stage, which is not really cost-effective given the relatively high price of a single porous tip capillary.

Apart from body fluids, the utility of CE-MS using a sheathless porous tip interface has also been recently evaluated for metabolic profiling of low number of mammalian cells, using HepG2 cells as a model system [64]. Given the nanomolar concentration sensitivity, the sheathless CE-MS method could be effectively used for obtaining metabolic profiles in HepG2 cells starting from 10,000 down to 500 cells. A typical profile obtained for cationic metabolites from a starting amount of 500 HepG2 cells only is shown in Fig. 5, in which more than 20 metabolites could be observed. Hence, these results suggest that the method has the sensitivity for performing single-cell mammalian metabolomics studies. Still, the long-term performance of

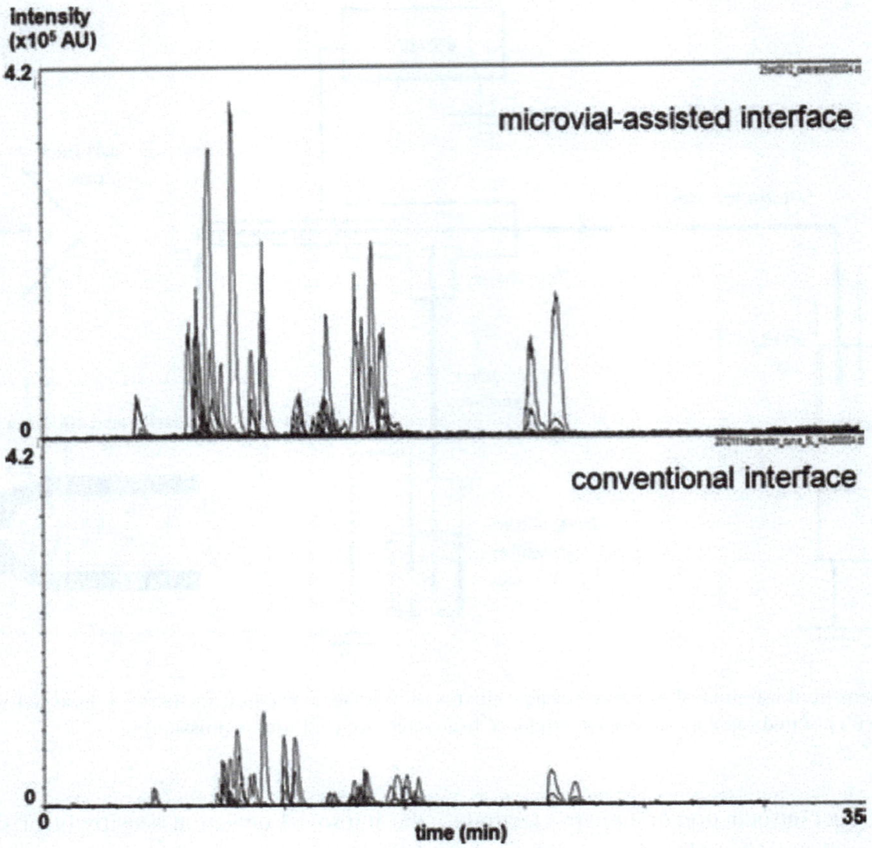

Fig. 3 Multiple extracted ion electropherograms of 35 cationic metabolites (25 μmol/L) obtained by CE-MS using a flow-through micro-vial interface (upper electropherogram) and a sheath-liquid interface (lower electropherogram). (Reproduced from reference [58] with permission)

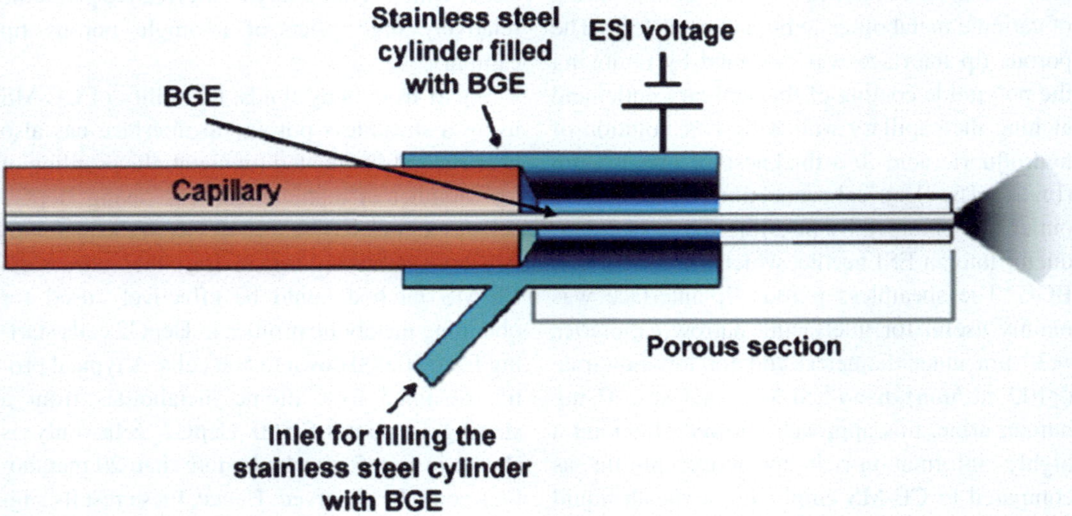

Fig. 4 Schematic of the high-sensitivity porous sprayer sheathless interface. (Reproduced from reference [59] with permission)

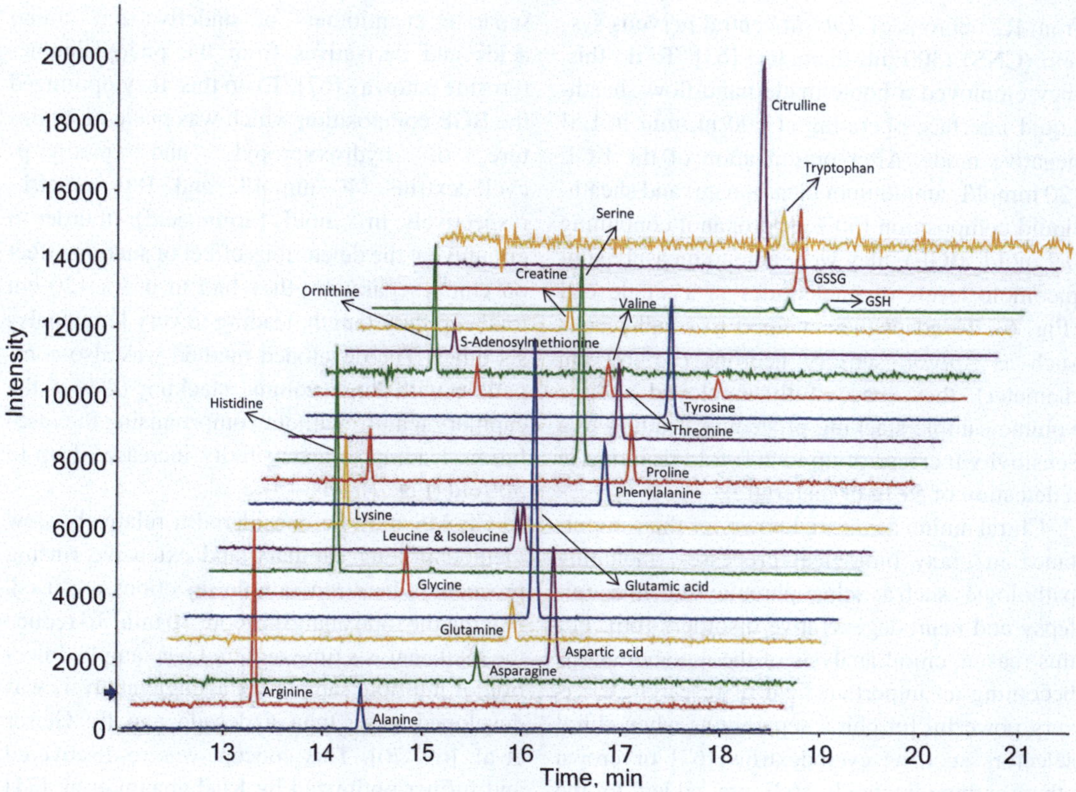

Fig. 5 Multiple extracted ion electropherograms for a selected number of metabolite peaks detected in an extract of 500 HepG2 cells by sheathless CE-MS in positive mode using a porous tip emitter. Separation conditions: BGE, 10% acetic acid (pH 2.2). Separation voltage: 30 kV. Sample injection: 6.0 psi for 60 s. (Reproduced from reference [64] with permission)

this approach needs to be assessed more comprehensively using substantially larger numbers of clinical samples.

3 Applications

In metabolomics, two different analytical approaches are typically used, targeted and untargeted analysis. Targeted metabolomics focuses on the quantification of a limited number of analytes. Therefore, accent is made on analysis throughput, sensitivity and robustness of the methods. In contrast, untargeted metabolomics focuses on every detectable feature in order to obtain a (unique) chemical fingerprint of the sample. Therefore, focus is made on unambiguous identification of the detected features. In the following sections, some state-of-the-art targeted and untargeted metabolomics studies performed by CE-MS are highlighted.

3.1 Targeted Applications

Nowadays, single-cell analysis remains a major challenge for current analytical platforms due to the large dilution required to handle the content of single cell, but also to selectively extract the relevant fraction for follow-up analysis. This kind of analysis is made even more complicated when the focus is on phosphorylated compounds, which are difficult to analyse by reversed-phase LC approaches due to their very high polarity. In this context, Liu et al. developed a CE-MS method for the quantification of 16 nucleotides from R2 neurons of *Aplysia* central nervous system (CNS) (300 μm diameter) [65]. To do this, they employed a home-made nano-flow sheath-liquid interface operating at 600 nL/min in ESI negative mode. After optimization of the BGE (20 mmol/L ammonium bicarbonate) and sheath-liquid composition (50% isopropanol containing 0.2 mol/L BGE), they were able to measure sub-picomole levels of nucleotides in a single cell (Fig. 6). To adapt their method to smaller cells such as *Aplysia* sensory neurons (i.e. 60 μm diameter), they successfully employed a large volume sample stacking procedure, leading to a sensitivity increase of up to 200-fold resulting in a detection of 51 fg of material.

Chiral amino acids are known for their importance in many biological processes, including pathologies such as schizophrenia, ischemia, epilepsy and neurodegenerative disorders [66]. For this reason, chiral analysis of the metabolome is becoming an important field of research. CE is very powerful for chiral separations when chiral selectors such as cyclodextrins [67] or crown ether carboxylic acids [68] are added to the BGE. However, these selectors are not MS compatible, leading to the necessity to develop alternative strategies. Recently, Sánchez-López et al. used a partial filing injection approach in order to separate enantiomers of underivatized amino acids and derivatives from the phenylalanine-tyrosine pathway [67]. To do this, they optimized the BGE composition which was made of a mixture of hydroxypropyl- and methyl-β-cyclodextrins (40 mmol/L and 180 mmol/L, respectively, in 2 mol/L formic acid). In order to circumvent the deleterious effect of suction effect on chiral separation, they had to use a 120 cm total capillary length, leading to very long analysis times. The developed method was also compatible with large volume stacking (7% of the capillary length) without compromising the resolution, leading to a sensitivity increase of up to 50-fold (Fig. 7).

CE-MS may be considered a relatively slow method if long capillary and extensive rinsing procedures between sample injections are used, making the total analysis about 30 min. To reduce the total analysis time required per sample, injection of multiple samples in a single analysis was developed more than a decade ago by Geiser et al. [69, 70]. This concept was re-discovered and further optimized by Kuehnbaum et al. [71] and is now known as multi-segment injection (MSI). MSI allows to inject up to 7 or more discrete sample zones in a single electrophoretic run, thereby increasing the throughput to only a

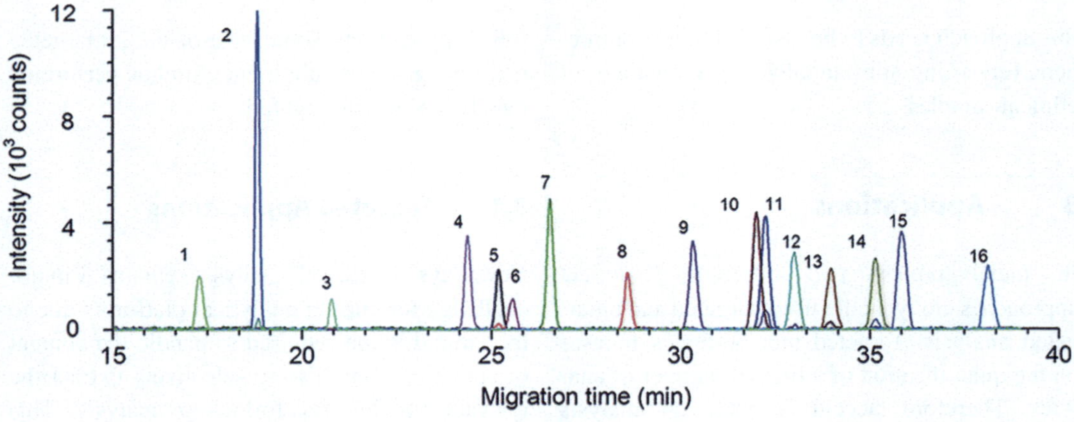

Fig. 6 Extracted ion electropherograms acquired from solutions containing 16 anionic analyte standards by CE-MS. Injection volume, 10 nL; BGE, 20 mmol/L ammonium bicarbonate (pH 10); separation voltage, 10 kV. Analysed compounds (each at 100 μg/L): 1 – NAD+ (*m/z* 622.102); 2 – cAMP (*m/z* 328.045); 3 – FAD (*m/z* 784.150); 4 – AMP (*m/z* 346.056); 5 – CMP (*m/z* 322.045); 6 – NADP+ (*m/z* 742.068); 7 – GMP (*m/z* 362.058); 8 – UMP (*m/z* 323.029); 9 – ADP (*m/z* 426.022); 10 – GDP (*m/z* 442.017); 11 – CDP (*m/z* 402.011); 12 – ATP (*m/z* 505.989); 13 – GTP (*m/z* 521.983); 14 – UDP (*m/z* 402.995); 15 – CTP (*m/z* 481.977); and 16 – UTP (*m/z* 482.961). (Reproduced from reference [65] with permission)

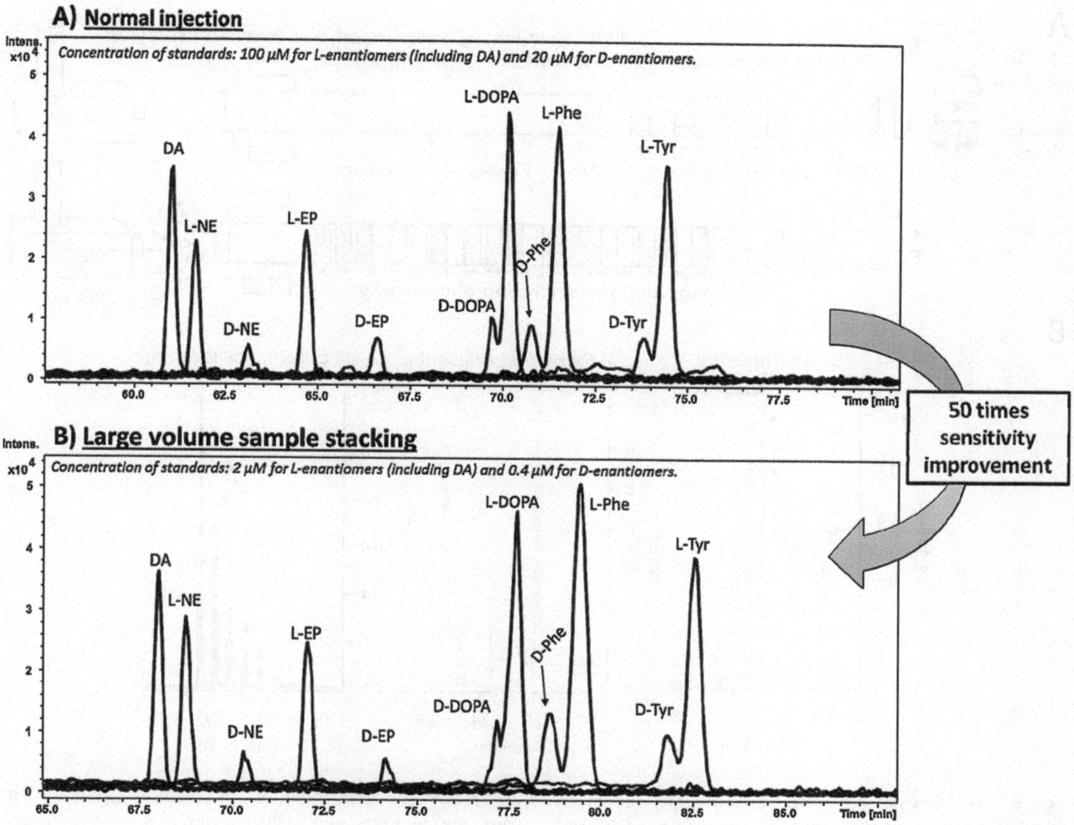

Fig. 7 Extracted ion electropherograms obtained by CE-MS for the chiral separation of the Phe-Tyr metabolic pathway constituents with injection time of 5 s (normal injection) in (**a**) and 250 s (injection with large volume sample stacking, LVSS) in (**b**). Peak identification: dopamine (DA), norepinephrine (NE), epinephrine (EP), 3,4-dihydroxyphenylalanine (DOPA), phenylalanine (Phe) and tyrosine (Tyr). (Reproduced from reference [67] with permission)

few minutes per sample. This method was successfully applied to various clinical studies, i.e. from the monitoring of 52 drugs of abuse [72] in human urine to the monitoring of inborn errors of metabolism from dried blood spot samples [73]. More recently, MSI was used in NACE-MS for the analysis of 18 fatty acids from C10 to C24 [74]. To reach a good separation of the metabolites, a complex BGE made of 35 mmol/L ammonium acetate in 70% v/v acetonitrile, 15% v/v methanol, 10% water and 5% v/v isopropanol with an apparent pH of 9.5 was used (Fig. 8). The developed NACE-MS method yielded a linear dynamic of at least two orders of magnitude for the tested compounds, with an average limit of quantification (LOQ) of 2.4 μmol/L. A further validation was performed by comparison of the newly developed MSI-NACE-MS with a standard GC-MS method, which revealed a similar sensitivity of both approaches; however, MSI-NACE-MS lacked isomeric resolution for some minor fatty acids.

As an alternative to using hydrodynamic separation plugs between the samples in MSI CE-MS, Drouin et al. proposed the use of electrokinetic plug to separate the samples [21]. This approach has the main advantage to circumvent the deleterious effect of multiple pressure applications, which can lead to peak diffusion and, as a result, to a loss of separation efficiency. Moreover, this approach also provides a longer effective capillary length for separation. With electrokinetic

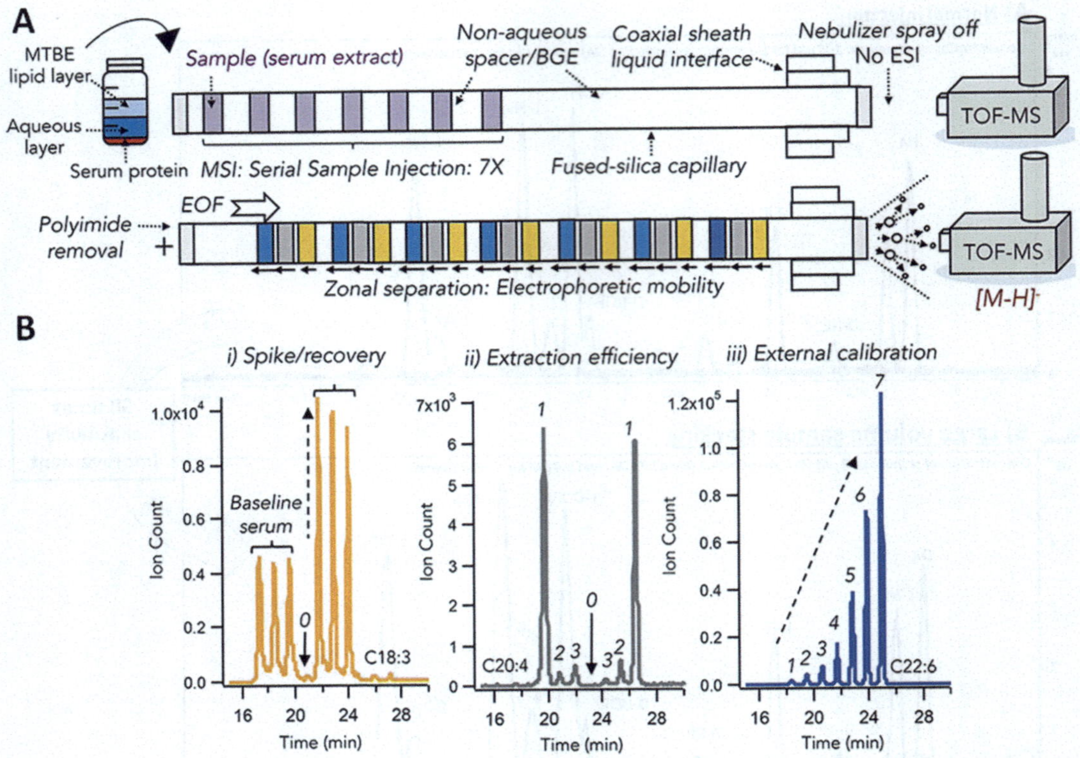

Fig. 8 (**a**) Multiplexed separation of fatty acids by NACE-MS using serial injection of seven or more discrete sample segments and their zonal electrophoretic separation following MTBE serum extraction with full-scan data acquisition under negative ion mode detection. (**b**) Customized serial injection configurations used in MSI-NACE-MS for FA quantification, including spike/recovery study for determination of method accuracy (in triplicate) and repeated MTBE serum extracts to evaluate extraction efficiency (in duplicate) along with seven-point external calibration curve over a 200-fold concentration range (1–200 μM). (Adapted from reference [74] with permission)

injection, optimization of space length and number of injected samples is simplified. However, the number of samples is limited by the migration time of the fastest compounds and the duration of the plugs.

Recently, Ouyang et al. developed a cIEF-MS method in order to achieve the separation of acidic oligosaccharides [75]. Using a set of commercial ampholytes with a pH range from 2.5 to 5.0, they were able to separate a mix of 16 disaccharides from different families. To do this, a reversed separation polarity was employed (Fig. 9). Under these conditions, the analyte solution (0.1–1% formic acid) was injected first, followed by the sample mixed with the ampholytes, and then the catholyte solution (0.2 mol/L ammonium hydroxide) was injected. Therefore, when a negative voltage of −30 kV was applied, a gradient pH was generated (from 5 at the inlet to 2.5 at the outlet) and compounds of interest migrated and were stacked in the zone where the pH was equal to their pI. Despite the presence of methanol in the solution, the gradient of apparent pH formed was demonstrated as linear based on the used pI markers. This method showed good repeatability as well as a good separation power for isomeric compounds. Indeed, it was possible to separate O- and N-sulfo isomers based on their slight pI difference. Finally, this method was successfully used for the separation of larger chondroitin sulphate oligosaccharides (tetra- and hexa-saccharides).

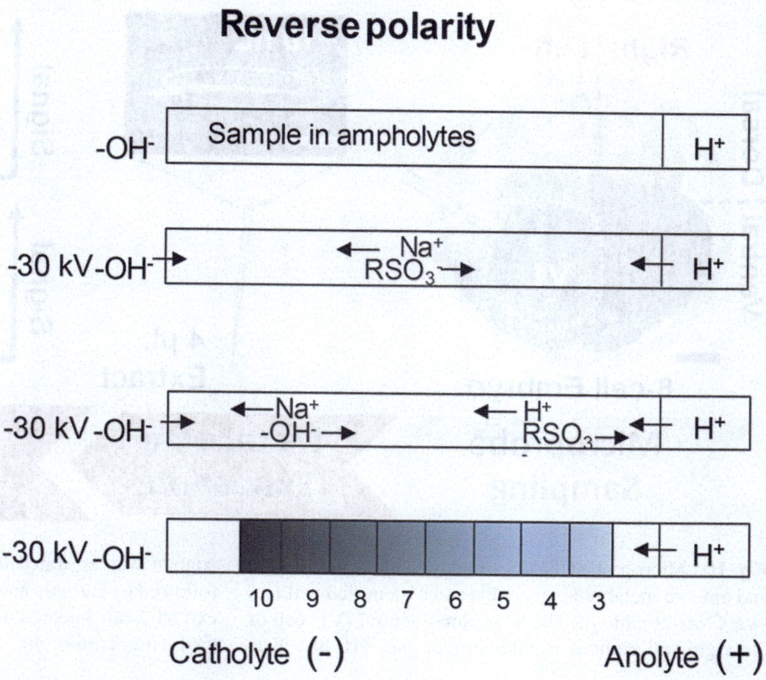

Fig. 9 Schematic of negative-ion mode cIEF-MS workflow showing reverse polarity separations. Anolyte solution (0.1–1% formic acid) was injected first, followed by sample mixed with the ampholytes (pH 2.5 to 5.0), and then catholyte solution (0.2 mol/L ammonium hydroxide) was injected. (Adapted from reference [75] with permission)

3.2 Untargeted Applications

Due to the lack of robustness and sensitivity often observed in CE-MS, LC-MS-based methods are generally used for untargeted metabolomics studies. However, recent developments in CE-MS indicate that this approach can reach the level of performance required for metabolomics. For example, in terms of robustness, a study involving the analysis of more than 8400 participant plasma samples was recently conducted by Harada et al. [11]. In this work, only 0.5% of the runs failed due to capillary current issues, highlighting the robustness of CE-MS.

In case of very limited sample amounts, such as single-cell profiling, CE-MS is extremely useful. Indeed, the extremely small injection volume limits the dilution factor required for injection and the nanospray in nano-sheath flow or in sheathless interfaces, leading to high detection sensitivities. In a recent work, Portero and Nemes were able to profile both anionic and the cationic metabolites of the left animal-ventral cell from a *Xenopus laevis*'s eight-cell embryo (Fig. 10) [76]. To do this, they designed a CE system which was capable to handle sample volumes below 1 μL. In order to avoid electrical discharge in ESI negative mode, the environment of the nanospray interface was saturated with acetonitrile gas. The LOD of the developed CE-MS method was on average around 5 nmol/L (~50 amol) for the tested compounds.

In another study, Sanchez-Lopez et al. took advantage of the CE-MS characteristics for profiling cationic metabolites in renal biopsies (20 μm thickness) using a sheathless porous tip interface [77]. However, in this study, only 5 metabolites were unambiguously identified using authentic standards and 21 compounds were putatively identified based on their accurate mass and MS/MS spectra. Indeed, one of the main challenges in untargeted metabolomics remains the annotation of the compounds. Criteria to achieve unambiguous identification (level 1) are extremely strict and require at least two independent and orthogonal parameters of a reference compound analysed under identical experimental conditions in the same laboratory [78–80]. Therefore, identification based on MS and MS/MS spectra search in libraries such as Metlin or HMDB can only reach the identification level 2 [2, 81]. In LC-MS-based methods, accurate mass

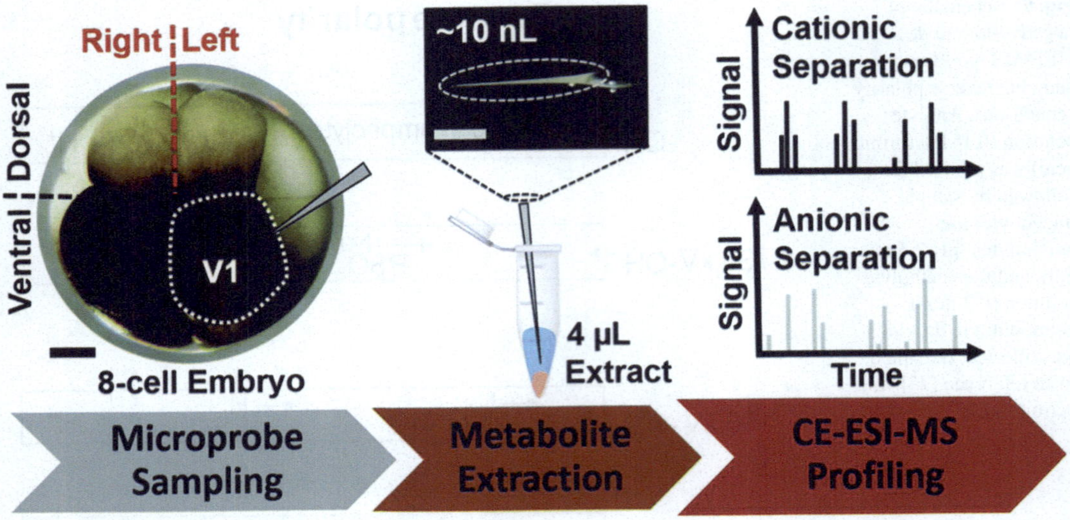

Fig. 10 Microprobe CE-MS strategy to measure cationic and anionic metabolites from the same identified cell in a live *X. laevis* embryo. The left animal-ventral (V1) cell of the eight-cell embryo was identified, and ~10 nL of its content was aspirated for one-pot metabolite extraction, followed by cationic and anionic profiling of the same cell extract. Scale bars = 250 μm. (Reproduced from reference [76] with permission)

and retention are commonly used for this purpose. However, the high variability of the migration times makes this strategy not very suitable in CE-MS-based analysis. To circumvent this problem, experimental relative migration times are commonly used [82, 83]. However, due to the cumulative effect of electrophoretic mobility and EOF on migration time and consequently on relative migration time, this approach is biased.

In comparison to migration time, under given conditions of BGE (pH and ionic strength) and temperature, electrophoretic mobility (μ_{ep}) can be considered as a physicochemical property of every compound (Eq. 1).

$$\mu_{ep} = \frac{z}{4.\pi.\eta.r} \quad (1)$$

where z is the charge of the molecule, η is the viscosity of the BGE and r is the hydrodynamic radius of the compounds. Therefore, μ_{ep} can be used for feature annotation. For this reason, Drouin et al. have created an experimental μ_{ep}-database [21]. To do this they used a two-step approach, with positive and negative CE polarity applied for cationic and anionic profiling, respectively. In both cases, a positive pressure was applied at the inlet of the capillary in order to reduce analysis time and generate an anodic flow in reverse CE polarity mode. Therefore, this approach allows global metabolic profiling of the sample in 20 min while maintaining a good peak shape and good separation of isomeric compounds, such as citric and isocitric acid (Fig. 11). In this study, 10% acetic acid buffer was used as BGE for both cationic and anionic metabolic profiling. Despite its low pH, this BGE was already reported earlier for cationic compounds as well as anionic species [84] and have as main advantages to be very easy and repeatable to prepare BGE. This method allows the detection of more than 450 compounds over a large variety of biochemical families (Fig. 12). To calculate the μ_{ep} of every compound, paracetamol was used as a neutral marker and was spiked in every sample. As expected, these strongly acidic conditions are highly suitable for basic compounds; however, as shown in Fig. 12, a large part of the acidic compounds are slightly charged and present a low μ_{ep} which can be detrimental for feature annotation. However, neutral compounds with the same exact mass, such as carbohydrates, cannot be separated or unambiguously identified with this method.

Interestingly, ammonium ions present in the sheath-liquid allowed the detection of a large

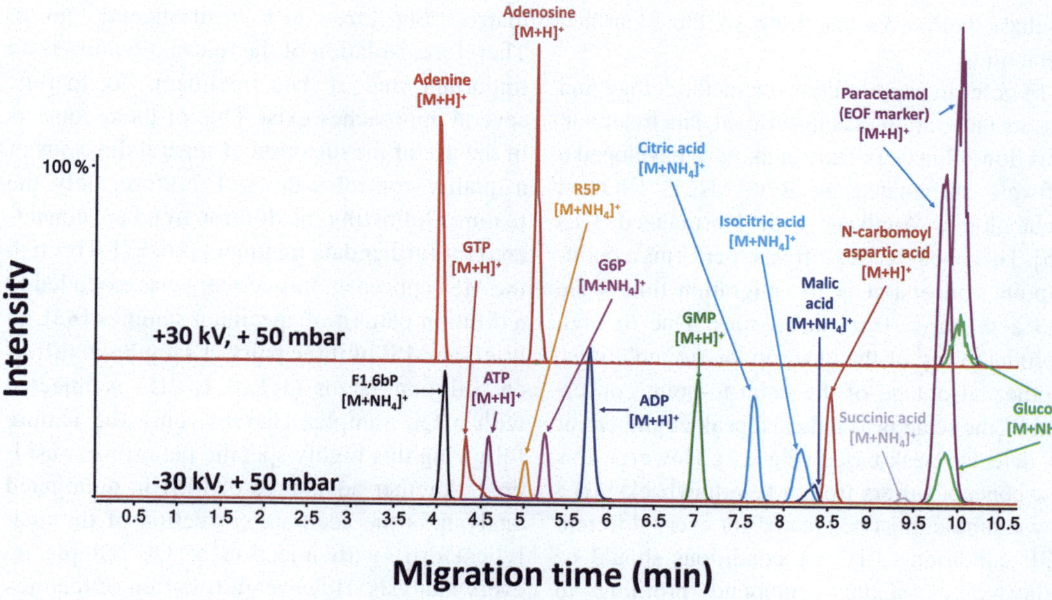

Fig. 11 Analytes observed in a metabolite test mixture by CE-MS in positive ESI mode using no nebulizer gas and an increased ESI voltage (5.5 kV). CE in normal polarity was used for cationic metabolites, while for anionic metabolites the polarity was reversed. (Reproduced from reference [21] with permission)

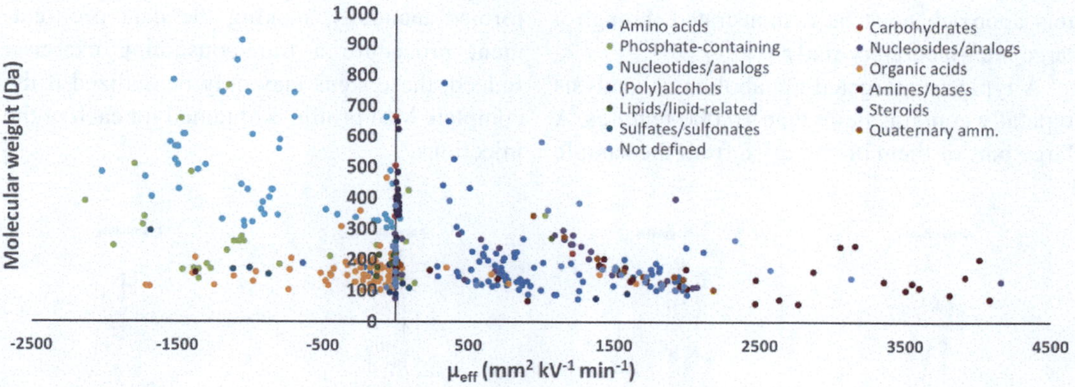

Fig. 12 Plot showing the detected compounds in a metabolite test mixture according to their effective mobilities and molecular weights. (Reproduced from Ref. [21] with permission)

variety of acidic metabolites in ESI positive mode by their ammonia adducts. However, the presence of ammonia in the sheath-liquid can also be considered as a drawback of this approach. Indeed, due to the reverse CE polarity applied during anionic metabolic profiling, ammonium ions, which have a very high but opposite mobility, migrate into the capillary, leading to the generation of a dynamic pH gradient in the capillary. In addition, the rising speed could be different between CE-MS platforms, generating a shift of the measured μ_{ep} from one CE-MS platform to another. However, for compounds with a sufficient μ_{ep}, this drawback was circumvented with fine-tuning of the positive pressure applied during the negative profiling, making the library transferable from one laboratory to another. Finally, after creation of the database in a first laboratory, it was successfully used for compound identification from cell culture extracts in a second laboratory, and 77 features were successfully annotated, with relative error to the database below 5% for 90% of the identified metabolites.

In order to support this new methodology and ease its integration in conventional data treatment workflow, Gonzàlez-Ruiz et al. have developed a software designated as ROMANCE (RObust Metabolomic Analysis with Normalized CE) [85]. This open-access software performs a point-to-point conversion of the migration time scale into a μ_{ep} scale from *mzml* files. Due to high reproducibility of the electrophoretic mobilities, another advantage of the point-to-point conversion of the scale is to induce a peak alignment of the detected peak based (Fig. 13). However, this new concept suffers from a few drawbacks: (i) a new database must be created for every different BGE conditions, (ii) new conditions should be dedicated to acidic compound profiling to improve robustness of the method and separation of acidic compounds, and (iii) this approach is only possible with robust and easy to prepare BGE systems. In addition, the universal aspect of this approach must be demonstrated through a large inter-laboratory study.

A typical untargeted metabolomics analysis typically contains more than 10,000 features. A large part of them do not arise from the sample matrix but are just instrumental noise. Therefore, isolation of the relevant feature is an important step of data treatment. To do this, several approaches exist. One of them consists of the use of the injection of several dilutions of a quality control sample. Therefore, only the features following the dilution trend are considered for further data treatment [86, 87]. Through the MSI approach, this concept was extended to a dilution pattern of specimen samples [88]. In brief, an MSI of three pairs of samples at different dilution factor (1:2; 1:1, 2:1) is injected with a QC sample. Thereby, only the feature following this highly specific pattern is considered. Another advantage of MSI in untargeted analysis is the accurate correction of the analytical drift with injection of QC sample on every analysis. However, attribution of features to a specific injection and their annotation is made extremely difficult in the case of metabolites which are detected near their LODs in the samples. In addition, this step can only be performed manually, making the data pre-treatment procedure a time-consuming exercise. Indeed, these steps may only be realized if the complete MSI profile is obtained for each of the injections.

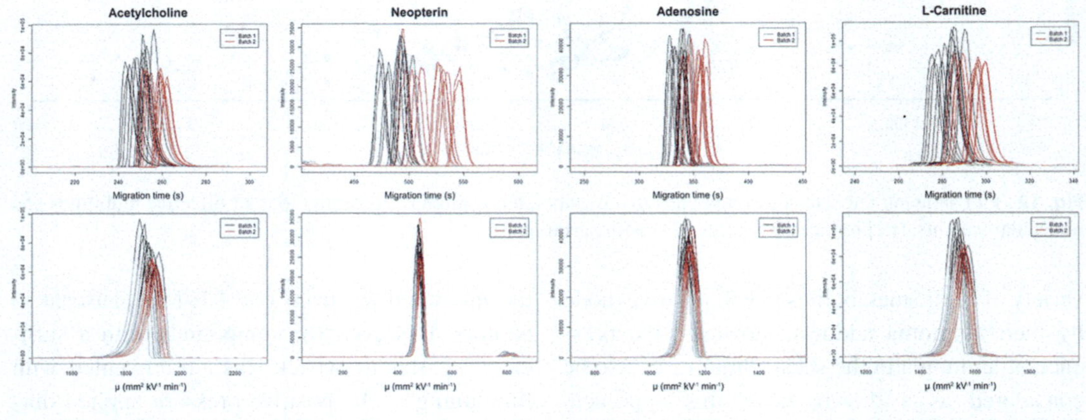

Fig. 13 Effect of the transformation into μ_{ep}-scale on a set of replicate analyses of a mix of standard compounds ($n = 12$), using two different batches of BGE ($j = 2$, red and black colours). The achieved pseudo-alignment is evident when top and bottom figures are compared. The upper electropherograms show the migration times of the compounds, while the lower electropherograms show the electrophoretic mobilities of the same compounds. (Reproduced from Ref. [85] with permission)

4 Conclusions and Perspectives

Over the past few decades, CE-MS has emerged as a strong analytical technique for the efficient profiling of polar and charged metabolites in the metabolomics field. Compared to chromatographic-based separation techniques, CE-MS provides complementary metabolic information. Recent developments in interfacing designs resulted in an increased interest in the CE-MS technique for metabolomics, as these interfaces provide a significantly improved metabolic coverage in comparison to the conventional interface. There are many more developments in interfacing designs, which have not been covered in this chapter as their utility for metabolomics still needs to be assessed. Concerning the use of novel interfacing designs, the long-term performance of these approaches for metabolomics studies needs to be demonstrated. One way to achieve this is to perform an inter-laboratory metabo-ring study using the (various) CE-MS approaches, and this is indeed a step that we have recently initiated. Next to such a study, standardized CE-MS-based analytical workflows need to be developed including data analysis procedures. The first steps into this direction have already been taken, for example, with the design of the open-access software ROMANCE for converting migration times into electrophoretic mobilities. Support from vendors would also be very helpful in this context.

We foresee the metabolomics study of the highly polar fraction of the metabolome and of volume- and biomass-limited samples as the main application areas for CE-MS, especially the approaches utilizing a new interfacing design. In this context, an important sweet spot for CE-MS will be metabolic profiling of primary cells, cell culture extracts from 3D microfluidic systems, spheroids and liquid biopsies. Chiral metabolic profiling of biological samples will be another key application field.

Acknowledgements Dr. Rawi Ramautar would like to acknowledge the financial support of the Vidi grant scheme of the Netherlands Organisation for Scientific Research (NWO Vidi 723.016.003).

References

1. Ramautar R et al (2013) Human metabolomics: strategies to understand biology. Curr Opin Chem Biol 17(5):841–846
2. Wishart DS et al (2018) HMDB 4.0: the human metabolome database for. Nucleic Acids Res 46(D1):D608–D617
3. Psychogios N et al (2011) The human serum metabolome. PLoS One 6(2):e16957
4. Miggiels P et al (2018) Novel technologies for metabolomics: more for less. TrAC Trends Anal Chem 120:115323
5. Kuehnbaum NL, Britz-McKibbin P (2013) New advances in separation science for metabolomics: resolving chemical diversity in a post-genomic era. Chem Rev 113(4):2437–2468
6. Kohler I, Giera M (2017) Recent advances in liquid-phase separations for clinical metabolomics. J Sep Sci 40(1):93–108
7. Gika HG et al (2014) Current practice of liquid chromatography-mass spectrometry in metabolomics and metabonomics. J Pharm Biomed Anal 87:12–25
8. Buscher JM et al (2009) Cross-platform comparison of methods for quantitative metabolomics of primary metabolism. Anal Chem 81(6):2135–2143
9. Tang DQ et al (2016) HILIC-MS for metabolomics: an attractive and complementary approach to RPLC-MS. Mass Spectrom Rev 35(5):574–600
10. Fukai K et al (2016) Metabolic profiling of total physical activity and sedentary behavior in community-dwelling men. PLoS One 11(10):e0164877
11. Harada S et al (2018) Reliability of plasma polar metabolite concentrations in a large-scale cohort study using capillary electrophoresis-mass spectrometry. PLoS One 13(1):e0191230
12. Soga T et al (2003) Quantitative metabolome analysis using capillary electrophoresis mass spectrometry. J Proteome Res 2(5):488–494
13. Kok MGM, Somsen GW, de Jong GJ (2014) The role of capillary electrophoresis in metabolic profiling studies employing multiple analytical techniques. TrAC Trends Anal Chem 61:223–235
14. Kok MG, Somsen GW, de Jong GJ (2015) Comparison of capillary electrophoresis-mass spectrometry and hydrophilic interaction chromatography-mass spectrometry for anionic metabolic profiling of urine. Talanta 132:1–7
15. Ramautar R (2017) Resolving volume-restricted metabolomics using Sheathless capillary electrophoresis-mass spectrometry. Lc Gc Europe 30(12):658–661
16. Ruta J et al (2010) A systematic investigation of the effect of sample diluent on peak shape in hydrophilic interaction liquid chromatography. J Chromatogr A 1217(52):8230–8240
17. Ramautar R et al (2011) Metabolic profiling of human urine by CE-MS using a positively charged capillary coating and comparison with UPLC-MS. Mol BioSyst 7(1):194–199

18. Naz S, Garcia A, Barbas C (2013) Multiplatform analytical methodology for metabolic fingerprinting of lung tissue. Anal Chem 85(22):10941–10948
19. Ibanez C et al (2012) CE/LC-MS multiplatform for broad metabolomic analysis of dietary polyphenols effect on colon cancer cells proliferation. Electrophoresis 33(15):2328–2336
20. Klein J et al (2014) Comparison of CE-MS/MS and LC-MS/MS sequencing demonstrates significant complementarity in natural peptide identification in human urine. Electrophoresis 35(7):1060–1064
21. Drouin N et al (2018) Effective mobility as a robust criterion for compound annotation and identification in metabolomics: toward a mobility-based library. Anal Chim Acta 1032:178–187
22. Ramautar R, Somsen GW, de Jong GJ (2009) CE-MS in metabolomics. Electrophoresis 30(1):276–291
23. Ramautar R et al (2011) CE-MS for metabolomics: developments and applications in the period 2008-2010. Electrophoresis 32(1):52–65
24. Ramautar R, Somsen GW, de Jong GJ (2013) CE-MS for metabolomics: developments and applications in the period 2010–2012. Electrophoresis 34(1):86–98
25. Ramautar R, Somsen GW, de Jong GJ (2015) CE-MS for metabolomics: developments and applications in the period 2012–2014. Electrophoresis 36(1):212–224
26. Ramautar R, Somsen GW, de Jong GJ (2017) CE-MS for metabolomics: developments and applications in the period 2014–2016. Electrophoresis 38(1):190–202
27. Ramautar R, Somsen GW, de Jong GJ (2019) CE-MS for metabolomics: developments and applications in the period 2016–2018. Electrophoresis 40(1):165–179
28. Ramautar R (2016) Capillary electrophoresis-mass spectrometry for clinical metabolomics. Adv Clin Chem 74:1–34
29. Jellum E, Thorsrud AK, Time E (1991) Capillary electrophoresis for diagnosis and studies of human disease, particularly metabolic disorders. J Chromatogr 559(1–2):455–465
30. Jellum E et al (1997) Diagnostic applications of chromatography and capillary electrophoresis. J Chromatogr B Biomed Sci Appl 689(1):155–164
31. Jellum E, Dollekamp H, Blessum C (1996) Capillary electrophoresis for clinical problem solving: analysis of urinary diagnostic metabolites and serum proteins. J Chromatogr B Biomed Appl 683(1):55–65
32. Barbas C et al (1998) Quantitative determination of short-chain organic acids in urine by capillary electrophoresis. Clin Chem 44(6 Pt 1):1340–1342
33. Mayboroda OA et al (2007) Amino acid profiling in urine by capillary zone electrophoresis – mass spectrometry. J Chromatogr A 1159(1–2):149–153
34. Soga T et al (2002) Simultaneous determination of anionic intermediates for Bacillus subtilis metabolic pathways by capillary electrophoresis electrospray ionization mass spectrometry. Anal Chem 74(10):2233–2239
35. Moreno-Gonzalez D et al (2013) Micellar electrokinetic chromatography-electrospray ionization mass spectrometry employing a volatile surfactant for the analysis of amino acids in human urine. Electrophoresis 34(18):2615–2622
36. Wu Q et al (2014) Pressurized CEC coupled with QTOF-MS for urinary metabolomics. Electrophoresis 35(17):2470–2478
37. Silvertand LH et al (2008) Recent developments in capillary isoelectric focusing. J Chromatogr A 1204(2):157–170
38. Huhn C et al (2010) Relevance and use of capillary coatings in capillary electrophoresis-mass spectrometry. Anal Bioanal Chem 396(1):297–314
39. Ramautar R (2018) Chapter 3 Capillary electrophoresis–mass spectrometry using non-covalently coated capillaries for metabolic profiling of biological samples. In Capillary electrophoresis–mass spectrometry for metabolomics 2018, The Royal Society of Chemistry, pp 53–65
40. Ramautar R (2018) Chapter 4 Capillary electrophoresis–mass spectrometry for metabolomics using new interfacing designs. In Capillary electrophoresis–mass spectrometry for metabolomics 2018, The Royal Society of Chemistry, pp 66–82
41. Zhao SS et al (2012) Capillary electrophoresis-mass spectrometry for analysis of complex samples. Proteomics 12(19–20):2991–3012
42. Hirayama A, Wakayama M, Soga T (2014) Metabolome analysis based on capillary electrophoresis-mass spectrometry. TrAC Trends Anal Chem 61:215–222
43. Drouin N, Rudaz S, Schappler J (2018) New supported liquid membrane for electromembrane extraction of polar basic endogenous metabolites. J Pharm Biomed Anal 159:53–59
44. Smith RD, Udseth HR (1988) Capillary zone electrophoresis-MS. Nature 331(6157):639–640
45. Klampfl CW (2008) Special issue: capillary electrophoresis-mass spectrometry. Electrophoresis 29(10):1955
46. Miksik I (2019) Coupling of CE-MS for protein and peptide analysis. J Sep Sci 42(1):385–397
47. Tycova A, Ledvina V, Kleparnik K (2017) Recent advances in CE-MS coupling: instrumentation, methodology, and applications. Electrophoresis 38(1):115–134
48. Bonvin G, Schappler J, Rudaz S (2012) Capillary electrophoresis-electrospray ionization-mass spectrometry interfaces: fundamental concepts and technical developments. J Chromatogr A 1267:17–31
49. Wenz C et al (2015) Interlaboratory study to evaluate the robustness of capillary electrophoresis-mass spectrometry for peptide mapping. J Sep Sci 38(18):3262–3270
50. Ramautar R et al (2008) Capillary electrophoresis-time of flight-mass spectrometry using noncovalently bilayer-coated capillaries for the analysis of amino acids in human urine. Electrophoresis 29(12):2714–2722
51. Chalcraft KR et al (2009) Virtual quantification of metabolites by capillary electrophoresis-electrospray ionization-mass spectrometry: predicting ionization

efficiency without chemical standards. Anal Chem 81(7):2506–2515
52. Baidoo EE et al (2008) Capillary electrophoresis-fourier transform ion cyclotron resonance mass spectrometry for the identification of cationic metabolites via a pH-mediated stacking-transient isotachophoretic method. Anal Chem 80(9):3112–3122
53. Ramautar R et al (2012) Enhancing the coverage of the urinary metabolome by Sheathless capillary electrophoresis-mass spectrometry. Anal Chem 84(2):885–892
54. Causon TJ et al (2014) Addition of reagents to the sheath liquid: a novel concept in capillary electrophoresis-mass spectrometry. J Chromatogr A 1343:182–187
55. Bonvin G, Rudaz S, Schappler J (2014) In-spray supercharging of intact proteins by capillary electrophoresis-electrospray ionization-mass spectrometry using sheath liquid interface. Anal Chim Acta 813:97–105
56. Lindenburg PW et al (2015) Developments in interfacing designs for CE-MS: towards enabling tools for proteomics and metabolomics. Chromatographia 78(5–6):367–377
57. Maxwell EJ et al (2010) Decoupling CE and ESI for a more robust interface with MS. Electrophoresis 31(7):1130–1137
58. Lindenburg PW et al (2014) Capillary electrophoresis-mass spectrometry using a flow-through microvial interface for cationic metabolome analysis. Electrophoresis 35(9):1308–1314
59. Busnel JM et al (2010) High capacity capillary electrophoresis-electrospray ionization mass spectrometry: coupling a porous Sheathless Interface with transient-Isotachophoresis. Anal Chem 82(22):9476–9483
60. Klampfl CW (2009) CE with MS detection: a rapidly developing hyphenated technique. Electrophoresis 30(Suppl 1):S83–S91
61. Simpson DC, Smith RD (2005) Combining capillary electrophoresis with mass spectrometry for applications in proteomics. Electrophoresis 26(7–8):1291–1305
62. Hirayama A et al (2018) Development of a sheathless CE-ESI-MS interface. Electrophoresis 39(11):1382–1389
63. Moini M (2007) Simplifying CE-MS operation. 2. Interfacing low-flow separation techniques to mass spectrometry using a porous tip. Anal Chem 79(11):4241–4246
64. Zhang W et al (2019) Utility of sheathless capillary electrophoresis-mass spectrometry for metabolic profiling of limited sample amounts. J Chromatogr B Analyt Technol Biomed Life Sci 1105:10–14
65. Liu JX et al (2014) Analysis of endogenous nucleotides by single cell capillary electrophoresis-mass spectrometry. Analyst 139(22):5835–5842
66. Genchi G (2017) An overview on D-amino acids. Amino Acids 49(9):1521–1533
67. Sánchez-López E et al (2016) Enantioseparation of the constituents involved in the phenylalanine-tyrosine metabolic pathway by capillary electrophoresis tandem mass spectrometry. J Chromatogr A 1467:372–382
68. Lee S et al (2019) Chiral separation of intact amino acids by capillary electrophoresis-mass spectrometry employing a partial filling technique with a crown ether carboxylic acid. J Chromatogr A 1586:128–138
69. Geiser L, Rudaz S, Veuthey J-L (2005) Decreasing analysis time in capillary electrophoresis: validation and comparison of quantitative performances in several approaches. Electrophoresis 26(12):2293–2302
70. Geiser L, Rudaz S, Veuthey J-L (2003) Validation of capillary electrophoresis – mass spectrometry methods for the analysis of a pharmaceutical formulation. Electrophoresis 24(17):3049–3056
71. Kuehnbaum NL, Kormendi A, Britz-McKibbin P (2013) Multisegment injection-capillary electrophoresis-mass spectrometry: a high-throughput platform for metabolomics with high data fidelity. Anal Chem 85(22):10664–10669
72. DiBattista A et al (2017) High throughput screening method for systematic surveillance of drugs of abuse by multisegment injection-capillary electrophoresis-mass spectrometry. Anal Chem 89(21):11853–11861
73. DiBattista A et al (2017) Temporal signal pattern recognition in mass spectrometry: a method for rapid identification and accurate quantification of biomarkers for inborn errors of metabolism with quality assurance. Anal Chem 89(15):8112–8121
74. Azab S, Ly R, Britz-McKibbin P (2019) Robust method for high-throughput screening of fatty acids by multisegment injection-nonaqueous capillary electrophoresis-mass spectrometry with stringent quality control. Anal Chem 91(3):2329–2336
75. Ouyang Y et al (2018) Negative-ion mode capillary isoelectric focusing mass spectrometry for charge-based separation of acidic oligosaccharides. Anal Chem 91(1):846–853
76. Portero EP, Nemes P (2019) Dual cationic-anionic profiling of metabolites in a single identified cell in a live Xenopus laevis embryo by microprobe CE-ESI-MS. Analyst 144(3):892–900
77. Sánchez-López E et al (2019) Sheathless CE-MS based metabolic profiling of kidney tissue section samples from a mouse model of Polycystic Kidney Disease. Sci Rep 9(1):806
78. Rochat B (2017) Proposed confidence scale and ID score in the identification of known-unknown compounds using high resolution MS data. J Am Soc Mass Spectrom 28(4):709–723
79. Sumner LW et al (2007) Proposed minimum reporting standards for chemical analysis Chemical Analysis Working Group (CAWG) Metabolomics Standards Initiative (MSI). Metabolomics 3(3):211–221
80. van Rijswijk M et al (2017) The future of metabolomics in ELIXIR [version 2; peer review: 3 approved]. F1000 Research 2017, 6(ELIXIR):1649. https://doi.org/10.12688/f1000research.12342.2

81. Guijas C et al (2018) METLIN: a technology platform for identifying knowns and unknowns. Anal Chem 90(5):3156–3164
82. Gonzalez-Pena D et al (2017) Metabolomic fingerprinting in the comprehensive study of liver changes associated with onion supplementation in Hypercholesterolemic Wistar rats. Int J Mol Sci 18(2):267. https://doi.org/10.3390/ijms18020267
83. Macedo AN et al (2017) The sweat metabolome of screen-positive cystic fibrosis infants: revealing mechanisms beyond impaired chloride transport. ACS Cent Sci 3(8):904–913
84. Gulersonmez MC et al (2016) Sheathless capillary electrophoresis-mass spectrometry for anionic metabolic profiling. Electrophoresis 37(7–8):1007–1014
85. Gonzalez-Ruiz V et al (2018) ROMANCE: a new software tool to improve data robustness and feature identification in CE-MS metabolomics. Electrophoresis 39(9–10):1222–1232
86. Gagnebin Y et al (2019) Toward a better understanding of chronic kidney disease with complementary chromatographic methods hyphenated with mass spectrometry for improved polar metabolome coverage. J Chromatogr B 1116:9–18
87. Gonzalez-Ruiz V et al (2017) Unravelling the effects of multiple experimental factors in metabolomics, analysis of human neural cells with hydrophilic interaction liquid chromatography hyphenated to high resolution mass spectrometry. J Chromatogr A 1527:53–60
88. DiBattista A et al (2019) Metabolic signatures of cystic fibrosis identified in dried blood spots for newborn screening without carrier identification. J Proteome Res 18(3):841–854

Liquid Chromatography-Mass Spectrometry for Clinical Metabolomics: An Overview

Izadora L. Furlani, Estéfane da Cruz Nunes, Gisele A. B. Canuto, Adriana N. Macedo, and Regina V. Oliveira

Abstract

Metabolomics is a discipline that offers a comprehensive analysis of metabolites in biological samples. In the last decades, the notable evolution in liquid chromatography and mass spectrometry technologies has driven an exponential progress in LC-MS-based metabolomics. Targeted and untargeted metabolomics strategies are important tools in health and medical science, especially in the study of disease-related biomarkers, drug discovery and development, toxicology, diet, physical exercise, and precision medicine. Clinical and biological problems can now be understood in terms of metabolic phenotyping. This overview highlights the current approaches to LC-MS-based metabolomics analysis and its applications in the clinical research.

I. L. Furlani · R. V. Oliveira (✉)
Núcleo de Pesquisa em Cromatografia (Separare), Department of Chemistry, Federal University of São Carlos, São Carlos, SP, Brazil
e-mail: oliveirarv@ufscar.br

E. da Cruz Nunes · G. A. B. Canuto
Department of Analytical Chemistry, Institute of Chemistry, Federal University of Bahia, Salvador, BA, Brazil

A. N. Macedo
Department of Chemistry, Federal University of Minas Gerais, Belo Horizonte, MG, Brazil

Keywords

Metabolomics · Mass analyzers · Sample preparation · Biomarkers · LC-MS · Clinical research

Abbreviations

ACN	Acetonitrile
CE	Capillary electrophoresis
DDA	Data-dependent acquisition
DIA	Data-independent acquisition
DIPEA	Diisopropylethylamine
DoE	Design of experiments
EIC	Extracted ion chromatogram
ESI	Electrospray
FA	Formic acid
FT-ICR	Fourier-transform ion cyclotron resonance
GC	Gas chromatography
HCD	Higher-energy collisional dissociation
$HDMS^E$	Ion mobility assisted-DIS
HFIP	1,1,1,3,3,3-hexafluoro-2-propanol
HILIC	Hydrophilic interaction liquid chromatography
HPLC	High-performance liquid chromatography
HRMS	High-resolution mass spectrometry
IC	Ionic chromatography

A. V. Colnaghi Simionato (ed.), *Separation Techniques Applied to Omics Sciences*, Advances in Experimental Medicine and Biology 1336, https://doi.org/10.1007/978-3-030-77252-9_10

ICP-MS	Inductively coupled plasma-mass spectrometry
IPC	Ion pair chromatography
LC	Liquid chromatography
LIT	Linear ion trap
LLE	Liquid-liquid extraction
LPME	Liquid-phase microextraction
LTQ-Orbitrap	Linear trap quadrupole-Orbitrap
m/z	Mass-to-charge ratio
MFA	Metabolic flux analysis
MIPs	Molecularly imprinted polymers
MP	Mobile phase
MS	Mass spectrometry
MS/MS	Tandem mass spectrometry
MS^3	Multistage fragmentation
MS^E	Elevated energy MS
MTBE	Methyl tert-butyl ether
NCEs	New chemical entities
NMR	Nuclear resonance magnetic
NP	Normal phase
OFAT	One-factor-at-a-time
OT	Orbitrap
PCA	Principal component analysis
PRM	Parallel reaction monitoring
QqQ	Triple quadrupole
QqTOF	Hybrid quadrupole time of flight
QTRAP	Quadrupole-ion trap
re-TOF	Ion reflector
RPLC	Reverse phase liquid chromatography
SALLE	Salting-out assisted liquid-liquid extraction
SBSE	Stir-bar sorptive extraction
SLE	Supported liquid extraction
SP	Stationary phase
SPE	Solid-phase extraction
SPME	Solid-phase microextraction
SRM	Selected reaction monitoring
SWATH-MS	Sequential Windowed Acquisition of All Theoretical Fragment Ion Mass Spectra
TBA	Tributylamine
TMA	Trimethylamine
TMAO	Trimethylamine N-oxide
TOF	Time of flight
UHPLC	Ultra-high-performance liquid chromatography
UPLC	Ultra-performance liquid chromatography

1 Introduction

In recent years, an increasing number of studies have been devoted to "omics" strategies, which mainly seek a great understanding of the cellular mechanism and their biological interaction and contributions. "Omic" technologies include studies of genes (genomics), mRNA (transcriptomics), proteins (proteomics), and metabolites (metabolomics) [1]. Metabolomics is the newest emerging field of "omics" research, with the aim of providing comprehensive identification and quantitation of small molecules, with typically low molecular weight from *m/z* 50–1500 Da, in biological samples, such as tissues, cells, and human or animal biological fluids [2, 3].

Since metabolomics refers to the study of endogenous metabolites, end products of cellular processes, the assessment of global metabolomics can reveal the metabolic profile of the metabolites and a complex network of changes in a biological system [4, 5]. In addition, metabolomics also benefits from the evaluation of exogenous metabolites, derived from environmental pollution, drug intake, food additives, toxins, and other xenobiotics, providing an increase in the complexity of the metabolome [2, 4]. In this context, metabolomics has become a powerful approach and it has been applied in multiple fields, such as natural products, food science and nutrition, environmental health and wellness, clinical, personalized medicine, and drug discovery, among others [6–9]. Studies describing the identification of metabolites have been reported for many years; however, it is important to mention that metabolomics rapidly expanded and has become a great resource for system biology and mechanistic insights into metabolic regulation [10].

Innovations in analytical platforms, including those hardware-based for mass spectrometers (MS) and nuclear magnetic resonance equipment (NMR), separation techniques, and statistical method for data analysis, have made possible the comprehensive profiling of metabolites. In addition, deconvolution, used to separate overlapping peaks based on mass spectral differences, offers a great advantage to process a large data set of raw mass spectral data, enabling as much information as possible about the metabolome [11]. These factors contributed to the advances in the field of metabolomics, which is reflected in the growing number of publications dedicated to this topic. Considering clinical metabolomics in the last decade, an exponential growth is evident, as shown in Fig. 1. According to the Web of Science database, approximately 5868 scientific articles were published between 2010 and 2020, *and the last 5 years (2016–2020) represented about 68% of this total*.

In the clinical approach, there is no doubt that the main application of metabolomics is in the discovery of disease-related biomarkers, which are altered metabolites when compared to control samples [12, 13]. These biomarkers can assist in clinical diagnosis, stratification of patients by disease type and/or response to treatment, selection of an adequate therapy response, and also improving monitoring disease progression. In drug discovery, biomarkers are critical to the rational development and discovery of new drugs as well as for target engagement. Moreover, in drug toxicology, metabolomics can be used in preclinical trials to verify the toxicity and effectiveness of new chemical entities (NCEs) [11].

Another important contribution of metabolomics is related to the study of cases of individuals with inborn error of metabolism (IEM) or other metabolic diseases [8, 9]. Clinical metabolomics has also been proving to be a useful and robust complement to genetic testing as an alternative for interpretation of genetic impacts on disease phenotypes [14, 15].

A map based on network data for the main diseases investigated by clinical metabolomics approaches was created using data extracted from the Web of Science database and VOSviewer [16] as a bibliometric tool (Fig. 2). In this context, different diseases have been benefited from metabolomics studies, where great efforts are concentrated in the area of oncology, including breast cancer, colorectal cancer, and prostate cancer, followed by research on inflammation and obesity. In addition, chronic diseases such as kidney disease, type 2 diabetes, and diabetes mellitus have shown an increasing number of cases over the years. Clinical metabolomics has also brought benefits to studies on Alzheimer's disease [17], the most common form of dementia, according to the World Health Organization [18], followed by other mental illness, such as schizo-

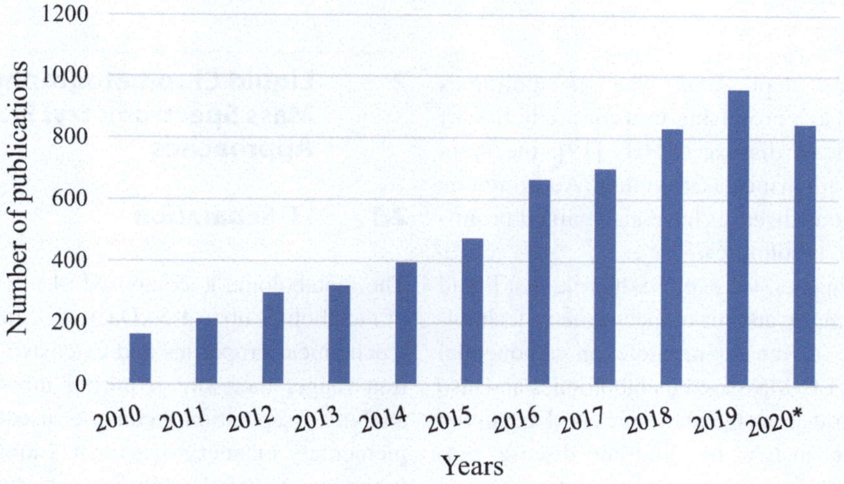

Fig. 1 Number of publications in the period 2010–2020* (up to October 27, 2020) on Web of Science database using ((*metabolom* OR *metabonom*) AND *clinical*) as keywords

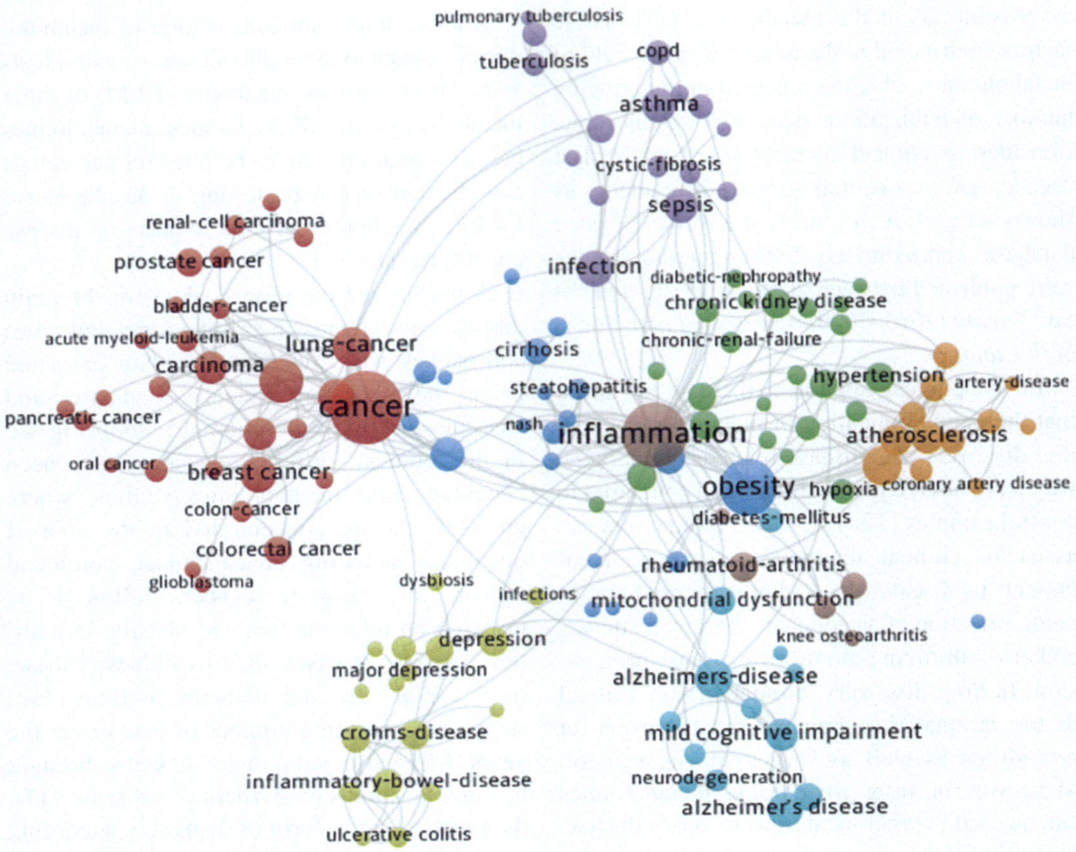

Fig. 2 Network-based data on diseases evaluated in clinical metabolomics. Minimum of 10 co-occurrences of all keywords present in titles and abstracts of articles published between 2010 and 2020* (up to October 27, 2020). (Data extracted from Web of Science using the search for the items ((*metabolom* OR *metabonom*) AND *clinical*). VOSviewer version 1.6.15)

phrenia and depression. The metabolomics profile is also a promising tool for prediction of coronary heart disease (CHD) [19], the main cause of death in men as in women. Autoimmune and infectious diseases have also gained prominence in metabolomics.

In this chapter, we explore the role that liquid chromatography and mass spectrometry technologies have driven to promote an exponential progress in LC-MS-based metabolomics and also to demonstrate how metabolomics is also gaining mechanistic insight of different disease processes. Therefore, the current approaches to metabolomics analysis based on LC-MS and their applications in the clinical research are highlighted.

2 Liquid Chromatography-Mass Spectrometry: Practical Approaches

2.1 LC Separation

The metabolome is composed of a wide variety of metabolites up to 1.5 kDa, with different physicochemical properties and extensive concentration range, therefore requiring more than one analytical separation technique, used in a complementary manner, to cover it. Liquid chromatography coupled with mass spectrometry (LC-MS) has been one the most used platform for metabolomics studies [20], mainly due to its high resolution, sensitivity, good repeatability, and robustness. The coupling between LC and MS is important for metabolomics analysis, since the previous separation of the metabolites reduces ion suppression, caused by the complexity of the biological matrices, increasing the sensibility [21]. Electrospray (ESI) ionization sources have been widely used in metabolomics studies because it produces intact molecular ions or few fragments, facilitating metabolite's identification [22]. Different mass analyzers are coupled with the LC system, depending on the chosen metabolomics approach, as discussed in the next section (Sect. 4).

The mechanism of separation in LC is based on the molecular characteristics and differential equilibrium of the solute between a stationary phase (SP) and a mobile phase (MP), according to polarity, hydrophobicity, electrical charge, and molecular size [23]. Typically, LC analyses for metabolomics assays are performed using an elution gradient to improve the resolution between analytes, especially in complex mixtures, such as biological samples, allowing shorter retention time for the later-eluting components so that faster separations and narrower chromatographic bands are achieved, resulting in more efficient separations. In the last decades, newer technology in LC column packings evolved with the development of smaller particle sizes, either totally porous (sub-3μm and sub-2μm) or fused-core particles, which offer higher throughput, higher resolution, and higher efficiency. Furthermore, ultra-high-performance liquid chromatography (UHPLC) instruments, capable of withstanding pressures greater than 10,000 psi, have demonstrated higher resolution and speed than the conventional high-pressure liquid chromatography (HPLC) [24, 25]. Nevertheless, both UHPLC and HPLC systems are frequently applied in metabolomics studies. In recent years, LC-MS has emerged as one of the most versatile analytical platforms for metabolomics studies [26], since it allows different modes of separation using the same instrument. This is achieved through the use of chromatographic columns with distinct SP and elution modes. Within the most applied modes are the reverse phase liquid chromatography (RPLC) for nonpolar and moderately polar compounds, and hydrophilic interaction liquid chromatography (HILIC), for highly polar metabolites. The ionic and polar fraction of the metabolome has been further investigated through the application of ion pair chromatography (IPC) [27, 28] and less often by ionic chromatography (IC) [29]. IPC and IC elution modes are discussed in more detail in Chap. 11.

RPLC mechanism of separation is based on the partition of the compounds between a nonpolar stationary phase and a polar mobile phase. The RPLC columns are typically composed of silica particle support chemically bonded with octadecylsilane (C18) or octylsilane (C8) carbon chains and eluted with MP composed of mixtures of polar (aqueous) and nonpolar (organic) solvents, e.g., methanol, tetrahydrofuran, and acetonitrile [20, 23]. The longer the linked chain, the more hydrophobic the stationary phase will be. Gradient elution is frequently used, especially for complex matrices, in which the MP starts with a high percentage of the aqueous phase, and the gradient strength increases with the increment of the organic phase along the chromatographic run. Differential retention is achieved according to the hydrophobicity of the analyte, in which more hydrophobic species are strongly retained [23]. Metabolites belonging to the class of organic and fatty acids, lipids and their derivatives (phosphocholines, sphingomyelins, glycerophospholipids, sterol lipids, etc.), sugars, steroids, etc. are frequently assessed by RPLC-MS. Although RPLC has been widely used in targeted and untargeted metabolomics assays, one of the major limitations of this separation technique is the low retention of ionic and highly polar compounds that elute near the void volume of the column, making accurate detection and identification difficult. Thus, HILIC emerged as a similar separation mode to normal phase (NP), for the resolution of polar analytes, and complementary to RPLC analyses [12, 24].

The application of HILIC-MS in the metabolomics field has been growing in recent years because the technique demonstrates many advantages over RPLC, such as increased analyte partition in the organic mobile phase, reduced backpressure due to low viscosity of the mobile phase, enhanced MS sensitivity due to better solvent desolvation, and better compatibility with organic extracts obtained from liquid-liquid extraction or protein precipitation [12, 20, 26]. Amino acids, amines, lipid derivatives, organic acids, nucleotides, and nucleosides are common classes of metabolites comprising critical biological pathways that are well detected by HILIC-MS [26]. Different silica supported-stationary phases have been developed for HILIC, like neutral, anionic, cationic, or zwitterionic compounds [13]. The separation occurs with hydrophobic mobile phases composed of aqueous (5–40% of water) and high contents of aprotic polar organic solvent mixtures (e.g., acetonitrile, tetrahydrofuran, and/or dioxane). HILIC elution mode is performed in isocratic mode either with a high and constant percentage of organic solvent or with gradients starting with high contents of organic solvents, followed by polarity increment, and finishing with a high percentage of the aqueous phase. Further information about the mechanism of separation, nature of the SP, and MP compositions for HILIC-based metabolomics studies is also covered in Chap. 11.

Some limitations of HILIC separation are related to poor retention times, batch-to-batch repeatability, and MS signal drift for long-term sample analysis. Otherwise, separations performed by RPLC are more reproducible, stable, and efficient, with less time needed for column equilibration [23, 26]. Despite these drawbacks, HILIC as a separation mode has been widely used in metabolomics studies. In general, HILIC analysis is performed to improve metabolome coverage to understand pathogenesis infection [30], to study cancer therapy [31], and to search biomarkers [32], among others. In the context of complementarity, Naser et al. [33] used only a single C18 column, and verified the inability to retain semipolar metabolites (logP range −2.0 to 1.5). Then, after testing four different columns, the authors optimized two orthogonal RPLC methods using a CORTECS T3 and a CORTECS C8 column, for analysis of semipolar compounds and lipids, respectively. Moreover, RPLC and HILIC (Luna NH_2, Phenomenex) methods have provided a good coverage of the metabolome without increasing the analysis time. The methods were developed using a mix of 47 analytical standards and applied to *E. coli* cells [33]. Some works have also integrated HILIC and RPLC in a multidimensional approach to improve metabolome coverage in a single run [34] such as the analysis of pharmaceutical formulations [35] and phenols in wine [36]. Another application merged metabolomics and lipidomics in a single method by using dual HILIC-RPLC columns [37]. The incompatibility between solvents in a multidimensional approach is known; however, to circumvent this problem, a new interface valve based on active solvent modulation (ASM) was developed. In this approach, it is possible to dilute the first column eluent with a weak solvent before transferring it to the second column [38]. Song et al. [39], who previously developed and validated a serial RPLC-HILIC system for targeted analysis [40], applied the serial separation system to a multiple reaction monitoring (MRM) system, called tailored MRM strategy, that was able to detect 164 metabolites of different classes in a mimic sample of medicinal bile [39].

2.2 Optimization in LC

2.2.1 Design of Experiments (DoE)

Despite the advancement in LC instrumentation and the development of efficient and robust chromatographic columns, there are no standard protocols for LC-MS metabolomics analysis, and this has been a topic of interest in recent years. In general, targeted methods present adequate analytical developments and validation for the quantification of selected metabolites [41]. On the other hand, there is a great challenge to detect unspecified metabolites in untargeted metabolomics studies, making it necessary to optimize comprehensive methods. The most common approach used is the one-factor-at-a-time (OFAT) optimization, in which each factor is varied individually, while the others are kept fixed [42]. A systematic optimization for untargeted metabolomics using HILIC and RPLC separations was conducted by Contrepois et al. [43], in order to find the best column to cover the metabolome profile of plasma and urine. Different stationary phases and mobile phases, combined with solvent extraction optimization, were univariately evaluated for metabolomics analysis in *E. coli* bacteria. A Imtakt Scherzo SM-C18 column (150 mm × 2.1 mm, i.d.; 3μm) proved to be useful for simultaneous polar and nonpolar metabolite separations, avoiding the use of HILIC and IPC, which can cause broader peak shapes and reagent incompatibility with MS, respectively. The mobile phase, composed of 1.0 mmol L^{-1} ammonium fluoride, showed good efficiency in ESI(−) ionization and reproducible results, enabling the identification of essential bacterial metabolic pathways [44]. In order to optimize the composition of MP for polar and nonpolar metabolites, Creydt and Fischer [45] investigated 17 additives for ESI(+) and 23 compositions for ESI(−), varying the pH between 3.5 and 6.5. Ammonium formate as a modifier (pH 6.5) proved to be more suitable for analysis in a positive ionization mode, whereas for ESI(−), acetic acid and formic acid (FA) showed better results for nonpolar and polar metabolites, respectively. After evaluating the influence of pH on the metabolite ionization and types of formed adducts, the authors concluded that ideally, an untargeted method should combine different MP compositions in order to improve metabolite coverage and detection. Therefore, this approach would make possible to detect polar and nonpolar metabolites, in both negative and positive ionization modes, resulting in a comprehensive metabolite response and improved performance of the analysis [45]. Despite being efficient, OFAT is a univariate approach whose response does not reflect the synergist and antagonist effects among the variables, which can lead to incorrect interpretations, in addition to being time-consuming and tedious [46, 47].

An interesting optimization strategy that has emerged is based on the design of experiments (DoE). DoE considers the impact of all evaluated factors on the final response, with a minimal number of experiments. In this way, the interpretation of the analytical parameters is facilitated, leading to a better estimation of the ideal condition of analysis [42]. Recently, a review highlighted the potential of DoE when applied to metabolomics and how this strategy could become an attractive choice to optimize sample preparation and instrumental method development [48].

Rhoades and Weljie [49] used the comprehensive optimization of LC-MS metabolomics methods using DoE (named COLMeD) to find the best conditions for polar metabolite analysis. Methods were optimized for acylcarnitine analysis by QqQ and untargeted metabolomics using a QTOF instrument. In this work, the authors systematically evaluated different gradient ramps and applied the full factorial design in three levels to set flow rate, gradient type, and column temperature. MS conditions, including cone voltage, desolvation and source temperature, and cone gas flow, were also optimized. According to the authors, the COLMeD method is an efficient approach to improve polar metabolite coverage and useful for improving optimizations from established LC-MS methods. Another example of application of DoE in an untargeted metabolome assay reports a rational and sequential optimization of organic mobile phase using modified central composition design and a 2^3 factorial design with central composition. This study was performed to improve metabolite resolution in MP composed of (A) water and (B) acetonitrile/methanol 68:32 v/v, both acidulated with 0.1% acetic acid, showing better metabolite separation. The negative effect of high temperatures was associated to changes in elution strength by decreasing viscosity, and the flow rate was related with changes in mass transfer properties. The injection volume inferred a positive effect related to decreased metabolite identification. Thus, column temperature, injection volume, and MP flow rate of 30 °C, 30μL, and 2.0 mL min^{-1}, respectively, were optimized for separation conditions [50].

In addition to the optimization of separation parameters, DoE has also been applied in the investigation of LC-MS metabolomics data processing [51] and the optimization of ionization source parameters of MS and funnel technology (ESI) with Jet Stream from Agilent Technologies [52]. MS parameters were systematically evaluated by a fractional design and varying LC flow rate and ESI conditions (source and desolvation temperature, sample cone voltage, desolvation gas, cone gas flow, and extraction voltage). This optimization was followed by a two-level full factorial design to set the ion mobility parameters in order to perform a reproducible and robust response for untargeted metabolomics [53]. The application of DoE for systematic studies of analytical methods development is very promising, but its use is still limited in the metabolomics field and should be encouraged. The choice of the design must be performed based on the study aim, and the factors that influence the process must be carefully planned [48].

3 Sample Preparation Strategies for LC-MS-Based Metabolomics

Metabolomics sample preparation is one of the most critical steps in the workflow and has been frequently revised [54–57]. Some pre-analytical factors must be taken into account. Rapid collection, transport, and storage of the biological samples at low temperatures are crucial for maintaining metabolome's integrity. Freeze and thaw cycles should be avoided; therefore, biofluids should be divided into aliquots and stored in separate freezers to minimize losses [58]. The extraction protocol depends on the biological sample, the analytical technique, and the metabolomics approach.

Ideally, sample preparation for untargeted metabolomics should be simple, rapid, and reproducible and requires minimum handling to avoid losses and degradation of the metabolites. Untargeted metabolomics methods should be as nonselective as possible, using procedures that do not introduce bias for specific classes of metabolites. In this sense, metabolome extraction involves protein precipitation, dilution, and cell/tissue lysis [59]. Protein removal is performed by precipitation using cold organic solvents acetonitrile, methanol, isopropanol, and acetone, and mixtures are often used [56]. In general, these methods are followed by centrifugation or ultrafiltration. Ultrafiltration is a preparation procedure in which the sample is filtered through a 1000–3000 Da filter to separate small metabolites from macromolecules (including the proteins) [55]. There is no consensus about the best method or better solvent for protein precipitation. For this reason, the evaluation of different methods, especially for plasma [60, 61] and serum samples [62, 63], is found important.

For urine preparation, which is a protein-free biofluid, simple centrifugation followed by dilution with water or buffer seems to be enough. However, it is important to emphasize the need for compatibility of the dilution solvent with the employed analytical system [64]. Cell and tissue samples require homogenization and lysis for metabolite extraction and protein precipitation, both performed in one step. Intracellular metabolomics requires culture medium removal by centrifugation or vacuum aspiration. Additionally, cell and tissue samples are frequently washed with cold phosphate-buffered saline (PBS) solution or deionized water to remove residues from extracellular medium or blood, respectively [65]. Furthermore, metabolic quenching is mandatory and must be fast in order to stop enzymatic activity [59]. Several methods have been applied for quenching, such as keeping samples at low temperatures, frozen in liquid nitrogen, and addition of organic solvents, acids, or bases. Special care must be taken when adding solvents to cell samples during quenching, to avoid rupture of the cellular membrane and, consequently, leaching the metabolites [66]. Metabolite extraction from tissue and cells is typically performed by adding the solvent and applying mechanical apparatus, including ball grinding, vortex mixing, and sonication [54, 66].

Liquid-liquid extraction (LLE) procedures are useful for polar and nonpolar metabolite extraction and have been used for sample cleanup and lipid removal. In general, the procedures use methanol, water, and chloroform in binary or ternary mixtures [54]. Yin and Xu [59] point out that the choice of the extracting solvent is associated with the class of desirable metabolites. Preferably, the extractant should not restrict classes of metabolites for untargeted analyses and must be compatible with the analytical technique to avoid additional solvent evaporation and resuspension steps. Some researchers have been suggesting substitution of chloroform by methyl tert-butyl ether (MTBE) [56], since MTBE has shown good recovery of metabolites against chloroform, with the advantage of lower toxicity [67]. An optimization of in-vial dual extraction was performed by Whiley et al. [68] to analyze polar and nonpolar metabolites. The extraction was conducted by adding MTBE to a methanol-water mixture after protein precipitation in plasma samples. This one-step LLE extraction was able to detect over 4500 molecular features by directly analyzing the organic and aqueous phase by LC-MS using two different LC methods. An advantage of this procedure is that the precipitated protein is deposited at the bottom of the vial. Otherwise, the solid is located between phases when using chloroform, making it difficult to recover the extract.

As discussed above, sample preparation is an important step in metabolomics workflow, and the choice of the method directly affects the data quality [55]. For this reason, it is necessary to optimize sample preparation protocols. An interesting strategy is based on DoE, which was previously described herein [48]. In this context, a mixture design based on triangle solvent was systematically employed to optimize a multiplatform metabolite extraction (for RPLC-MS and HILIC-MS, both ESI(+) and ESI(−), and also for CE-MS) using pooled tissue samples (*Fasciola hepatica* parasites) [69]. The sample preparation for fecal samples using different analytical platforms, including HILIC-MS, was also optimized by various combinations of organic solvents, pH, and sample weight/extractant volume ratio [70].

Solid-phase extraction (SPE) is another sample preparation technique applied to metabolomics studies and has been used to remove interferences and promote metabolite concentration. SPE mechanism is similar to liquid chromatography, in which the analytes interact differently with a liquid mobile phase and sorbent materials [71]. The SPE applied to the preparation of urine samples has aided in the removal of salts, reducing the effects of ion suppression, commonly observed in LC-MS methods [72]. Despite being popular in metabolomics, SPE is more suitable for targeted analyses since its extraction sorbents (such as RPLC-based, amino or phenol groups, and ion-exchange materials) are highly selective [56, 73]. SPE has also been applied using online configuration (online SPE-LC). Online SPE-LC systems have advantages such as miniaturization (microliter-scale injections), ease of automation, and reduced manual steps, which are more prone to errors. The online SPE-LC system also allows good selectivity with the combination of different chromatographic columns (e.g., RPLC and HILIC) and adequate robustness [74]. In this sense, a dual-cartridge SPE-RPLC-MS/MS method for human serum was optimized by Calderón-Santiago et al. [75]. Different sorbents, based on C8 and C18 phases, resin phases (GP, polymeric polydivinylbenzene, and SH, strong hydrophobic-modified polystyrene-divinylbenzene), and mixed-mode ionic phases (anionic and cationic), were evaluated. After individual analysis of the SPE phases, the authors concluded that the capacity to retain metabolites would have been increased using two combined cartridges, unless the chemical nature and physicochemical properties of them. Therefore, the best extractions were obtained by coupling a mixed-mode cationic phase with a C18 sorbent, when using ESI(+), and a SH resin in ESI(−). Combining these chemically different phases allowed an efficient extraction from polar to nonpolar metabolites, with high sensitivity and selectivity [75].

Still regarding sample treatment for biofluids, there are other techniques derived from conventional extraction methods (LLE and SPE) that are also available, such as salting-out, supported liquid extraction (SLE), microextraction, molecularly imprinted polymers (MIPs), and stir-bar sorptive extraction (SBSE), among others [71]. Salting-out assisted liquid-liquid extraction (SALLE) combines the advantages of LLE and conventional protein precipitation. The SALLE procedure includes the addition of an organic solvent (such as ACN) followed by a high concentrated salt solution (usually alkali or alkaline earth metal chlorides, citrates, or acetates). The separation of the organic and aqueous phases allows good extraction with fewer interferents than the protein precipitation procedure. The extracts can be injected directly into the LC-MS system without solvent evaporation or reconstitution [76]. Another alternative to LLE is the SLE, in which porous sorbents, such as modified diatomaceous earth, are employed to extract polar analytes from biological samples. The SLE methods provide good recovery and the absence of emulsion, unlike SALLE. Besides, as it is a flow-through technique, it is easier to be automated than LLE. When properly optimized, the SLE is useful for removing phospholipids. Microextraction techniques, such as liquid-phase microextraction (LPME) and solid-phase microextraction (SPME), have also been developed as alternatives to LLE and SPE, respectively. These techniques have arisen to reduce the amount of toxic solvents used by LLE and the high costs associated with SPE cartridges. The LPME is the miniaturized form of LLE, in which microliters of solvents are used, providing excellent concentration and cleanup of the target analytes. The SPME is a technique established in 1989, in which a fiber (polymeric coatings) or coated capillary tubes are used as a stationary phase. After stirring to increase the extraction process, the analytes are removed from the fiber by heating and then injected into the separation system. Although it was initially developed to analyze volatile compounds by GC-MS, the SPME has been interfaced with LC-based methods [77, 78]. Overall, these advanced methodologies are more selective and focus on analytes or classes of analytes, constituting interesting approaches for applications in targeted metabolomics. Further details on these extraction techniques can be found in specialized literature [71, 79].

3.1 Immunocapture for Targeted and Semi-targeted Metabolomics: New Perspectives

Despite their usefulness, metabolomics approaches still present multiple analytical challenges. In particular, the samples are very complex, containing a diverse set of compounds in a wide range of polarity. In addition to the separation and detection step, the extraction of metabolites during the sample preparation prior to analysis by LC-MS is of fundamental importance. Although different extractions protocols are applied to the same fractionated biological sample to expand the coverage of the analyzed metabolome for polar and nonpolar metabolites, it is not a straightforward task. Therefore, the main disadvantage of this strategy is the time spent for sample preparation plus the analysis time required for samples obtained through different extraction protocols. To overcome this issue, researchers are always looking for new and better alternatives to improve the efficiency of the sample preparation step. Typically, metabolomics studies employ sample preparation techniques that are based on liquid-liquid extraction, protein precipitation, and, in some cases, solid-phase extraction, as discussed earlier.

Proteomic analysis faces similar challenges, where relevant biomarkers are often constituted of low concentration proteins [80]. In order to overcome this problem, immunoaffinity enrichment techniques (immunocapture) are used, which can be employed before or after protein digestion, resulting in the interaction of the target protein or the surrogate peptide with the capture antibody, respectively. This strategy offers high selectivity and, consequently, high sensitivity, resulting in a highly efficient process for detection of target proteins or peptides [15, 81, 82].

Although the sample preparation protocols by immunocapture have been increasingly used for the analysis of proteins, peptides, and biotherapeuticals of clinical relevance [83], this strategy has not been extensively explored for the analysis of metabolites in targeted and semi-targeted metabolomics assays. Therefore, we think it might be worthwhile to explore it for metabolomics analysis. As an example of the potential application of immunocapture for metabolites, Chen et al. described a method for determination of mitochondrial metabolites and its potential metabolite profiling to other mammalian systems. According to the authors, mitochondria constitute a small fraction of cellular contents;

therefore, the whole-cell profiling is likely inadequate for monitoring changes within the mitochondrial matrix [84]. Several challenges were described for metabolite profiling when using subcellular organelles such as long hours for matrix purification, commercially available kits constituted by multiple immunocapture steps to complete, and traditional buffers used for organelle extraction, which severely interfere with the MS ionization process. By using epitope-tagged recombinant protein for immunocapture, it was possible to evaluate the dynamics of mitochondrial metabolism by quantification of more than 100 metabolites across different states of the respiratory chain. The sample preparation protocol took only 12 min from the homogenization to the extraction step, demonstrating its potential for metabolomics assays.

4 LC-MS/MS-Based Metabolomics and Data Acquisition

Following the genome era, omics analysis has been promoting an essential shift in the study of biomedical sciences to support a better understanding of biological systems and the development of successful precision medicine. As a part of the omics strategies, metabolomics is responsible for the analysis of end products of biochemical reactions underlying how phenotypic, disease, and lifestyle traits contribute to a meaningful interpretation of metabolic associations of an individual's metabolic phenotype [20, 21].

Therefore, the primary goal when using metabolomics strategies is to reveal system-wide alterations of unexpected biochemistry pathways in response to a complex interplay between an organism's genes and its environment. The result is a more comprehensive understanding of biological systems as a whole [85].

Nuclear magnetic resonance (NMR) and mass spectrometry (MS) are the primary analytical techniques that have often been used to perform metabolomics studies [86–88]. In brief, NMR has the advantages of simple sample preparation, nondestructive sample analysis, chemical characterization of complex mixtures, excellent reproducibility, high throughput of analysis, and inherently quantitative metabolite levels; however, a major weakness of NMR is its relatively low sensitivity, detecting only the most abundant compounds (typically >1μmol/L) in a sample, and small spectral databases [89]. Mass spectrometry-based metabolomics (MS) provides a great depth of information by offering high selectivity, sensitivity, high accuracy for low- and high-molecular-mass determination, and a greater structural identification and characterization capability. The technique continues to advance and it is considered a key analytical technology on which the emerging "-omics" approaches are mostly based. Mass spectrometry and NMR are complementary techniques and many reviews have comprehensively discussed how each of these techniques works and how each of them can be used in metabolomics [85, 87, 88].

Every year MS systems are launched by manufacturers and many new applications continue to grow, with the remarkable rise in system biology analysis. The combination of mass spectrometers with separation methods such as gas chromatography (GC-MS) [90], liquid chromatography (LC-MS) [85, 91], capillary electrophoresis (CE-MS) [92], and even inductively coupled plasma (ICP-MS) [93] for organometallic metabolites maximizes the power of mass spectrometry. The separation techniques offer metabolite separation in a time dimension, including isobar separation, decreasing the mass spectra complexity for better analysis and interpretation of challenging clinical samples. Moreover, MS provides a rapid measurement of hundreds of metabolites (small organic molecules, generally <1500 Da) in a single sample and run and requires minimal volumes of sample for injection (1–5μL), fast data acquisition, detection levels down to low picomolar concentrations, and relative short analysis time (20–30 min) [1, 2].

High-performance liquid chromatography (HPLC) and ultra-performance and ultra-high-performance liquid chromatography (UPLC and UHPLC) are considered more comprehensive techniques than GC-MS, since no derivatization is required and nonvolatile polar compounds may be easily analyzed, which simplifies sample preparation and allows the analysis of a broader range of metabolites, i.e., from peptides to hydrophilic organic acids, even to hydrophobic lipids [5, 94]. Furthermore, it is important to emphasize that in order to maximize metabolite coverage and to get a deeper metabolome characterization, the integration of different analytical techniques as a multiplatform metabolomics approach is crucial, as previously discussed (Sect. 2).

The successful interfacing of LC with MS was possible due to the invention of electrospray ionization (ESI). ESI is a soft ionization technique, which means that relatively small amounts of energy are retained by the analytes upon ionization, with minimal or no fragmentation occurrence [95]. As a result, cations or anions in solution-phase are converted into gas phase before entering into the mass spectrometer analyzer. Small molecules (<1500 Da) will produce mainly monocharged ions, whereas large molecules with several ionization sites will produce multiple charged ions. Protonated molecule $[M + H]^+$, $[M + nH]^{n+}$ and deprotonated molecule $[M - H]^-$, $[M - nH]^{n-}$ are the molecular species generated by addition of a proton or by removal of a proton from a molecule M, respectively [95, 96].

A diversity of mass analyzers are used in mass spectrometers. High-resolution mass spectrometry, such as time of flight (TOF), quadrupole time of flight (QTOF), Fourier-transform ion cyclotron resonance, and orbital ion traps, confers great specificity and sensitivity, thereby improving the quality of metabolome data. These advances successfully allow LC-MS to be used for routine separation and measurement of thousands of discrete chemical features for long-term or large-scale sample sets by using target, semi-targeted, or untargeted analytical assays, which are three possible metabolomics approaches to explore [97–102].

Untargeted Metabolomic: Also called global metabolomic, it is recommended to analyze as many metabolites as possible, either identified or unidentified, to achieve a comprehensive answer to a particular problem. The use of multiple analytical platforms is highly recommended, such as NMR, LC-MS, CE-MS, and GC-MS, thereby allowing the greatest metabolite information as possible [97–99].

Semi-targeted Metabolomic: Also known as pseudo-target metabolomic, this strategy quantifies a large number of previously identified metabolites. Typically, the method uses one single calibration curve to quantify a number of related compounds within a known compound class, e.g., phospholipids. For this reason, these compounds often show a characteristic fragmentation pattern, so that neutral loss scans, precursor ion scans, data-dependent analysis (DDA), and data-independent analysis (DIA) help to identify the structural related compounds. Therefore, it is possible to quantify hundreds of metabolites by using only 10–20 calibration curves. Due to metabolite's chemical similarity, the quantification is still precise and accurate [85, 97, 100].

Targeted Metabolomic: Targeted metabolomics approaches are used for the absolute quantitation of chemically characterized and biochemically annotated set of metabolites, for example, the quantification of pre-identified biomarkers using a particular analytical platform [100, 101].

Semi-targeted and untargeted approaches are designated for evaluating a hypothesis-generating studies, while targeted metabolomics approaches are used for validation and translation of novel discoveries of a specific hypothesis study [85].

Therefore, the number of metabolites, the aimed chemical coverage, and the chemical characteristic of the metabolites are crucial aspects in determining the type of metabolomics experiment to be used. These considerations will also define the sample preparation method and the correct choice of instrumentation [20–22].

4.1 Mass Analyzers

The mass analyzers used in metabolomics when coupled with separation techniques (LC-MS, GC-MS, and CE-MS) can either be of low resolution, such as triple quadrupole (QqQ) and quadrupole-ion trap (QTRAP), which are indicated for targeted and semi-targeted metabolomics analysis, or of high resolution, based on systems such as time of flight (TOF), hybrid quadrupole time of flight (QqTOF), Fourier-transform ion cyclotron resonance (FT-ICR), and Orbitrap (OT), which are the main choice for untargeted metabolomics assays; however, QqTOF and Orbitrap analyzers can also be successfully used for semi-targeted (QqTOF) and large-scale targeted metabolomics (Orbitrap) by parallel reaction monitoring (PRM) approaches [91, 102]. The latter is a targeted method in which all product ions originated from the precursor ion are monitored.

4.1.1 Triple Quadrupole (QqQ)

When it comes to targeted metabolomics assays, there is no doubt that triple quadrupole (*QqQ*) MS systems provide fast and reliable results for absolute quantitation. A tandem quadrupole mass spectrometer, called triple quadrupole, consists of two quadrupole (*Q1* and *Q3*) mass analyzers, separated by a collision cell (q_2). *Q1* and *Q3* act as mass filters, while q_2 is a RF-only quadrupole collision cell. Quadrupoles are the most used mass analyzers in mass spectrometers, which functions by scanning, transmission, isolation, and fragmentation of ions. Some of the key features of triple quadrupole MS systems are high sensitivity, high selectivity, accuracy, and reproducibility in absolute quantitative analysis of known compounds. Moreover, *QqQ* MS systems have an excellent MS/MS capability [91, 101, 103].

The benefits of triple quadrupole technology rely on its high detection sensibility when operated in a tandem MS/MS assay, designated as selected reaction monitoring (SRM). In this type of experiment, Q_1 is used for selecting and isolating a specific ion (precursor ion) measured by its mass-to-charge ratio (*m/z*), which is then directed to the collision cell (q_2) and undergoes fragmentation by collision of the precursor ion with an inert gas (argon or nitrogen). This process is called collision-induced dissociation (CID) and produces product ions. After that, the product ions are directed to Q_3 and then sorted according to their *m/z* and recorded by the detector [101]. The SRM experiment is widely used in absolute quantitative analysis. Because only ions with an exact transaction are detected, SRM scan mode greatly increases the selectivity by decreasing the background noise and increasing the signal-to-noise ratio, yielding lower detection and quantification limits, which is difficult with less sensitive methods, such as NMR [101, 104]. Therefore, SRM experiments can be applied to measure a significant number of chemically and biologically relevant metabolites with relatively high throughput, in a tandem-in-space manner. In a full-scan mode, ions based on their *m/z* are transmitted from the ionization source through the quadrupoles that are in radio frequency (RF) only mode [105]. Full-scan MS mode offers more information about the mass spectrum because it shows all ions that are formed in the ion source. A full-scan MS is a very useful mode of operation for the screening and identification of unknown compounds in complex matrices. On the other hand, full-scan MS does not yield the sensitivity that can be achieved with SRM experiments.

Besides SRM acquisition and full-scan MS, triple quadrupole mass spectrometer is designated to work under other three scan modes: precursor ion scan, neutral loss scan, and product ion scan.

A product ion scan can provide structural information of a given precursor ion that is selected in the Q_1 and fragmented in the collision cell (q_2), then all resultant product ions are scanned in the Q_3. This is a common experiment carried out to identify which transitions should be selected for quantification in the SRM scan mode. Precursor ion scan and neutral loss scan are very useful for screening analysis of compounds with similar chemical structures. In the precursor ion scan, Q_1 is scanned across a defined *m/z* range, and ions are fragmented in the q_2, and then Q_3 is set to transmit to the detector only a diagnostic charged fragment at a fixed *m/z*. The neutral loss scan monitors a constant difference between *m/z* set for Q_1 and Q_3. In this scan mode, the diagnostic fragment is a neutral fragment [101, 105].

Although *QqQ* systems are not outstanding for mass resolving power and mass range, it has several advantages such as high efficiency and specificity, robustness, fast acquisition rate, and simplicity to operate for users of all expertise levels.

Hybrid mass spectrometer systems such as QTRAP are similar to the ion path of a *QqQ* mass spectrometer. However, the third quadrupole (Q_3) can be configured as a linear ion trap (LIT), which offers many advantages due to the LIT high ion storage capacity. As a result, QTRAP mass spectrometers show excellent performance and flexibility through enhanced scan functions that allow simultaneous quantification and qualitative information for compound identification, without sensitivity losses and with higher data selectivity than *QqQ*, mainly when using multistage fragmentation (MS^3) [105].

4.1.2 Time of Flight (TOF) and Hybrid Quadrupole Time of Flight (QqTOF)

Time of flight (TOF) is one of the most preferred high-resolution mass spectrometry (HRMS) analyzers when it comes to untargeted metabolomics analysis, especially for studies using isotope labeling [106]. The operation of the TOF mass analyzers is based on the principle that lower *m/z* ions accelerate faster than the higher ones in an electric field, reaching the detector earlier than large *m/z* ions. As a temporally dispersive technique, the TOF mass analyzer is used in a linear configuration, which suffers from poor resolution, or as a ion reflector (re-TOF). The latter uses a constant electrostatic field (lenses) to reflect the ion beam toward the detector, promoting a fundamental upgrade in TOF's resolving power, without losing sensitivity. The re-TOF configuration increases the length of the flight (or drift) when compared to the length of the TOF instruments, without changes on peak width, leading to an optimal performance [106–108].

TOF mass analyzers have very high data acquisition sensitivity for an extensive mass range, which makes it adequate for the detection of a detailed and comprehensive view of expected and unexpected metabolites, at a high acquisition rate [109].

The hybrid quadrupole time of flight (QqTOF) combines quadrupole technology with a time-of-flight mass analyzer. The quadrupoles (Q_1 *and* q_2) function in the same manner as in the *QqQ* mass spectrometers. In summary, Q_1 can operate either as a mass filter, selecting ions based on their *m/z*, or as an ion transmission device, directing all íons into the TOF mass analyzer. Orthogonal acceleration was introduced to overcome the challenges of coupling a continuous ion source (ESI) and a pulsed mass analyzer. In this way, the beam of ions first passes into an ion pulser, which will accelerate orthogonally to their original direction and acquire the same level of energy. The ions leave the ion pulse and are injected into the flight tube [107–109].

The main features of the QqTOF systems are their high mass resolving power ($R > 30{,}000$ at full width at half maximum) and mass accuracy (<5 ppm deviation), which makes it useful for reliable identification of new compounds. Besides, QqTOF shows high data acquisition speed (≥5 Hz) and an attractive cost-benefit. It is worth mentioning that the high data acquisition of the QqTOF mass spectrometer ensures a sufficient number of points across narrow chromatographic bands, when UPLC and UHPLC systems are used as separation platforms [109].

4.1.3 Orbitrap

Orbitrap mass spectrometers are the last development in mass analyzers. It is a modified ion trap mass analyzer that consists of two electrodes – a barrel-shaped outer surface and an inner spindle-shaped electrode. A static electrostatic field is used to trap the ions radially around the center electrode, and they oscillate harmonically on elliptical trajectories with a frequency characteristic to its *m/z* value. There is no ion mass dependence with the orbit radius; thus, the amplitude of the orbit is the same for all ions independent of the corresponding *m/z*. A detector records the oscillation frequencies as images of the axial velocity of the ions, and the signal is converted by a Fourier transform into a high-resolution mass spectrum [110, 111].

In 2005, the Orbitrap analyzer was commercially introduced in the hybrid version. The combination of a low-resolution linear ion trap with the high-resolution Orbitrap mass analyzer (LTQ-Orbitrap) became widely used in the omics strategies. Thus, new configurations have been often released to provide more versatility and possibilities for quantification, identification, and characterization of small and large molecules [111, 112].

The operating mechanism of the LTQ-Orbitrap represents a multistage trap combination. In the full-scan mode, the linear ion trap directs all ions to the Orbitrap after passing through a C-Trap. The C-Trap is responsible for storage and compression of the ion population, followed by efficient transmission into the Orbitrap mass analyzer. The main advantage of the geometry used in the C-Trap is the radial ejection, which provides less spatial dispersion of the ions. Additionally, it allows the determination of product ions with high accuracy [111].

Different configurations and recent technology innovations are available for Orbitrap mass spectrometers. For example, the combination of a quadrupole precursor ion selection with the high-resolution, accurate-mass Orbitrap (QExactive™ systems) [113] provided high performance for targeted and untargeted applications, while the Orbitrap Fusion™ Tribrid™ mass spectrometer combines the power of quadrupole, ion trap, and Orbitrap mass analyzers in a single system [114]. Another innovation is the use of higher-energy collisional dissociation (HCD) as new fragmentation technique. The HDC, a multipole device directly interfaced with a C-Trap, promotes precursor ion acceleration into the gas-filled cell, collision with neutral gas, and cleavage to generate product ions and neutral species, which are then pushed back toward the C-Trap in order to focus ions, and after they are pushed toward the Orbitrap mass analyzer. Therefore, ions are produced in a collision cell rather than in an ion trap, and, because it employs higher-energy dissociation than those used in ion trap CID, the HCD fragmentation can enable a wider range of fragmentation pathways [115, 116]. Nowadays, the Orbitrap Fusion Lumos Tribrid™ can provide a high resolving power up to 1,000,000 (at *m/z* 200), mass accuracy <1 ppm, and a dynamic range up to five orders of magnitude. Additionally, the Orbitrap analyzers offer the capacity of enabling efficient multiplexed scan modes with high resolution, due to their ability to separate "in space" and analyze MS and MS/MS ranges [113, 115].

5 Typical HRMS Data Acquisition

Following the advances in high-resolution mass spectrometers, the development of integrated approaches using data-dependent acquisition (DDA) and data-independent acquisition (DIA), along with multiple data processing tools, has been used in the identification and characterizations of compounds. Basically, both acquisition modes acquire full-scan MS for precursor ions, simultaneously followed by MS/MS experiments to obtain product ions. The selection of parameters for the MS/MS experiments is the main difference between DDA and DIA. Consequently, the data generated and data processing are also different, as summarized in Fig. 3. These differences will be described in more detail on the following topics.

5.1 Full-Scan MS HRMS Data

The full-scan MS monitoring provides a full mass spectrum of each analyte, thus a qualitative picture of the sample composition. This operation involves the mass analyzer scanning from a low to a high *m/z*, as defined in the mass range interval that was set in the MS method. Typically the *m/z* range interval would be from *m/z* 50 to 1500, depending on what the users expect in the sample [118]. The full-scan MS will trigger MS/

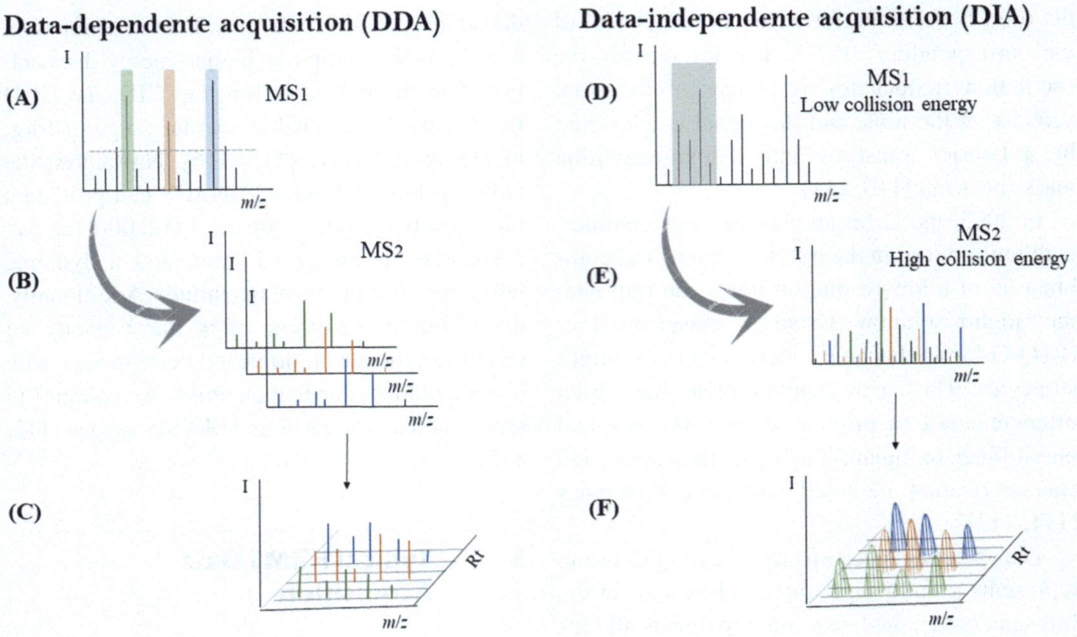

Fig. 3 Comparison of data-dependent acquisition and data-independent acquisition. (**a–c**) Data-dependent acquisition (DDA). The data acquisition automatically switches from full-scan MS to MS/MS when precursor ions exceed a predefined threshold. Then, the MS/MS data are acquired taking into account retention time, *m/z*, and ion intensity. (**d–f**) Data-independent acquisition (DIA). Acquires MS/MS scans with wide isolation windows that do not select any particular precursor ion. Several precursor ions are fragmented together in a single MS2 event. The product ion information can be extracted over time and used for quantification owing to the repetitive MS/MS sampling cycle of DIA, and the data generated is more complex than data generated from DDA acquisition. (Adapted from Pappireddi et al. [117])

MS acquisition of ions that are detected above a preset intensity – so-called intensity-dependent data acquisition method. From the full-scan MS analysis, extracted ion chromatogram (EIC) are used to obtain one or more selected *m/z* values of interest, using a *m/z* tolerance with parts per million (ppm), to visualize the chromatographic peak quality [119, 120]. In general, data acquisition using full-scan is the first step performed in any type of MS data acquisition method, as a way to obtain a general and fast profiling of the studied sample.

5.1.1 Data-Dependent Acquisition (DDA)

The analysis of complex samples is a difficult task and a time-consuming process. Additional experiments are often needed to assist the detection and characterization of compounds that often coelute with large amounts of interfering ions and may still be present in trace levels. Taking these drawbacks into account, new approaches to data acquisition were developed to maximize the qualitative mass spectral information, as is the case of data-dependent acquisition (DDA) [120, 121]. With DDA MS, the mass analyzer acquires a survey (full-scan MS) of precursor ions, and the DDA MS/MS is triggered by the detection of a number of the most intense ions recorded in MS^1 (e.g., top N ions), producing product ion spectra data sets. In terms of fragmentation, DDA MS/MS is promoted by charge-state optimized collision energy. Moreover, a mass-dependent ramp collision energy of DDA can enable more balanced MS/MS fragmentation for unknown metabolites. Thus, both total ion scan spectrum and product ion scan spectrum are acquired in the same analysis, combining comprehensive qualitative data information [112, 121].

DDA methods with HRMS QqTOF have been developed to maximize the qualitative information that may be recorded in a single run and to increase the throughput. The survey scans are acquired in accordance with parameters predefined by the analyst. Usually, for targeted and semi-targeted analysis, inclusion of *m/z* list and pseudo-neutral loss-dependent are used as data acquisition methods to trigger the MS/MS mode. On the other hand, for untargeted metabolomics, exposome (or xenometabolome), and drug metabolism studies, the most appropriate survey scans for data acquisition are full-scan, product ion-dependent, precursor ion intensity-dependent, isotope pattern-dependent, pseudo-neutral loss-dependent, and mass defect-dependent [119, 120].

Although DDA is a flexible and powerful approach, it suffers from some disadvantages, namely: (i) greater chemical background interference compromises the method sensitivity and, consequently, the selection of the precursor ion for further MS/MS experiments; (ii) the dynamic exclusion approach prevents the mass spectrometer from acquiring, in a short time period (typically 30 s), more than one MS/MS of the same precursor ions, which impacts mainly peptide and protein analysis; and (iii) there is poor reproducibility for repeated runs of the same sample. This is caused by the semi-random selection of the precursor ions, which is dependent on the elution time, peak intensity, and dynamic exclusion time lists [118, 121, 122].

5.1.2 Data-Independent Acquisition (DIA)

As an alternative to conventional full-scan acquisition tools, the data-independent acquisition (DIA) is another mass spectrometric technique based on nonspecific collision-induced dissociation (CID) [118, 119]. Differently from DDA, DIA provides information about fragment ions in a nonselective manner, i.e., all ionized compounds (even those coeluting) of a given sample that fall within a specified precursor *m/z* range are fragmented in a systematic and unbiased way [122].

The DIA acquisition method has been developed to improve the collection of information from MS/MS experiments and has been applied in different mass spectrometer systems. Certain QqTOF instruments perform MS^E (elevated energy MS) (Waters Corp., QTof) or MS/MS^{ALL} with SWATH™ acquisition (Sciex, TripleTOF®), while Orbitrap™ systems (Thermo Scientific) use "all-ion fragmentation" (AIF) experiments to collect untargeted data [118, 121, 122].

5.2 MS^E Data Acquisition Using QTOF Systems

MS^E (elevated energy MS) acquisition mode acquires two full-scan acquisitions (MS1 and MS2), one at low collision energy and the other at a higher collision energy through dynamic switch between energies in the collision cell. This approach allows users to obtain both precursor ions and full-scan MS information in parallel, maximizing the data collection efficiency, mainly for a large sample set. Plum et al. described some experiments exploring MS^E data acquisition for molecular fragment information for biomarker identification, and they reported that MS^E approach maximizes the duty cycle of the QTOF instrument and therefore did not affect the quality of the chromatographic data since seven data points were still acquired across each peak in both collision energy settings [123].

Another main advantage of MS^E data acquisition is that MS/MS experiments are carried out without *m/z* pre-selection in Q_1. Thus, all exact-mass precursor ions from every detectable sample component are fragmented in the collision cell, followed by the TOF analyzers. The MS^E experiments have been applied as the first step to improve the acquisition data throughput by generating multiplexed spectra. It is worth mentioning that since there is no pre-selection of precursor ions, data processing can be complicated, due also to product ions from coeluting compounds or interferences from the matrix. However, this can be minimized by enhanced chromatographic separation promoted by the UPLC systems (Waters). Because of data complexity, different algorithms are applied during data preprocessing, requiring data evaluation by a database searching algorithm [124]. In general, a particular set of product ions are grouped to a specific subset of precursors based on chromatographic elution and ion mobility profiles. Ion mobility enhances the peak capacity and it is referred as ion mobility-assisted DIA or HDMS^E [125].

5.3 MS/MS^{ALL} with SWATH™ Acquisition Using TripleTOF™ Systems

Sequential Windowed Acquisition of All Theoretical Fragment Ion Mass Spectra (SWATH-MS) is a variant from DIA. In SWATH-MS data acquisition method, successive predefined ranges of precursor *m/z* values are isolated and subjected to cofragmentation. Quadrupole (Q_1) selects the precursor ions using a small *m/z* range of typically 25 *m/z* each; within this range, all ions are transmitted to the collision cell where they are fragmented by CID, and the product ions are analyzed by the TOF mass analyzer [112]. Note that the MS/MS^{ALL} with SWATH™ acquisition mode combines the benefit of the fast scan speed of TOF analyzers with data-independent acquisition. The main advantage is attributed to increased selectivity due to the multiple selections of precursor ions in a narrow range of *m/z*. Thus, the obtained data is relatively simple when compared to MS^E and AIF acquisition modes.

5.4 All-Ion Fragmentation Using Orbitrap

All-ion fragmentation (AIF) acquisition mode in Orbitrap systems fragments all precursor ions in a similar manner to MS^E. First, the data is acquired in full-scan MS, then AIF occurs in a higher-energy collisional dissociation (HCD) cell for the entire *m/z* range of coeluting molecules. The energy values are set considering values around the chosen middle energy and the ion characteristics [113, 120].

In untargeted metabolomics assays by HRMS, full-scan MS, DDA, and DIA (MS^E, all-ion fragmentation, MS/MS^{ALL} with SWATH™) are the most used data acquisition modes. The work published by Chaleckis et al. [126] describes a metabolomics assay for plasma and/or serum in which AIF was used in combination with a data processing method using an in-house spectral library. Selectivity for the coeluting compounds and interferences was improved using AIF. Geiger et al. [127] demonstrated the potential of MS and AIF data acquisition for the characterization of protein mixtures. The technique was able to identify proteins with more than 100-fold abundance differences in a high dynamic range standard.

6 Informatic Tools and Databases

Due to the high number of data generated during DIA and DDA experiments, as well as their complexity, the visual inspection of over a thousand of peaks detected is impractical. Thus, numerous efforts have been made to offer data analysis options that enable a relatively faster and less laborious step.

Within the metabolomics community, there is a variety of software that provide a method for peak picking, nonlinear retention time alignment, relative quantitation, extensive visualization of results, and statistical analysis. Most of these tools are implemented with R packages [104, 128, 129]. For processing metabolomics data, the most used software are XCMS and MZmine [130, 131]. These two tools are free and open source, allowing users to upload data and perform processing through a friendly interface [129].

In addition, other free software intended for data preprocessing has also been widely reported, such as OpenMS [132], MetAlign [133], and MS-DIAL [134]. The latter promotes the deconvolution of spectral data from ion fragments acquired by DIA [134].

Currently, a growing number of users are taking a workflow-based approach for processing LC-MS data. This strategy provides several interconnected tools and includes preprocessing, annotation, and statistical analysis in a single tool. This approach allows users to carry out analysis using only one software, instead of having to use several tools for each stage of the data processing [128, 135]. Some examples include XCMS Online [136], Metabolomic Analysis and Visualization Engine (MAVEN) [137], MZmine2 [131], MetaboAnalyst [138], and Workflow4metabolomics [139].

Indeed, the most important and difficult part of untargeted metabolomics studies is identifying the metabolites. Only limited structural information can be obtained from mass spectrometry, so it is a challenge to identify unknown features. It is important to note that many software are available and have the promise of providing the metabolite identification. However, they provide only a table of features with p-values and identify differences in relative ion intensity between samples [104, 128].

In the first step of metabolite identification, the exact measured mass of the investigated molecule is searched in metabolite databases such as METLIN [140], Human Metabolome Database (HMDB) [141], or MassBank [142]. The correlation with databases represents only a putative identification. Therefore, it is necessary to confirm the proposed molecule by comparing its retention time and MS/MS data with a model compound. For this purpose, software to match the MS/MS spectra with in silico results can be used, as is the case with MetFrag [140, 142, 143]. This tool generates in silico fragmentation and allows users to search a wider selection of databases to identify candidate molecules generated from topological fragments. In addition, users can select filtering criteria, such as inclusion or exclusion of substructures and elements [143]. It is worth mentioning that in this section, we only list some tools used to assist in the identification of metabolites. Many other software are available for free and can be found in the literature.

As previously mentioned, the metabolite identification process represents the limiting step in the metabolomics workflow, despite recent advances in data acquisition and processing. Moreover, even with the growth of metabolomics databases, a large number of metabolite features detected in biological samples do not return any matches. These factors make it difficult to comprehensively identify all metabolites present in the sample, and a broad response is not yet possible [104, 128]. Further discussion about this topic will be presented in Chap. 12 of this book.

7 Metabolomics for Biomarkers Discovery and Clinical Applications

Biomarkers are “any substance, structure, or process that can be measured in the body or its products, which influence or predict the incidence of an outcome or disease,” according to the definition of the World Health Organization [144]. Other broader definitions include indicators of exposure to environmental pollution, dietary intake, and effects of treatment interventions [145]. The majority of clinical tests used in modern medicine are based on biomarkers. For instance, blood pressure can be considered as a biomarker for cardiovascular health, whereas blood glucose is a biomarker for diabetes and insulin resistance. A variety of clinical roles can be performed by biomarkers, including disease risk prediction, screening, diagnosis, stratification, prognosis, and treatment monitoring [146]. Regardless of its role, a biomarker should desirably fulfill a series of requirements, which include the following: (1) be significantly associated to a certain clinical outcome or disease in a way that can be biochemically understood; (2) preferably, be found in biological specimens that can be collected with minimal intervention (e.g., urine, saliva, blood, feces); and (3) be reliably analyzed by a simple, rapid, accurate, robust, and cost-effective method that can be adopted by clinical laboratories [147].

The majority of biomarkers currently used in the clinical practice were discovered based on serendipitous observations or insights that inspired hypothesis-driven studies, which subsequently identified their link to a disease pathophysiology and confirmed their usefulness as a biomarker. The “omics” sciences have initiated a new era of data-driven, hypothesis-generating studies for biomarker discovery. This approach allows the active search of putative biomarkers among thousands of molecular features measured in biological specimens, which does not rely solely on spontaneous insights, as previously, and has a great potential to streamline the discovery of new biomarkers [146, 148]. In this regard, metabolomics can be highlighted as the "omics" approach that reflects more closely the phenotype of an organism, as metabolites represent a downstream product of genes and proteins, which incorporate the effects of epigenetic and post-translational modifications, as well as perturbations from diet, lifestyle, and the environment [104, 146].

To our knowledge, no biomarkers discovered by metabolomics have been translated into clinical practice yet, although some have been the focus of several recent confirmatory studies with promising results. After analyzing a discovery/training set to first detect and identify a candidate biomarker, the next step is an internal validation that should be performed using a similar population and similar analytical conditions [149]. This step tests whether the proposed biomarker can distinguish the disease group in a similar cohort, allowing researchers to exclude inconsistent metabolites, for example. Frequently, initial studies utilize what is called a hold-out set, originated from splitting the available data into a discovery/training set and a test/hold-out set, which is used to evaluate overfitting in multivariate models [150]. The next step is an external validation, which is often performed as a separate study using an independently collected and analyzed set of samples from a similar population [149]. If still consistent, the candidate biomarker can then be evaluated to characterize its clinical effectiveness and relevance, a step that is sometimes called "clinical validation," although the term "evaluation" is preferred to highlight that it refers to an ongoing process, rather than a definitive conclusion [145]. According to Pepe et al. [151], the biomarker evaluation step comprises a multistage process, which includes mainly (1) the development and validation of a clinical assay to measure the biomarker; (2) the assessment of factors that can be associated to biomarker status (e.g., gender, age) to identify the need for subpopulation thresholds; (3) retrospective studies using repository samples to evaluate the ability of detecting early stages of a disease; (4) prospective studies to evaluate the biomarker utility and practical feasibility, including an assessment of false-positive and false-negative cases; and (5) control studies to report the real impact of the biomarker in improving the clinical outcomes in the population. Instead of "evaluation," regulatory agencies, such as the Food and Drug Administration (FDA) and the European Medicines Agency (EMA), adopt the word "qualification" to refer to the formal process of submission and evaluation of candidate biomarkers [152, 153].

Suggestions on how to further advance in this long road toward clinical application have been proposed by several researchers working with metabolomics biomarkers. For instance, Marchand et al. [149] highlight the importance of engaging all the stakeholders (clinicians, clinical biochemists, organizations, and policy makers) in the process of biomarker translation. Xia et al. [150] have published a tutorial with several recommendations for metabolomics biomarker research, in order to help improve the consistency and quality of the data generated and reported. This tutorial also mentions the importance of clear communication with the medical community, in terms of biomarker statistical analysis and reporting. The importance of standardizing the design and execution of biomarker studies is also a topic discussed by Monteiro et al. [148], considering the importance of making results more comparable across different studies. In addition, an aspect raised by Lindahl et al. [154] is that metabolomics studies can easily identify metabolites that are not specific to a disease, but rather common compounds related to inflammation in general, which will likely be irrelevant as disease biomarkers. For this reason, they suggest the inclusion of groups of different diseases when performing the discovery study, in addition to the disease of interest and the non-diseased group. Ioannidis and Bossuyt [155] presented a summary of key reasons of failure in the metabolomics biomarker "pipeline," including each stage of the process (discovery, validation, evaluation, and clinical implementation), with suggestions of possible solutions for each one of them. Therefore, although omics studies are susceptible to false discoveries and many putative biomarkers may still fail to reach real clinical application, the chances of having true discoveries can be considerably increased by following the recommended practices while performing and reporting the study.

An example of potential biomarker discovered by metabolomics is the compound trimethylamine N-oxide (TMAO), which has been extensively studied over the last decade, due to the association between elevated plasma TMAO levels and an increased risk for major cardiovascular events in humans [89, 156]. In 2011, Wang et al. [157] reported a study performed with the goal of identifying metabolites to predict risk for cardiovascular diseases by untargeted metabolomics, in which human plasma samples were analyzed using LC-MS. The subjects included in the study were stable patients undergoing elective cardiac evaluation, who subsequently had a major cardiovascular event (myocardial infarction, stroke, or death) after up to 3 years later, compared to age- and gender-matched patients who did not have these events. Initially, the test cohort (n = 100) identified 40 candidate metabolites, a number that was reduced to 18 after analyzing the validation cohort (n = 50). The clinical prognostic utility of the metabolites was then confirmed in a larger prospective cohort (n = 1876), as well as with a series of more focused follow-up human- and animal-based experiments to confirm the links between TMAO, liver metabolism, gut microflora, and dietary habits [157–161]. In summary, the proposed hypothesis is that some dietary nutrients (phosphatidylcholine, choline, carnitine, betaine) are metabolized to trimethylamine (TMA) by the gut microbiota, and then TMA is subsequently converted to TMAO by hepatic enzymes. TMAO, in turn, contributes to the rapid development of atherosclerotic plaques in animal models.

Several other studies have helped to strengthen the evidences about the role of gut microbiota-derived TMAO as a potential marker for risk of major adverse cardiovascular events in humans. For instance, Guasti et al. [156] have recently published a systematic review and meta-analysis of cohort studies that focused on further evaluation of the role of TMAO in cardiovascular diseases. The data included in the meta-analysis comprised three targeted metabolomics studies, with patients from the USA, Swiss, and the UK, in which TMAO was quantified by LC-MS/MS using stable isotope dilution. A total of 3807 patients with a high/very high cardiovascular risk were included in the meta-analysis, 1907 of which with a high baseline plasmatic TMAO level and 1900 with a low baseline TMAO level, which were followed for 1 month to 7 years for major cardiovascular events and/or all-cause death. The study found that patients with a high baseline TMAO level had a two-time higher risk for major cardiovascular events and a three-time higher risk for all-cause mortality than patients with a low TMAO baseline level. The authors highlight the potential use of TMAO to better stratify individual cardiovascular risk, as well as to design new prevention strategies, which points to the importance of conducting new prospective studies about TMAO [156].

Despite all studies that have already helped researchers and clinicians to understand the source of plasmatic TMAO and its association with atherosclerosis and cardiovascular diseases, more studies are still required to explain the exact mechanism of gut dysbiosis that increases the levels of microflora-derived TMAO and the influence of dietary habits [162–164]. For instance, diets rich in TMAO and TMAO precursors, like seafood, are important sources of protein and vitamins, which have been considered beneficial for the cardiovascular health in a number of studies [165]. Therefore, multiple mechanisms are likely involved and must be evaluated in future studies to further elucidate the roles of TMAO as a potential early biomarker in cardiovascular disease and a target for disease prevention. This example highlights that several follow-up studies are often necessary to establish the knowledge about a putative biomarker before proceeding to clinical qualification/evaluation studies as a step toward approval for clinical application.

Another example of potential biomarkers discovered by metabolomics is given by some compounds that have been associated to the risk of developing type 2 diabetes mellitus. In a first study, Wang et al. [166] performed an LC-MS/MS-based targeted metabolomics study, which identified three branched-chain amino acids (i.e., leucine, isoleucine, and valine) and two aromatic amino acids (i.e., phenylalanine and tyrosine) in fasting human plasma as metabolites associated to an increased risk for type 2 diabetes up to 12 years prior to the onset of the disease. This study involved initially 189 individuals who developed new-onset diabetes during the 12-year follow-up period, along with 189 matched controls who did not develop diabetes in this period. The initial findings were supported by validation using an independent cohort [166].

Subsequently, several other targeted and untargeted metabolomics studies were conducted, and, despite some variations in the reported differentiating metabolites reported, many studies confirmed the elevated plasma levels of branched-chain amino acids and aromatic amino acids prior to type 2 diabetes manifestation [167, 168]. In addition, some meta-analyses have also compiled results from different studies, which corroborate to strengthen the evidences about the association of these metabolites with early identification of type 2 diabetes and/or prediabetes. For instance, Guasch-Ferré et al. [169] meta-analyzed eight prospective studies including 8000 individuals and found a higher risk of developing diabetes in individuals with higher blood levels of these amino acids (leucine, isoleucine, valine, phenylalanine, and tyrosine), as well as a lower risk associated with higher levels of glycine and glutamine. In this study, other metabolites were also associated to diabetes and prediabetes, including hexoses, phospholipids, and triglycerides. Other recent meta-analyses [170, 171] also support the previous findings, indicating a consistent measurement of higher levels of branched-chain and aromatic amino acids associated to a later development of diabetes. Therefore, many studies point to these metabolites as potential early biomarkers for monitoring the risk of type 2 diabetes, which may allow for early medical intervention a few years or even a decade prior to the first symptoms.

Currently, the only biomarkers approved for patient diagnosis and monitoring in diabetes are blood glucose levels and glycosylated hemoglobin [172]. New potential biomarkers discovered by metabolomics are certainly promising, although actual translation into clinical practice still requires further research. Future studies should, for instance, help to elucidate the impact of diet and gut microflora on these metabolites and in diabetes/prediabetes, considering that medical intervention in diabetes is mostly focused on diet and lifestyle [169, 172]. Further studies should also comprise different age groups, ethnicities, and both genders, in order to generate results that can be more easily extended to a variety of individuals in heterogeneous populations, helping to achieve robust replications and improve the power of the predictive risk assessment [168, 170, 171].

Overall, the translation of metabolomics research into clinical applications is making clear progress, supported by the contributions of many researchers in the field, who work in different stages of biomarker discovery and translation studies. Even though no metabolomics biomarker has been translated yet, some candidates remain promising even after almost a decade of follow-up studies involving different cohorts, as well as different analysts and laboratories/instruments.

8 Applications in Metabolic Flux Analysis (MFA)

The determination of one or more metabolites' level produces information about the biochemical changes of these compounds in complex systems. Biological processes are dynamic, in which metabolites participate in several metabolic pathways [173, 174]. The combination of the transformations of these chemical species works together, forming a real network [175]. In this sense, the metabolic flux analysis (MFA) with stable isotopic markers enables the dynamic understanding of the metabolome of a living system [173, 174]. MFA, also called fluxomics, is based on the idea of tracing marked atoms incorporated into the intracellular metabolites via the administration of biochemical reactions [176]. In metabolomics, MFA can be explored in three ways: (i) by the discrimination of metabolic variants (isotopic profiling), (ii) by the specific flow measurement (targeted flow analysis), and (iii) through the investigation of the entire biological system (global fluxomics) [177].

As with all metabolomics experiments, MFA also requires a careful experimental design to maximize the information obtained through comparative experiments using a group containing marked compounds and another with unmarked ones. MFA requires highly sensitive analytical techniques to measure metabolite levels. MS and NMR are the most used techniques for detecting and quantifying isotopically labeled metabolites [173]. However, one of the problems faced when working with MFA assays is the detection of compounds of low abundance; thus, MS is more attractive compared to NMR due to its higher sensitivity [173, 176]. Separation techniques such as GC and LC hyphenated to MS contribute to increased metabolome coverage and facilitate the determination of metabolites based on isotope isomers (isotopomers) [174, 178].

^{13}C is usually the stable isotope-labeled of choice because it can be easily supplied in the form of pure carbon sources to the synthetic culture medium, in addition to being easily transferred during organic reactions [179, 180]. Overall, labeled amino acids are incorporated into the system since they are important intermediaries of central metabolic pathways, providing crucial information about the changes observed in central carbon metabolism [177]. Incorporation of the isotope-labeled compound may occur in several ways and different positions in the molecules, causing the resulting metabolite to present different labeling states from different isotopomers [181].

The determination of data workflow and interpretation requires robust computational tools and adequate mathematical methods, which generate models that describe isotopic propagation in the studied organism [182]. Unsupervised statistical methods (such as PCA) and classification methods (clustering analysis) have been employed. The flow ratio estimation is finally based on stoichiometric and isotopic mass balances to estimate flux models [177].

Understanding metabolic flows is important for developing therapies and diagnostic methods for metabolic diseases [182]. A survey carried out in the databases showed that from 2018 to 2020, LC-MS (HILIC and RPLC) has been one of the most used techniques in MFA. Table 1 presents a summary highlighting the analytical conditions of these clinical metabolomics applications. Due to straightforward experimental planning, almost all researches are performed in cells or tissue samples. Overall, the isotopic markers are incorporated into the medium culture to study diseases that cause severe problems for the world population, such as cancer and neglected diseases (Table 1).

In order to study reductive glutamine metabolism in mitochondria and cytosol of cancer cells, Lee et al. [183] used a LC-MS-based flux analysis. In this work, ^{13}C-glucose and ^{13}C-glutamine were incorporated into derived cervical cancer cells (HeLa cells). Flow analysis by HILIC-MS enabled the mapping of important metabolites of compartment-specific metabolic fluxes, demonstrating the contribution of reductive glutamine metabolism to fatty acid production. The results presented important insights that could support advances in identifying tumor vulnerabilities and the discovery of target pathways for the development of new therapies against cancer. Studying compartmentalization is a challenge for metabolomics since the metabolites may come from different organelles or microcompartments [176].

Another interesting study performed by Wang et al. [184] used MFA to assess substrates' contributions to gluconeogenesis in fasted mice. The experimental design was carried out by infusing $^{13}C_6$-glucose, $^{13}C_3$-glycerol or $^{13}C_3$-pyruvate, and $^{13}C_3$-lactate in mice for 6 h. Mice serum samples were analyzed by isocratic elution in HILIC-MS. The flow analysis suggests that, although lactate serves as an important substrate for gluconeogenesis, glycerol seems to contribute

Table 1 Applications of LC-MS in fluxomics from 2018 to 2020

Sample matrix	Disease	Metabolomics approach	System	Elution mode	Isotopic marker	Ref.
Mice serum	Diabetes mellitus	Untargeted and targeted	HILIC-MS	MP: (A) 20 mM NH_4Ac + 20 mM NH_4OH in 95:5 water/ACN and (B) 20 mM NH4Ac + 20 mM NH_4OH in 20:80 water/ACN. Elution: isocratic, 73% B for 6 min	^{13}C	[184]
Tumor tissue	Cancer	Untargeted and targeted	RPLC-QqTOF	Elution: 3–100% methanol with 5 mM tributylamine over 22 min	^{15}N and ^{13}C	[186]
Human cells	Cancer	Targeted	RPLC-QqTOF	MP: (A) 0.1% FA in water and (B) 0.1% FA in ACN. Elution: 2–95% B over 44 min	^{13}C	[187]
Human cells	Toxoplasmosis	Untargeted and targeted	RPLC-Orbitrap	MP: (A) 97:3 water/methanol + 9 mM NH_4Ac + 10 mM TBA and (B) 100% methanol. Elution: 5% B for 2.5 min, 5–95% B over 14.5 min	^{15}N	[188]
Human cells	Cancer	Untargeted and targeted	HILIC- and RPLC-Orbitrap	*HILIC*: MP: (A) 10 mM NH_4Ac in water and (B) ACN. Elution: isocratic, 10% B for 12 min *RP*: MP: (A) 0.2% F.A. in water and (B) methanol. Elution: gradient 30–80% B in 8 min	^{13}C and ^{15}N	[189]
Mouse cells	Cancer	Targeted	HILIC-Orbitrap	MP: (A) 20 mM $(NH_4)_2CO_3$ and 0.01% NH_4OH and (B) ACN. Elution: 80% to 20% B over 15 min	^{13}C	[183]
Lymphoma cells	Cancer	Untargeted and targeted	IPC-QqQ	MP: (A) 400 mM HFIP and 10 mM DIPEA in water and (B) 300 mM HFIP and 10 mM DIPEA in methanol. Elution: 2–95% B over 31 min	^{13}C and ^{15}N	[190]
Fibroblast cells	Acyl-CoA dehydrogenase deficiency	Untargeted and targeted	RPLC-Orbitrap	MP: (A) 0.025% heptafluorobutyric acid with 0.1% F.A. in water and (B) ACN. Elution: NI	^{13}C and ^{15}N	[191]
T. brucei bloodstream form	African trypanosomiasis	Targeted	HILIC-Orbitrap	MP: (A) 20 mM $(NH_4)_2CO_3$ in water and (B) ACN. Elution: 80% to 5% B over 15 min	^{13}C	[192]
Cancer cells	Cancer	Targeted	RPLC- and HILIC-QqTOF	*RP*: MP: (A) water with 0.1% F.A. and (B) 75:25 can/isopropanol with 0.1% F.A. Elution: min 0: 0.1% B, min 2: 10% B, min 7: 99% B, min 9: 99% B *HILIC*: MP: (A) 50 mM NH_4HCO_2 in water and (B) 90:10 (ACN): (water + 50 mM NH_4HCO_2). Elution: min 0: 90% B; min 4: 85% B, min 5: 70% B, min 7: 55% B, min 10: 20% B	^{13}C	[193]

(continued)

Table 1 (continued)

Sample matrix	Disease	Metabolomics approach	System	Elution mode	Isotopic marker	Ref.
Human cells	Cancer	Targeted	HILIC-QqQ	MP: (A) 1 mM NH_4Ac in water and (B) ACN with 0.5% NH_4OH. Elution: 5% B at 15 min, 10% B at 22 min, 90% B over 23 min	^{13}C	[194]
Human cells	Cancer	Targeted	IPC-QqQ	MP: (A) 10 mM NH_4HCO_2 and 1.5 mM hexylamine in water (pH 7.0, adjusted with FA) and (B) ACN. Elution: 1% B at 3.5 min, 1–15% B at 4.5 min, 15% B over 10 min	^{13}C	[195]
Mouse cells	Obesity	Targeted	HILIC-Orbitrap	MP: (A) 9/1 ACN/water with 5 mM NH_4Ac and (B) 1/9 ACN/water with 5 mM NH_4Ac. Elution: 100% to 36% A at 20 min, 36% to 20% A at 24 min, 20% A at 27 min	^{13}C	[196]
T. b. brucei cells	African trypanosomiases	Targeted	HILIC-Orbitrap	MP: (A) 20 mM $(NH_4)_2CO_3$ in water and (B) ACN. Elution: 80% to 20% B at 30 min, 20% to 5% B at 32 min	^{13}C	[197]
L. mexicana cells	Leishmaniasis	Untargeted and targeted	HILIC-Orbitrap	NI	^{13}C	[198]
Human cells	Cancer	Targeted	RPLC-Orbitrap	MP: (A) 95% water, 5% ACN, 20 mM NH_4OH, and 20 mM NH_4Ac and (B) 95% ACN and 5% water. Elution: 100% to 0% B over 45 min	^{13}C	[199]
T. b. brucei cells	African trypanosomiases	Untargeted and targeted	HILIC-Orbitrap	MP: (A) 20 mM $(NH_4)_2CO_3$ in water and (B) ACN. Elution: NI	^{13}C	[200]
Skin fibroblast cells	Barth syndrome	Targeted	HILIC-Orbitrap	NI	^{13}C and ^{15}N	[201]

NI no information, *ACN* acetonitrile, *NH_4Ac* ammonium acetate, *NH_4OH* ammonium hydroxide, *TBA* tributylamine, *$(NH_4)_2CO_3$* ammonium carbonate, *HFIP* 1,1,1,3,3,3-hexafluoro-2-propanol, *DIPEA* diisopropylethylamine, *NH_4HCO_2* ammonium formate

mostly to this process during short and long fasting periods. Since glycerol has been shown to be a biomarker for the development of type II diabetes, the authors have pointed out this metabolite as a potential source for treating this chronic disease.

Isotope labeling experiments are an interesting approach to decipher how complex metabolic pathways work, discovering hidden functions in living cells [185]. Although there are some significant technical barriers in these experiments, including insufficient measurements of isotopic labeling information in the metabolites, recent mathematical modeling developments for solving metabolic fluxes help circumvent these problems and determine the dynamic picture of the phenotype.

9 Conclusions and Perspectives

The rapid development of UHPLC systems and the high mass resolving power and mass accuracy of the modern HRMS instruments have provided an opportunity to dramatically promote the growth of metabolomics strategies for a better understanding of metabolic signatures related to a given biological condition. In this chapter, we presented classical sample treatment approaches and how they can be improved to promote a high selectivity by means of immunocapture extraction procedures. The use of complementary chromatographic modes (RPLC and HILIC) was highlighted in order to demonstrate the importance of using orthogonal separation strategies to obtain as much responses as possible from complex matrices used in clinical metabolomics. Moreover, DoE has also emerged as an attractive alternative for sample preparation and instrumental method development in a fast and fashionable way. For biomarker discovery and flux metabolomics, objective descriptions and examples were presented to demonstrate their contributions for understanding disease-related biomarkers and the importance of metabolic flows for the development of new therapies and diagnostic methods for metabolic diseases. Metabolomics is the newest technique among "omics" approaches and we believe it will continue to evolve in different areas of the life science. Advances in the instrumental methods, software, hardware, and databases will still provide major contributions to drug discovery and clinical science.

References

1. Horgan RP, Kenny LC (2011) "Omic" technologies: genomics, transcriptomics, proteomics and metabolomics. Obstet Gynaecol 13:189–195. https://doi.org/10.1576/toag.13.3.189.27672
2. Bedia C (2018) Experimental approaches in omic sciences. In: Comprehensive analytical chemistry. Elsevier B.V., pp 13–36. https://doi.org/10.1016/bs.coac.2018.07.002
3. Godzien J, Gil de la Fuente A, Otero A, Barbas C (2018) Metabolite annotation and identification. In: Comprehensive analytical chemistry. Elsevier B.V., pp 415–445. https://doi.org/10.1016/bs.coac.2018.07.004
4. Sindelar M, Patti GJ (2020) Chemical discovery in the era of metabolomics. J Am Chem Soc 142:9097–9105. https://doi.org/10.1021/jacs.9b13198
5. Tebani A, Afonso C, Bekri S (2018) Advances in metabolome information retrieval: turning chemistry into biology. Part I: analytical chemistry of the metabolome. J Inherit Metab Dis 41:379–391. https://doi.org/10.1007/s10545-017-0074-y
6. Tebani A, Bekri S (2019) Paving the way to precision nutrition through metabolomics. Front Nutr 6:41. https://doi.org/10.3389/fnut.2019.00041
7. Sakaguchi CA, Nieman DC, Signini EF, Abreu RM, Catai AM (2019) Metabolomics-based studies assessing exercise-induced alterations of the human metabolome: a systematic review. Metabolites 9:164. https://doi.org/10.3390/metabo9080164
8. Ismail IT, Showalter MR, Fiehn O (2019) Inborn errors of metabolism in the era of untargeted metabolomics and lipidomics. Metabolites 9:242. https://doi.org/10.3390/metabo9100242
9. Mordaunt D, Cox D, Fuller M (2020) Metabolomics to improve the diagnostic efficiency of inborn errors of metabolism. Int J Mol Sci 21:1195. https://doi.org/10.3390/ijms21041195
10. Kell DB (2004) Metabolomics and systems biology: making sense of the soup. Curr Opin Microbiol 7:296–307. https://doi.org/10.1016/j.mib.2004.04.012
11. Lazar AG, Romanciuc F, Socaciu MA, Socaciu C (2015) Bioinformatics tools for metabolomic data processing and analysis using untargeted liquid chromatography coupled with mass spectrometry. Bull Univ Agric Sci Vet Med Cluj-Napoca Anim Sci Biotechnol 72:1843–1536. https://doi.org/10.15835/buasvmcn-asb:11536
12. Kuehnbaum NL, Britz-Mckibbin P (2013) New advances in separation science for metabolomics: resolving chemical diversity in a post-genomic era. Chem Rev 113:2437–2468. https://doi.org/10.1021/cr300484s
13. Rojo D, Barbas C, Rupérez FJ (2012) LC-MS metabolomics of polar compounds. Bioanalysis 4:1235–1243. https://doi.org/10.4155/bio.12.100
14. Troisi J, Cavallo P, Colucci A, Pierri L, Scala G, Symes S, Jones C, Richards S (2020) Metabolomics in genetic testing. In: Advances in clinical chemistry. Academic Press Inc, pp 85–153. https://doi.org/10.1016/bs.acc.2019.07.009
15. Neubert H, Shuford CM, Olah TV, Garofolo F, Schultz GA, Jones BR, Amaravadi L, Laterza OF, Xu K, Ackermann BL (2020) Protein biomarker quantification by immunoaffinity liquid chromatography-tandem mass spectrometry: current state and future vision. Clin Chem 66:282–301. https://doi.org/10.1093/CLINCHEM/HVZ022
16. van Eck NJ, Waltman L (2010) Software survey: VOSviewer, a computer program for bibliometric

mapping. Scientometrics 84:523–538. https://doi.org/10.1007/s11192-009-0146-3
17. Wilkins JM, Trushina E (2018) Application of metabolomics in Alzheimer's disease. Front Neurol 8:1. https://doi.org/10.3389/fneur.2017.00719
18. WHO (n.d.) Dementia. https://www.who.int/news--room/fact-sheets/detail/dementia. Accessed 29 Oct 2020
19. Malakar AK, Choudhury D, Halder B, Paul P, Uddin A, Chakraborty S (2019) A review on coronary artery disease, its risk factors, and therapeutics. J Cell Physiol 234:16812–16823. https://doi.org/10.1002/jcp.28350
20. Gika H, Virgiliou C, Theodoridis G, Plumb RS, Wilson ID (2019) Untargeted LC/MS-based metabolic phenotyping (metabonomics/metabolomics): the state of the art. J Chromatogr B 1117:136–147
21. Ortmayr K, Causon TJ, Hann S, Koellensperger G (2016) Increasing selectivity and coverage in LC-MS based metabolome analysis. TrAC – Trends Analyt Chem 82:358–366. https://doi.org/10.1016/j.trac.2016.06.011
22. Zhou B, Xiao JF, Tuli L, Ressom HW (2012) LC-MS-based metabolomics. Mol BioSyst 8:470–481. https://doi.org/10.1039/c1mb05350g
23. Snyder LR, Kirkland JJ, Glajch JL (1997) Practical HPLC method development. Wiley, New York. https://doi.org/10.1002/9781118592014
24. Saurina J, Sentellas S (2017) Strategies for metabolite profiling based on liquid chromatography. J Chromatogr B Anal Technol Biomed Life Sci 1044–1045:103–111. https://doi.org/10.1016/j.jchromb.2017.01.011
25. Fekete S, Oláh E, Fekete J (2012) Fast liquid chromatography: the domination of core-shell and very fine particles. J Chromatogr A 1228:57–71. https://doi.org/10.1016/j.chroma.2011.09.050
26. Tang DQ, Zou L, Yin XX, Ong CN (2016) HILIC-MS for metabolomics: an attractive and complementary approach to RPLC-MS. Mass Spectrom Rev 35:574–600. https://doi.org/10.1002/mas.21445
27. Knee JM, Rzezniczak TZ, Barsch A, Guo KZ, Merritt TJS (2013) A novel ion pairing LC/MS metabolomics protocol for study of a variety of biologically relevant polar metabolites. J Chromatogr B Anal Technol Biomed Life Sci 936:63–73. https://doi.org/10.1016/j.jchromb.2013.07.027
28. Seifar RM, Ras C, Deshmukh AT, Bekers KM, Suarez-Mendez CA, da Cruz ALB, van Gulik WM, Heijnen JJ (2013) Quantitative analysis of intracellular coenzymes in Saccharomyces cerevisiae using ion pair reversed phase ultra high performance liquid chromatography tandem mass spectrometry. J Chromatogr A 1311:115–120. https://doi.org/10.1016/j.chroma.2013.08.076
29. Schwaiger M, Rampler E, Hermann G, Miklos W, Berger W, Koellensperger G (2017) Anion-exchange chromatography coupled to high-resolution mass spectrometry: a powerful tool for merging targeted and non-targeted metabolomics. Anal Chem 89:7667–7674. https://doi.org/10.1021/acs.analchem.7b01624
30. Diop F, Vial T, Ferraris P, Wichit S, Bengue M, Hamel R, Talignani L, Liegeois F, Pompon J, Yssel H, Marti G, Missé D (2018) Zika virus infection modulates the metabolomic profile of microglial cells. PLoS One 13(10):e0206093. https://doi.org/10.1371/journal.pone.0206093
31. Licha D, Vidali S, Aminzadeh-Gohari S, Alka O, Breitkreuz L, Kohlbacher O, Reischl RJ, Feichtinger RG, Kofler B, Huber CG (2019) Untargeted metabolomics reveals molecular effects of ketogenic diet on healthy and tumor xenograft mouse models. Int J Mol Sci 20:3873. https://doi.org/10.3390/ijms20163873
32. D'Elia RV, Goodchild SA, Winder CL, Southam AD, Weber RJM, Stahl FM, Docx C, Patel V, Green AC, Viant MR, Lukaszewski RA, Dunn WB (2019) Multiple metabolic pathways are predictive of ricin intoxication in a rat model. Metabolomics 15:102. https://doi.org/10.1007/s11306-019-1547-9
33. Naser FJ, Mahieu NG, Wang L, Spalding JL, Johnson SL, Patti GJ (2018) Two complementary reversed-phase separations for comprehensive coverage of the semipolar and nonpolar metabolome. Anal Bioanal Chem 410:1287–1297. https://doi.org/10.1007/s00216-017-0768-x
34. Haggarty J, Oppermann M, Dalby MJ, Burchmore RJ, Cook K, Weidt S, Burgess KEV (2015) Serially coupling hydrophobic interaction and reversed-phase chromatography with simultaneous gradients provides greater coverage of the metabolome. Metabolomics 11:1465–1470. https://doi.org/10.1007/s11306-014-0770-7
35. Louw S, Pereira AS, Lynen F, Hanna-Brown M, Sandra P (2008) Serial coupling of reversed-phase and hydrophilic interaction liquid chromatography to broaden the elution window for the analysis of pharmaceutical compounds. J Chromatogr A 1208:90–94. https://doi.org/10.1016/j.chroma.2008.08.058
36. Greco G, Grosse S, Letzel T (2013) Serial coupling of reversed-phase and zwitterionic hydrophilic interaction LC/MS for the analysis of polar and nonpolar phenols in wine. J Sep Sci 36:1379–1388. https://doi.org/10.1002/jssc.201200920
37. Schwaiger M, Schoeny H, El Abiead Y, Hermann G, Rampler E, Koellensperger G (2019) Merging metabolomics and lipidomics into one analytical run. Analyst 144:220–229. https://doi.org/10.1039/c8an01219a
38. Stoll DR, Shoykhet K, Petersson P, Buckenmaier S (2017) Active solvent modulation: a valve-based approach to improve separation compatibility in two-dimensional liquid chromatography. Anal Chem 89:9260–9267. https://doi.org/10.1021/acs.analchem.7b02046
39. Song Q, Liu W, Chen X, Li J, Li P, Yang F, Wang Y, Song Y, Tu P (2018) Serially coupled reversed phase-hydrophilic interaction liquid chromatography–tailored multiple reaction monitoring, a fit-for-

purpose tool for large-scale targeted metabolomics of medicinal bile. Anal Chim Acta 1037:119–129. https://doi.org/10.1016/j.aca.2017.11.072
40. Yan Y, Song Q, Chen X, Li J, Li P, Wang Y, Liu T, Song Y, Tu P (2017) Simultaneous determination of components with wide polarity and content ranges in Cistanche tubulosa using serially coupled reverse phase-hydrophilic interaction chromatography-tandem mass spectrometry. J Chromatogr A 1501:39–50. https://doi.org/10.1016/j.chroma.2017.04.034
41. Klassen A, Faccio AT, Canuto GAB, da Cruz PLR, Ribeiro HC, Tavares MFM, Sussulini A (2017) Metabolomics: definitions and significance in systems biology. In: Metabolomics: from fundamentals to clinical applications. Advances in experimental medicine and biology. Springer, Cham. https://doi.org/10.1007/978-3-319-47656-8_1
42. Montgomery DC (2012) Design and analysis of experiments, 8th edn. Wiley, New York. https://doi.org/10.1198/tech.2006.s372
43. Contrepois K, Jiang L, Snyder M (2015) Optimized analytical procedures for the untargeted metabolomic profiling of human urine and plasma by combining hydrophilic interaction (HILIC) and reverse-phase liquid chromatography (RPLC)–mass spectrometry. Mol Cell Proteomics 14:1684–1695. https://doi.org/10.1074/mcp.M114.046508
44. Yanes O, Tautenhahn R, Patti GJ, Siuzdak G (2011) Expanding coverage of the metabolome for global metabolite profiling. Anal Chem 83:2152–2161. https://doi.org/10.1021/ac102981k
45. Creydt M, Fischer M (2017) Plant metabolomics: maximizing metabolome coverage by optimizing mobile phase additives for nontargeted mass spectrometry in positive and negative electrospray ionization mode. Anal Chem 89:10474–10486. https://doi.org/10.1021/acs.analchem.7b02592
46. Leardi R (2009) Experimental design in chemistry: a tutorial. Anal Chim Acta 652:161–172. https://doi.org/10.1016/j.aca.2009.06.015
47. Ebrahimi-Najafabadi H, Leardi R, Jalali-Heravi M (2014) Experimental design in analytical chemistry – Part I: Theory. J AOAC Int 97:3–11. https://doi.org/10.5740/jaoacint.SGEEbrahimi1
48. Jacyna J, Kordalewska M, Markuszewski MJ (2019) Design of experiments in metabolomics-related studies: an overview. J Pharm Biomed Anal 164:598–606. https://doi.org/10.1016/j.jpba.2018.11.027
49. Rhoades SD, Weljie AM (2016) Comprehensive optimization of LC–MS metabolomics methods using design of experiments (COLMeD). Metabolomics 12:183. https://doi.org/10.1007/s11306-016-1132-4
50. Pilon AC, Carnevale Neto F, Freire RT, Cardoso P, Carneiro RL, Da Silva Bolzani V, Castro-Gamboa I (2016) Partial least squares model and design of experiments toward the analysis of the metabolome of Jatropha gossypifolia leaves: extraction and chromatographic fingerprint optimization. J Sep Sci 39:1023–1030. https://doi.org/10.1002/jssc.201500892
51. Zheng H, Clausen MR, Dalsgaard TK, Mortensen G, Bertram HC (2013) Time-saving design of experiment protocol for optimization of LC-MS data processing in metabolomic approaches. Anal Chem 85:7109–7116. https://doi.org/10.1021/ac4020325
52. Peris-Díaz MD, Rodak O, Sweeney SR, Krężel A, Sentandreu E (2019) Chemometrics-assisted optimization of liquid chromatography-quadrupole-time-of-flight mass spectrometry analysis for targeted metabolomics. Talanta 199:380–387. https://doi.org/10.1016/j.talanta.2019.02.075
53. Tebani A, Schmitz-Afonso I, Rutledge DN, Gonzalez BJ, Bekri S, Afonso C (2016) Optimization of a liquid chromatography ion mobility-mass spectrometry method for untargeted metabolomics using experimental design and multivariate data analysis. Anal Chim Acta 913:55–62. https://doi.org/10.1016/j.aca.2016.02.011
54. Álvarez-Sánchez B, Priego-Capote F, Luque de Castro MD (2010) Metabolomics analysis II. Selection of biological samples and practical aspects preceding sample preparation. TrAC – Trends Analyt Chem 29:120–127
55. Vuckovic D (2012) Current trends and challenges in sample preparation for global metabolomics using liquid chromatography-mass spectrometry. Anal Bioanal Chem 403:1523–1548. https://doi.org/10.1007/s00216-012-6039-y
56. Raterink RJ, Lindenburg PW, Vreeken RJ, Ramautar R, Hankemeier T (2014) Recent developments in sample-pretreatment techniques for mass spectrometry-based metabolomics. TrAC – Trends Analyt Chem 61:157–167. https://doi.org/10.1016/j.trac.2014.06.003
57. Tsakelidou E, Virgiliou C, Valianou L, Gika HG, Raikos N, Theodoridis G (2017) Sample preparation strategies for the effective quantitation of hydrophilic metabolites in serum by multi-targeted HILIC-MS/MS. Metabolites 7:13. https://doi.org/10.3390/metabo7020013
58. Stevens VL, Hoover E, Wang Y, Zanetti KA (2019) Pre-analytical factors that affect metabolite stability in human urine, plasma, and serum: a review. Metabolites 9:156. https://doi.org/10.3390/metabo9080156
59. Yin P, Xu G (2014) Current state-of-the-art of nontargeted metabolomics based on liquid chromatography-mass spectrometry with special emphasis in clinical applications. J Chromatogr A 1374:1–13. https://doi.org/10.1016/j.chroma.2014.11.050
60. Sánchez-López E, Crego AL, Marina ML (2017) Design of strategies to study the metabolic profile of highly polar compounds in plasma by reversed-phase liquid chromatography–high resolution mass spectrometry. J Chromatogr A 1490:156–165. https://doi.org/10.1016/j.chroma.2017.02.029
61. Wawrzyniak R, Kosnowska A, Macioszek S, Bartoszewski R, Markuszewski MJ (2018) New plasma preparation approach to enrich metabolome coverage in untargeted metabolomics: plasma pro-

tein bound hydrophobic metabolite release with proteinase K. Sci Rep 8:9541. https://doi.org/10.1038/s41598-018-27983-0
62. Chan W, Zhao Y, Zhang J (2019) Evaluating the performance of sample preparation methods for ultra-performance liquid chromatography/mass spectrometry based serum metabonomics. Rapid Commun Mass Spectrom 33:561–568. https://doi.org/10.1002/rcm.8381
63. Want EJ, O'Maille G, Smith CA, Brandon TR, Uritboonthai W, Qin C, Trauger SA, Siuzdak G (2006) Solvent-dependent metabolite distribution, clustering, and protein extraction for serum profiling with mass spectrometry. Anal Chem 78:743–752. https://doi.org/10.1021/ac051312t
64. Ivanisevic J, Want EJ (2019) From samples to insights into metabolism: uncovering biologically relevant information in LC-HRMS metabolomics data. Metabolites 9:308. https://doi.org/10.3390/metabo9120308
65. Chetwynd AJ, Dunn WB, Rodriguez-Blanco G (2017) Collection and preparation of clinical samples for metabolomics. In: Metabolomics: from fundamentals to clinical applications. Advances in experimental medicine and biology. Springer, Cham, pp 19–44. https://doi.org/10.1007/978-3-319-47656-8_2
66. Lu W, Su X, Klein MS, Lewis IA, Fiehn O, Rabinowitz JD (2017) Metabolite measurement: pitfalls to avoid and practices to follow. Annu Rev Biochem 86:277–304. https://doi.org/10.1146/annurev-biochem-061516-044952
67. Chen S, Hoene M, Li J, Li Y, Zhao X, Häring HU, Schleicher ED, Weigert C, Xu G, Lehmann R (2013) Simultaneous extraction of metabolome and lipidome with methyl tert-butyl ether from a single small tissue sample for ultra-high performance liquid chromatography/mass spectrometry. J Chromatogr A 1298:9–16. https://doi.org/10.1016/j.chroma.2013.05.019
68. Godzien J, Ciborowski M, Whiley L, Legido-Quigley C, Ruperez FJ, Barbas C (2013) In-vial dual extraction liquid chromatography coupled to mass spectrometry applied to streptozotocin-treated diabetic rats. Tips and pitfalls of the method. J Chromatogr A 1304:52–60. https://doi.org/10.1016/j.chroma.2013.07.029
69. Saric J, Want EJ, Duthaler U, Lewis M, Keiser J, Shockcor JP, Ross GA, Nicholson JK, Holmes E, Tavares MFM (2012) Systematic evaluation of extraction methods for multiplatform-based metabotyping: application to the Fasciola hepatica metabolome. Anal Chem 84:6963–6972. https://doi.org/10.1021/ac300586m
70. Deda O, Chatziioannou AC, Fasoula S, Palachanis D, Raikos N, Theodoridis GA, Gika HG (2017) Sample preparation optimization in fecal metabolic profiling. J Chromatogr B Anal Technol Biomed Life Sci 1047:115–123. https://doi.org/10.1016/j.jchromb.2016.06.047
71. Lee MS (2019) Wiley series on pharmaceutical science and biotechnology: practices, applications, and methods. Series Editor. http://www.wiley.com/go/permissions
72. Michopoulos F, Gika H, Palachanis D, Theodoridis G, Wilson ID (2015) Solid phase extraction methodology for UPLC-MS based metabolic profiling of urine samples. Electrophoresis 36:2170–2178. https://doi.org/10.1002/elps.201500101
73. Álvarez-Sánchez B, Priego-Capote F, Luque de Castro MD (2010) Metabolomics analysis I. Selection of biological samples and practical aspects preceding sample preparation. TrAC – Trends Analyt Chem 29:111–119
74. Rogeberg M, Malerod H, Roberg-Larsen H, Aass C, Wilson SR (2014) On-line solid phase extraction-liquid chromatography, with emphasis on modern bioanalysis and miniaturized systems. J Pharm Biomed Anal 87:120–129. https://doi.org/10.1016/j.jpba.2013.05.006
75. Calderón-Santiago M, Priego-Capote F, de Castro MDL (2015) Enhancing detection coverage in untargeted metabolomics analysis by solid-phase extraction on-line coupled to LC-MS/MS. Electrophoresis 36:2179–2187. https://doi.org/10.1002/elps.201500060
76. Tang YQ, Weng N (2013) Salting-out assisted liquid-liquid extraction for bioanalysis. Bioanalysis 5:1583–1598. https://doi.org/10.4155/bio.13.117
77. Bojko B, Reyes-Garcés N, Bessonneau V, Goryński K, Mousavi F, Souza Silva EA, Pawliszyn J (2014) Solid-phase microextraction in metabolomics. TrAC – Trends Analyt Chem 61:168–180. https://doi.org/10.1016/j.trac.2014.07.005
78. Reyes-Garcés N, Gionfriddo E (2019) Recent developments and applications of solid phase microextraction as a sample preparation approach for mass-spectrometry-based metabolomics and lipidomics. TrAC – Trends Analyt Chem 113:172–181. https://doi.org/10.1016/j.trac.2019.01.009
79. Hemmati M, Nix C, Crommen J, Servais AC, Fillet M (2020) Benefits of microsampling and microextraction for metabolomics studies. TrAC – Trends Analyt Chem 127:115899. https://doi.org/10.1016/j.trac.2020.115899
80. Levernæs MCS, Farhat B, Oulie I, Abdullah SS, Paus E, Reubsaet L, Halvorsen TG (2019) Immunocapture sample clean-up in determination of low abundant protein biomarkers-a feasibility study of peptide capture by anti-protein antibodies. RSC Adv 9:34902–34911. https://doi.org/10.1039/c9ra05071j
81. Chen L (2018) Current status of anti-drug antibody analysis using immunocapture-liquid chromatography/mass spectrometry. J Appl Bioanal 4:74–80. https://doi.org/10.17145/jab.18.011
82. Fredolini C, Byström S, Pin E, Edfors F, Tamburro D, Iglesias MJ, Häggmark A, Hong MG, Uhlen M, Nilsson P, Schwenk JM (2016) Immunocapture strategies in translational proteomics. Expert Rev

Proteomics 13:83–98. https://doi.org/10.1586/14789450.2016.1111141
83. Liu A (2019) Immunocapture in LC-MS bioanalysis. In: Sample preparation in LC-MS bioanalysis. Wiley, pp 85–97. https://doi.org/10.1002/9781119274315.ch7
84. Chen WW, Freinkman E, Wang T, Birsoy K, Sabatini DM (2016) Absolute quantification of matrix metabolites reveals the dynamics of mitochondrial metabolism. Cell 166:1324–1337.e11. https://doi.org/10.1016/j.cell.2016.07.040
85. Liu X, Locasale JW (2017) Metabolomics: a primer. Trends Biochem Sci 42:274–284. https://doi.org/10.1016/j.tibs.2017.01.004
86. Emwas AH, Roy R, McKay RT, Tenori L, Saccenti E, Nagana Gowda GA, Raftery D, Alahmari F, Jaremko L, Jaremko M, Wishart DS (2019) Nmr spectroscopy for metabolomics research. Metabolites 9:123. https://doi.org/10.3390/metabo9070123
87. Liu X, Zhou L, Shi X, Xu G (2019) New advances in analytical methods for mass spectrometry-based large-scale metabolomics study. TrAC – Trends Analyt Chem 121:115665. https://doi.org/10.1016/j.trac.2019.115665
88. Wishart DS (2019) NMR metabolomics: a look ahead. J Magn Reson 306:155–161. https://doi.org/10.1016/j.jmr.2019.07.013
89. Wishart DS (2016) Emerging applications of metabolomics in drug discovery and precision medicine. Nat Rev Drug Discov 15:473–484. https://doi.org/10.1038/nrd.2016.32
90. Beale DJ, Pinu FR, Kouremenos KA, Poojary MM, Narayana VK, Boughton BA, Kanojia K, Dayalan S, Jones OAH, Dias DA (2018) Review of recent developments in GC–MS approaches to metabolomics-based research. Metabolomics 14:152. https://doi.org/10.1007/s11306-018-1449-2
91. Werner E, Heilier JF, Ducruix C, Ezan E, Junot C, Tabet JC (2008) Mass spectrometry for the identification of the discriminating signals from metabolomics: current status and future trends. J Chromatogr B Anal Technol Biomed Life Sci 871:143–163. https://doi.org/10.1016/j.jchromb.2008.07.004
92. Ramautar R (2018) Chapter 1: Capillary electrophoresis-mass spectrometry for metabolomics – from metabolite analysis to metabolic profiling. In: Capillary electrophoresis–mass spectrometry for metabolomics. New developments in mass spectrometry. Royal Society of Chemistry, pp 1–20. https://doi.org/10.1039/9781788012737-00001
93. Henry R, Cassel T (2012) Metabolomic applications of inductively coupled plasma-mass spectrometry (ICP-MS). Methods Pharmacol Toxicol 17:99–125. https://doi.org/10.1007/978-1-61779-618-0_5
94. Haggarty J, Burgess KE (2017) Recent advances in liquid and gas chromatography methodology for extending coverage of the metabolome. Curr Opin Biotechnol 43:77–85. https://doi.org/10.1016/j.copbio.2016.09.006
95. Banerjee S, Mazumdar S (2012) Electrospray ionization mass spectrometry: a technique to access the information beyond the molecular weight of the analyte. Int J Anal Chem 2012:1–40. https://doi.org/10.1155/2012/282574
96. Tugizimana F, Steenkamp PA, Piater LA, Dubery IA (2018) Mass spectrometry in untargeted liquid chromatography/mass spectrometry metabolomics: electrospray ionisation parameters and global coverage of the metabolome. Rapid Commun Mass Spectrom 32:121–132. https://doi.org/10.1002/rcm.8010
97. Zhou J, Yin Y (2016) Strategies for large-scale targeted metabolomics quantification by liquid chromatography-mass spectrometry. Analyst 141:6362–6373. https://doi.org/10.1039/c6an01753c
98. Rochat B (2016) From targeted quantification to untargeted metabolomics: why LC-high-resolution-MS will become a key instrument in clinical labs. TrAC – Trends Analyt Chem 84:151–164. https://doi.org/10.1016/j.trac.2016.02.009
99. Gorrochategui E, Jaumot J, Lacorte S, Tauler R (2016) Data analysis strategies for targeted and untargeted LC-MS metabolomic studies: overview and workflow. TrAC – Trends Analyt Chem 82:425–442. https://doi.org/10.1016/j.trac.2016.07.004
100. Che N, Ma Y, Ruan H, Xu L, Wang X, Yang X, Liu X (2018) Integrated semi-targeted metabolomics analysis reveals distinct metabolic dysregulation in pleural effusion caused by tuberculosis and malignancy. Clin Chim Acta 477:81–88. https://doi.org/10.1016/j.cca.2017.12.003
101. Roberts LD, Souza AL, Gerszten RE, Clish CB (2012) Targeted metabolomics. Curr Protoc Mol Biol 1:Unit30.2. https://doi.org/10.1002/0471142727.mb3002s98
102. Ren JL, Zhang AH, Kong L, Wang XJ (2018) Advances in mass spectrometry-based metabolomics for investigation of metabolites. RSC Adv 8:22335–22350. https://doi.org/10.1039/c8ra01574k
103. Gu H, Zhang P, Zhu J, Raftery D (2015) Globally optimized targeted mass spectrometry: reliable metabolomics analysis with broad coverage. Anal Chem 87:12355–12362. https://doi.org/10.1021/acs.analchem.5b03812
104. Patti GJ, Yanes O, Siuzdak G (2012) Innovation: metabolomics: the apogee of the omics trilogy. Nat Rev Mol Cell Biol 13(4):263–269. https://doi.org/10.1038/nrm3314
105. Hopfgartner G (2007) Mass spectrometry in bioanalysis – methods, principles and instrumentation. In: Mass spectrometry in medicinal chemistry. Wiley Blackwell, pp 1–62. https://doi.org/10.1002/9783527610907.ch1
106. Boesl U (2017) Time-of-flight mass spectrometry: introduction to the basics. Mass Spectrom Rev 36:86–109. https://doi.org/10.1002/mas.21520
107. Allen D, McWhinney B (2019) Quadrupole time-of-flight mass spectrometry: a paradigm shift in

toxicology screening applications. Clin Biochem Rev 40:135–146. https://doi.org/10.33176/aacb-19-00023
108. Ens W, Standing KG (2005) Hybrid quadrupole/time-of-flight mass spectrometers for analysis of biomolecules. Methods Enzymol 402:49–78. https://doi.org/10.1016/S0076-6879(05)02002-1
109. Campbell JL, Le Blanc JCY (2012) Using high-resolution quadrupole TOF technology in DMPK analyses. Bioanalysis 4:487–500. https://doi.org/10.4155/bio.12.14
110. Makarov A (2000) Electrostatic axially harmonic orbital trapping: a high-performance technique of mass analysis. Anal Chem 72:1156–1162. https://doi.org/10.1021/ac991131p
111. Scigelova M, Makarov A (2006) Orbitrap mass analyzer – overview and applications in proteomics. Proteomics 6(Suppl 2):16–21. https://doi.org/10.1002/pmic.200600528
112. Ludwig C, Gillet L, Rosenberger G, Amon S, Collins BC, Aebersold R (2018) Data-independent acquisition-based SWATH – MS for quantitative proteomics: a tutorial. Mol Syst Biol 14:e8126. https://doi.org/10.15252/msb.20178126
113. Michalski A, Damoc E, Hauschild JP, Lange O, Wieghaus A, Makarov A, Nagaraj N, Cox J, Mann M, Horning S (2011) Mass spectrometry-based proteomics using Q exactive, a high-performance benchtop quadrupole orbitrap mass spectrometer. Mol Cell Proteomics 10:M111.011015. https://doi.org/10.1074/mcp.M111.011015
114. Saint Hilaire PB, Rousseau K, Seyer A, Dechaumet S, Damont A, Junot C, Fenaille F (2020) Comparative evaluation of data dependent and data independent acquisition workflows implemented on an orbitrap fusion for untargeted metabolomics. Metabolites 10:158. https://doi.org/10.3390/metabo10040158
115. Hecht ES, Scigelova M, Eliuk S, Makarov A (2019) Fundamentals and advances of orbitrap mass spectrometry. In: Encyclopedia of analytical chemistry. Wiley, pp 1–40. https://doi.org/10.1002/9780470027318.a9309.pub2
116. Frese CK, Altelaar AFM, Hennrich ML, Nolting D, Zeller M, Griep-Raming J, Heck AJR, Mohammed S (2011) Improved peptide identification by targeted fragmentation using CID, HCD and ETD on an LTQ-Orbitrap Velos. J Proteome Res 10:2377–2388. https://doi.org/10.1021/pr1011729
117. Pappireddi N, Martin L, Wühr M (2019) A review on quantitative multiplexed proteomics. ChemBioChem 20:1210–1224. https://doi.org/10.1002/cbic.201800650
118. Guo J, Huan T (2020) Comparison of full-scan, data-dependent, and data-independent acquisition modes in liquid chromatography-mass spectrometry based untargeted metabolomics. Anal Chem 92:8072–8080. https://doi.org/10.1021/acs.analchem.9b05135
119. Xie C, Zhong D, Yu K, Chen X (2012) Recent advances in metabolite identification and quantitative bioanalysis by LC-Q-TOF MS. Bioanalysis 4:937–959. https://doi.org/10.4155/bio.12.43
120. Ma S, Chowdhury SK (2013) Data acquisition and data mining techniques for metabolite identification using LC coupled to high-resolution MS. Bioanalysis 5:1285–1297. https://doi.org/10.4155/bio.13.103
121. Bern M, Finney G, Hoopmann MR, Merrihew G, Toth MJ, MacCoss MJ (2010) Deconvolution of mixture spectra from ion-trap data-independent-acquisition tandem mass spectrometry. Anal Chem 82:833–841. https://doi.org/10.1021/ac901801b
122. Pino LK, Just SC, MacCoss MJ, Searle BC (2020) Acquiring and analyzing data independent acquisition proteomics experiments without spectrum libraries. Mol Cell Proteomics 19:1088–1103. https://doi.org/10.1074/mcp.P119.001913
123. Plumb RS, Johnson KA, Rainville P, Smith BW, Wilson ID, Castro-Perez JM, Nicholson JK (2006) UPLC/MSE; a new approach for generating molecular fragment information for biomarker structure elucidation. Rapid Commun Mass Spectrom 20:1989–1994. https://doi.org/10.1002/rcm.2550
124. Li GZ, Vissers JPC, Silva JC, Golick D, Gorenstein MV, Geromanos SJ (2009) Database searching and accounting of multiplexed precursor and product ion spectra from the data independent analysis of simple and complex peptide mixtures. Proteomics 9:1696–1719. https://doi.org/10.1002/pmic.200800564
125. Shliaha PV, Bond NJ, Gatto L, Lilley KS (2013) Effects of traveling wave ion mobility separation on data independent acquisition in proteomics studies. J Proteome Res 12:2323–2339. https://doi.org/10.1021/pr300775k
126. Chaleckis R, Naz S, Meister I, Wheelock CE (2018) LC-MS-based metabolomics of biofluids using all-ion fragmentation (AIF) acquisition. In: Methods in molecular biology. Humana Press Inc, pp 45–58. https://doi.org/10.1007/978-1-4939-7592-1_3
127. Geiger T, Cox J, Mann M (2010) Proteomics on an orbitrap benchtop mass spectrometer using all-ion fragmentation. Mol Cell Proteomics 9:2252–2261. https://doi.org/10.1074/mcp.M110.001537
128. Spicer R, Salek RM, Moreno P, Cañueto D, Steinbeck C (2017) Navigating freely-available software tools for metabolomics analysis. Metabolomics 13:106. https://doi.org/10.1007/s11306-017-1242-7
129. Fenaille F, Barbier Saint-Hilaire P, Rousseau K, Junot C (2017) Data acquisition workflows in liquid chromatography coupled to high resolution mass spectrometry-based metabolomics: where do we stand? J Chromatogr A 1526:1–12. https://doi.org/10.1016/j.chroma.2017.10.043
130. Smith CA, Want EJ, O'Maille G, Abagyan R, Siuzdak G (2006) XCMS: processing mass spectrometry data for metabolite profiling using nonlinear peak alignment, matching, and identification. Anal Chem 78:779–787. https://doi.org/10.1021/ac051437y
131. Pluskal T, Castillo S, Villar-Briones A, Orešič M (2010) MZmine 2: modular framework for process-

ing, visualizing, and analyzing mass spectrometry-based molecular profile data. BMC Bioinformatics 11:395. https://doi.org/10.1186/1471-2105-11-395
132. Bertsch A, Gröpl C, Reinert K, Kohlbacher O (2011) OpenMS and TOPP: open source software for LC-MS data analysis. Methods Mol Biol 696:353–367. https://doi.org/10.1007/978-1-60761-987-1_23
133. Lommen A, Kools HJ (2012) MetAlign 3.0: performance enhancement by efficient use of advances in computer hardware. Metabolomics 8:719–726. https://doi.org/10.1007/s11306-011-0369-1
134. Tsugawa H, Cajka T, Kind T, Ma Y, Higgins B, Ikeda K, Kanazawa M, Vandergheynst J, Fiehn O, Arita M (2015) MS-DIAL: data-independent MS/MS deconvolution for comprehensive metabolome analysis. Nat Methods 12:523–526. https://doi.org/10.1038/nmeth.3393
135. O'Shea K, Misra BB (2020) Software tools, databases and resources in metabolomics: updates from 2018 to 2019. Metabolomics 16:36. https://doi.org/10.1007/s11306-020-01657-3
136. Tautenhahn R, Patti GJ, Rinehart D, Siuzdak G (2012) XCMS online: a web-based platform to process untargeted metabolomic data. Anal Chem 84:5035–5039. https://doi.org/10.1021/ac300698c
137. Melamud E, Vastag L, Rabinowitz JD (2010) Metabolomic analysis and visualization engine for LC – MS data. Anal Chem 82:9818–9826. https://doi.org/10.1021/ac1021166
138. Xia J, Sinelnikov IV, Han B, Wishart DS (2015) MetaboAnalyst 3.0-making metabolomics more meaningful. Nucleic Acids Res 43:W251–W257. https://doi.org/10.1093/nar/gkv380
139. Giacomoni F, Le Corguillé G, Monsoor M, Landi M, Pericard P, Pétéra M, Duperier C, Tremblay-Franco M, Martin JF, Jacob D, Goulitquer S, Thévenot EA, Caron C (2015) Workflow4Metabolomics: a collaborative research infrastructure for computational metabolomics. Bioinformatics 31:1493–1495. https://doi.org/10.1093/bioinformatics/btu813
140. Smith CA, O'Maille G, Want EJ, Qin C, Trauger SA, Brandon TR, Custodio DE, Abagyan R, Siuzdak G (2005) METLIN: a metabolite mass spectral database. Ther Drug Monit 27:747–751. https://doi.org/10.1097/01.ftd.0000179845.53213.39
141. Wishart DS, Jewison T, Guo AC, Wilson M, Knox C, Liu Y, Djoumbou Y, Mandal R, Aziat F, Dong E, Bouatra S, Sinelnikov I, Arndt D, Xia J, Liu P, Yallou F, Bjorndahl T, Perez-Pineiro R, Eisner R, Allen F, Neveu V, Greiner R, Scalbert A (2013) HMDB 3.0—the human metabolome database in 2013. Nucleic Acids Res 41:D801–D807. https://doi.org/10.1093/nar/gks1065
142. Horai H, Arita M, Kanaya S, Nihei Y, Ikeda T, Suwa K, Ojima Y, Tanaka K, Tanaka S, Aoshima K, Oda Y, Kakazu Y, Kusano M, Tohge T, Matsuda F, Sawada Y, Hirai MY, Nakanishi H, Ikeda K, Akimoto N, Maoka T, Takahashi H, Ara T, Sakurai N, Suzuki H, Shibata D, Neumann S, Iida T, Tanaka K, Funatsu K, Matsuura F, Soga T, Taguchi R, Saito K, Nishioka T (2010) MassBank: a public repository for sharing mass spectral data for life sciences. J Mass Spectrom 45:703–714. https://doi.org/10.1002/jms.1777
143. Ruttkies C, Schymanski EL, Wolf S, Hollender J, Neumann S (2016) MetFrag relaunched: incorporating strategies beyond in silico fragmentation. J Cheminform 8:3. https://doi.org/10.1186/s13321-016-0115-9
144. WHO, International Programme on Chemical Safety (2001) Biomarkers in risk assessment: validity and validation. Environmental health criteria series, vol 222. WHO, Geneva. https://apps.who.int/iris/handle/10665/42363
145. Strumbu K, Tavel JA (2010) What are biomarkers? Curr Opin HIV AIDS 5:463–466. https://doi.org/10.1097/COH.0b013e32833ed177.What
146. Puntmann VO (2009) How-to guide on biomarkers: biomarker definitions, validation and applications with examples from cardiovascular disease. Postgrad Med J 85:538–545. https://doi.org/10.1136/pgmj.2008.073759
147. Jain KK (2010) The handbook of biomarkers. Springer Science and Business Media, Basel. https://doi.org/10.1007/978-1-60761-685-6
148. Monteiro MS, Carvalho M, Bastos ML, Guedes de Pinho P (2013) Metabolomics analysis for biomarker discovery: advances and challenges. Curr Med Chem 20:257–271. https://doi.org/10.2174/092986713804806621
149. Marchand CR, Farshidfar F, Rattner J, Bathe OF (2018) A framework for development of useful metabolomic biomarkers and their effective knowledge translation. Metabolites 8:59. https://doi.org/10.3390/metabo8040059
150. Xia J, Broadhurst DI, Wilson M, Wishart DS (2013) Translational biomarker discovery in clinical metabolomics: an introductory tutorial. Metabolomics 9:280–299. https://doi.org/10.1007/s11306-012-0482-9
151. Pepe MS, Etzioni R, Feng Z, Potter JD, Lou M, Thornquist M, Winget M, Yasui Y (2001) Phases of biomarker development for early detection of cancer. J Natl Cancer Inst 93:1054–1061. https://doi.org/10.1093/jnci/93.14.1054
152. Food and Drug Administration (2018) Biomarker qualification program. https://www.fda.gov/drugs/drug-development-tool-ddt-qualification-programs/biomarker-qualification-program. Accessed 13 Oct 2020
153. European Medicines Agency (2020) Biomarker qualification. https://www.ema.europa.eu/en/human-regulatory/research-development/scientific-advice-protocol-assistance/novel-methodologies-biomarkers/opinions-letterssupport-qualification-novel-methodologies-medicine-development. Accessed 13 Oct 2020
154. Lindahl A, Forshed J, Nordström A (2016) Overlap in serum metabolic profiles between non-related diseases: implications for LC-MS metabolomics biomarker discovery. Biochem Biophys Res

Commun 478:1472–1477. https://doi.org/10.1016/j.bbrc.2016.08.155
155. Ioannidis JPA, Bossuyt PMM (2017) Waste, leaks, and failures in the biomarker pipeline. Clin Chem 63:963–972. https://doi.org/10.1373/clinchem.2016.254649
156. Guasti L, Galliazzo S, Molaro M, Visconti E, Pennella B, Gaudio GV, Lupi A, Grandi AM, Squizzato A (2021) TMAO as a biomarker of cardiovascular events: a systematic review and meta-analysis. Intern Emerg Med 16(1):201–207. https://doi.org/10.1007/s11739-020-02470-5
157. Wang Z, Klipfell E, Bennett BJ, Koeth R, Levison BS, Dugar B, Feldstein AE, Britt EB, Fu X, Chung Y-M, Wu Y, Schauer P, Smith JD, Allayee H, Tang WHW, DiDonato JA, Lusis AJ, Hazen SL (2011) Gut flora metabolism of phosphatidylcholine promotes cardiovascular disease. Nature 472:57–63. https://doi.org/10.1038/nature09922
158. Koeth RA, Wang Z, Levison BS, Buffa JA, Org E, Sheehy BT, Britt EB, Fu X, Wu Y, Li L, Smith JD, Didonato JA, Chen J, Li H, Wu GD, Lewis JD, Warrier M, Brown JM, Krauss RM, Tang WHW, Bushman FD, Lusis AJ, Hazen SL (2013) Intestinal microbiota metabolism of l-carnitine, a nutrient in red meat, promotes atherosclerosis. Nat Med 19:576–585. https://doi.org/10.1038/nm.3145
159. Wang Z, Tang WHW, Buffa JA, Fu X, Britt EB, Koeth RA, Levison BS, Fan Y, Wu Y, Hazen SL (2014) Prognostic value of choline and betaine depends on intestinal microbiota-generated metabolite trimethylamine-N-oxide. Eur Heart J 35:904–910. https://doi.org/10.1093/eurheartj/ehu002
160. Gregory JC, Buffa JA, Org E, Wang Z, Levison BS, Zhu W, Wagner MA, Bennett BJ, Li L, DiDonato JA, Lusis AJ, Hazen SL (2015) Transmission of atherosclerosis susceptibility with gut microbial transplantation. J Biol Chem 290:5647–5660. https://doi.org/10.1074/jbc.M114.618249
161. Warrier M, Shih DM, Burrows AC, Ferguson D, Gromovsky AD, Brown AL, Marshall S, McDaniel A, Schugar RC, Wang Z, Sacks J, Rong X, Vallim TA, Chou J, Ivanova PT, Myers DS, Brown HA, Lee RG, Crooke RM, Graham MJ, Liu X, Parini P, Tontonoz P, Lusis AJ, Hazen SL, Temel RE, Brown JM (2015) The TMAO-generating enzyme flavin monooxygenase 3 is a central regulator of cholesterol balance. Cell Rep 10:326–338. https://doi.org/10.1016/j.celrep.2014.12.036
162. Janeiro MH, Ramírez MJ, Milagro FI, Martínez JA, Solas M (2018) Implication of trimethylamine n-oxide (TMAO) in disease: potential biomarker or new therapeutic target. Nutrients 10:1398. https://doi.org/10.3390/nu10101398
163. Zhu Y, Li Q, Jiang H (2020) Gut microbiota in atherosclerosis: focus on trimethylamine N-oxide. APMIS 128:353–366. https://doi.org/10.1111/apm.13038
164. Canyelles M, Tondo M, Cedó L, Farràs M, Escolà-Gil JC, Blanco-Vaca F (2018) Trimethylamine N-oxide: a link among diet, gut microbiota, gene regulation of liver and intestine cholesterol homeostasis and HDL function. Int J Mol Sci 19:3228. https://doi.org/10.3390/ijms19103228
165. Ufnal M, Zadlo A, Ostaszewski R (2015) TMAO: a small molecule of great expectations. Nutrition 31:1317–1323. https://doi.org/10.1016/j.nut.2015.05.006
166. Wang TJ, Larson MG, Vasan RS, Cheng S, Rhee EP, McCabe E, Lewis GD, Fox CS, Jacques PF, Fernandez C, O'Donnell CJ, Carr SA, Mootha VK, Florez JC, Souza A, Melander O, Clish CB, Gerszten RE (2011) Metabolite profiles and the risk of developing diabetes. Nat Med 17:448–453. https://doi.org/10.1038/nm.2307
167. Klein MS, Shearer J (2016) Metabolomics and type 2 diabetes: translating basic research into clinical application. J Diabetes Res 2016:3898502. https://doi.org/10.1155/2016/3898502
168. Pallares-Méndez R, Aguilar-Salinas CA, Cruz-Bautista I, Del Bosque-Plata L (2016) Metabolomics in diabetes, a review. Ann Med 48:89–102. https://doi.org/10.3109/07853890.2015.1137630
169. Guasch-Ferré M, Hruby A, Toledo E, Clish CB, Martínez-González MA, Salas-Salvadó J, Hu FB (2016) Metabolomics in prediabetes and diabetes: a systematic review and meta-analysis. Diabetes Care 39:833–846. https://doi.org/10.2337/dc15-2251
170. Satheesh G, Ramachandran S, Jaleel A (2020) Metabolomics-based prospective studies and prediction of type 2 diabetes mellitus risks. Metab Syndr Relat Disord 18:1–9. https://doi.org/10.1089/met.2019.0047
171. Sun Y, Gao HY, Fan ZY, He Y, Yan YX (2020) Metabolomics signatures in type 2 diabetes: a systematic review and integrative analysis. J Clin Endocrinol Metab 105:1000–1008. https://doi.org/10.1210/clinem/dgz240
172. Arneth B, Arneth R, Shams M (2019) Metabolomics of type 1 and type 2 diabetes. Int J Mol Sci 20:1–14. https://doi.org/10.3390/ijms20102467
173. Shi X, Xi B, Jasbi P, Turner C, Jin Y, Gu H (2020) Comprehensive isotopic targeted mass spectrometry: reliable metabolic flux analysis with broad coverage. Anal Chem 92:11728–11738. https://doi.org/10.1021/acs.analchem.0c01767
174. Badur MG, Metallo CM (2018) Reverse engineering the cancer metabolic network using flux analysis to understand drivers of human disease. Metab Eng 45:95–108. https://doi.org/10.1016/j.ymben.2017.11.013
175. Cascante M, Benito A, Marín de Mas I, Centelles JJ, Miranda A, de Atauri P (2014) Fluxomics. In: A systems biology approach to study metabolic syndrome. Springer International Publishing, pp 237–250. https://doi.org/10.1007/978-3-319-01008-3_12
176. Basler G, Fernie AR, Nikoloski Z (2018) Advances in metabolic flux analysis toward genome-scale profiling of higher organisms. Biosci Rep

38:BSR20170224. https://doi.org/10.1042/BSR20170224
177. Heux S, Bergès C, Millard P, Portais JC, Létisse F (2017) Recent advances in high-throughput 13C-fluxomics. Curr Opin Biotechnol 43:104–109. https://doi.org/10.1016/j.copbio.2016.10.010
178. Antoniewicz MR (2018) A guide to 13C metabolic flux analysis for the cancer biologist. Exp Mol Med 50:19. https://doi.org/10.1038/s12276-018-0060-y
179. Birkemeyer C, Luedemann A, Wagner C, Erban A, Kopka J (2005) Metabolome analysis: the potential of in vivo labeling with stable isotopes for metabolite profiling. Trends Biotechnol 23:28–33. https://doi.org/10.1016/j.tibtech.2004.12.001
180. Klein S, Heinzle E (2012) Isotope labeling experiments in metabolomics and fluxomics. Wiley Interdiscip Rev Syst Biol Med 4:261–272. https://doi.org/10.1002/wsbm.1167
181. Cascante M, Marin S (2008) Metabolomics and fluxomics approaches. Essays Biochem 45:67–81. https://doi.org/10.1042/bse0450067/853546/bse0450067.pdf
182. Cheah YE, Young JD (2018) Isotopically nonstationary metabolic flux analysis (INST-MFA): putting theory into practice. Curr Opin Biotechnol 54:80–87. https://doi.org/10.1016/j.copbio.2018.02.013
183. Lee WD, Mukha D, Aizenshtein E, Shlomi T (2019) Spatial-fluxomics provides a subcellular-compartmentalized view of reductive glutamine metabolism in cancer cells. Nat Commun 10:1351. https://doi.org/10.1038/s41467-019-09352-1
184. Wang Y, Kwon H, Su X, Wondisford FE (2020) Glycerol not lactate is the major net carbon source for gluconeogenesis in mice during both short and prolonged fasting. Mol Metab 31:36–44. https://doi.org/10.1016/j.molmet.2019.11.005
185. Allen DK, Young JD (2020) Tracing metabolic flux through time and space with isotope labeling experiments. Curr Opin Biotechnol 64:92–100. https://doi.org/10.1016/j.copbio.2019.11.003
186. Lemberg KM, Zhao L, Wu Y, Veeravalli V, Alt J, Aguilar JMH, Dash RP, Lam J, Tenora L, Rodriguez C, Nedelcovych MT, Brayton C, Majer P, Blakeley JO, Rais R, Slusher BS (2020) The novel glutamine antagonist prodrug JHU395 has antitumor activity in malignant peripheral nerve sheath tumor. Mol Cancer Ther 19:397–408. https://doi.org/10.1158/1535-7163.MCT-19-0319
187. Bian X, Qian Y, Tan B, Li K, Hong X, Wong CC, Fu L, Zhang J, Li N, Wu JL (2020) In-depth mapping carboxylic acid metabolome reveals the potential biomarkers in colorectal cancer through characteristic fragment ions and metabolic flux. Anal Chim Acta 1128:62–71. https://doi.org/10.1016/j.aca.2020.06.064
188. Olson WJ, Di Genova BM, Gallego-Lopez G, Dawson AR, Stevenson D, Amador-Noguez D, Knoll LJ (2020) Dual metabolomic profiling uncovers Toxoplasma manipulation of the host metabolome and the discovery of a novel parasite metabolic capability. PLoS Pathog 16:e1008432. https://doi.org/10.1371/journal.ppat.1008432
189. Sweeney SR, Collins M, Pandey R, Chiou J, Lodi A, Tiziani S (2020) Identification of a synergistic combination of dimethylaminoparthenolide and shikonin alters metabolism and inhibits proliferation of pediatric precursor-B cell acute lymphoblastic leukemia. Mol Carcinog 59:399–411. https://doi.org/10.1002/mc.23163
190. Lee SC, Shestov AA, Guo L, Zhang Q, Roman JC, Liu X, Wang HY, Pickup S, Nat K, Lu P, Hofbauer S, Mesaros C, Wang YL, Nelson DS, Schuster SJ, Blair IA, Glickson JD, Wasik MA (2019) Metabolic detection of Bruton's tyrosine kinase inhibition in mantle cell lymphoma cells. Mol Cancer Res 17:1365–1377. https://doi.org/10.1158/1541-7786.MCR-18-0256
191. McCalley S, Pirman D, Clasquin M, Johnson K, Jin S, Vockley J (2019) Metabolic analysis reveals evidence for branched chain amino acid catabolism crosstalk and the potential for improved treatment of organic acidurias. Mol Genet Metab 128:57–61. https://doi.org/10.1016/j.ymgme.2019.05.008
192. Johnston K, Kim DH, Kerkhoven EJ, Burchmore R, Barrett MP, Achcar F (2019) Mapping the metabolism of five amino acids in bloodstream form Trypanosoma brucei using U-13C-labelled substrates and LC–MS. Biosci Rep 39:BSR20181601. https://doi.org/10.1042/BSR20181601
193. Muthu M, Kumar R, Khaja ASS, Gilthorpe JD, Persson JL, Nordström A (2019) Glul ablation can confer drug resistance to cancer cells via a malate-aspartate shuttle-mediated mechanism. Cancers (Basel) 11:1945. https://doi.org/10.3390/cancers11121945
194. Vantaku V, Amara CS, Piyarathna DWB, Donepudi SR, Ambati CR, Putluri V, Tang W, Rajapakshe K, Estecio MR, Terris MK, Castro PD, Ittmann MM, Williams SB, Lerner SP, Sreekumar A, Bollag R, Coarfa C, Kornberg MD, Lotan Y, Ambs S, Putluri N (2019) DNA methylation patterns in bladder tumors of African American patients point to distinct alterations in xenobiotic metabolism. Carcinogenesis 40:1332–1340. https://doi.org/10.1093/carcin/bgz128
195. Kim J, Jonus HC, Zastre JA, Bartlett MG (2019) Development of an IPRP-LC-MS/MS method to determine the fate of intracellular thiamine in cancer cells. J Chromatogr B Anal Technol Biomed Life Sci 1124:247–255. https://doi.org/10.1016/j.jchromb.2019.05.037
196. Held NM, Kuipers EN, Van Weeghel M, Van Klinken JB, Denis SW, Lombès M, Wanders RJ, Vaz FM, Rensen PCN, Verhoeven AJ, Boon MR, Houtkooper RH (2018) Pyruvate dehydrogenase complex plays a central role in brown adipocyte energy expenditure and fuel utilization during short-term beta-adrenergic activation. Sci Rep 8:9562. https://doi.org/10.1038/s41598-018-27875-3

197. Kovářová J, Nagar R, Faria J, Ferguson MAJ, Barrett MP, Horn D (2018) Gluconeogenesis using glycerol as a substrate in bloodstream-form Trypanosoma brucei. PLoS Pathog 14:e1007475. https://doi.org/10.1371/journal.ppat.1007475
198. Kovářová J, Pountain AW, Wildridge D, Weidt S, Bringaud F, Burchmore RJS, Achcar F, Barrett MP (2018) Deletion of transketolase triggers a stringent metabolic response in promastigotes and loss of virulence in amastigotes of Leishmania mexicana. PLoS Pathog 14:e1006953. https://doi.org/10.1371/journal.ppat.1006953
199. Gelman SJ, Naser F, Mahieu NG, McKenzie LD, Dunn GP, Chheda MG, Patti GJ (2018) Consumption of NADPH for 2-HG synthesis increases pentose phosphate pathway flux and sensitizes cells to oxidative stress. Cell Rep 22:512–522. https://doi.org/10.1016/j.celrep.2017.12.050
200. Steketee PC, Vincent IM, Achcar F, Giordani F, Kim DH, Creek DJ, Freund Y, Jacobs R, Rattigan K, Horn D, Field MC, MacLeod A, Barrett MP (2018) Benzoxaborole treatment perturbs S-adenosyl-L-methionine metabolism in Trypanosoma brucei. PLoS Negl Trop Dis 12:e0006450. https://doi.org/10.1371/journal.pntd.0006450
201. Chatzispyrou IA, Guerrero-Castillo S, Held NM, Ruiter JPN, Denis SW, IJlst L, Wanders RJ, van Weeghel M, Ferdinandusse S, Vaz FM, Brandt U, Houtkooper RH (2018) Barth syndrome cells display widespread remodeling of mitochondrial complexes without affecting metabolic flux distribution. Biochim Biophys Acta Mol basis Dis 1864:3650–3658. https://doi.org/10.1016/j.bbadis.2018.08.041

Analytical Platforms for Mass Spectrometry-Based Metabolomics of Polar and Ionizable Metabolites

Adriana N. Macedo, Andrea T. Faccio, Tatiana S. Fukuji, Gisele A. B. Canuto, and Marina F. M. Tavares

Abstract

Metabolomics studies rely on the availability of suitable analytical platforms to determine a vast collection of chemically diverse metabolites in complex biospecimens. Liquid chromatography-mass spectrometry operated under reversed-phase conditions is the most commonly used platform in metabolomics, which offers extensive coverage for nonpolar and moderately polar compounds. However, complementary techniques are required to obtain adequate separation of polar and ionic metabolites, which are involved in several fundamental metabolic pathways. This chapter focuses on the main mass-spectrometry-based analytical platforms used to determine polar and/or ionizable compounds in metabolomics (GC-MS, HILIC-MS, CE-MS, IPC-MS, and IC-MS). Rather than comprehensively describing recent applications related to GC-MS, HILIC-MS, and CE-MS, which have been covered in a regular basis in the literature, a brief discussion focused on basic principles, main strengths, limitations, as well as future trends is presented in this chapter, and only key applications with the purpose of illustrating important analytical aspects of each platform are highlighted. On the other hand, due to the relative novelty of IPC-MS and IC-MS in the metabolomics field, a thorough compilation of applications for these two techniques is presented here.

Keywords

Metabolomics · Analytical platforms · Polar compounds · Ionic compounds · Ionizable compounds · HILIC · CE-MS · GC-MS · Ion pairing chromatography · Ion chromatography

A. N. Macedo (✉)
Department of Chemistry, Federal University of Minas Gerais, Belo Horizonte, MG, Brazil
e-mail: adrianamacedo@qui.ufmg.br

A. T. Faccio · T. S. Fukuji · M. F. M. Tavares
Institute of Chemistry, University of Sao Paulo, Sao Paulo, SP, Brazil
e-mail: mfmtavar@iq.usp.br

G. A. B. Canuto
Department of Analytical Chemistry, Institute of Chemistry, Federal University of Bahia, Salvador, BA, Brazil

Abbreviations

BGE	Background electrolyte
C18	Octadecyl-derived silica
CE	Capillary electrophoresis
DIPEA	Diisopropylethylamine
EI	Electron ionization
EOF	Electroosmotic flow
ESI	Electrospray ionization

A. V. Colnaghi Simionato (ed.), *Separation Techniques Applied to Omics Sciences*, Advances in Experimental Medicine and Biology 1336, https://doi.org/10.1007/978-3-030-77252-9_11

GC×GC	Two-dimensional gas chromatography
GC	Gas chromatography
HFBA	Heptafluorobutyric acid
HILIC	Hydrophilic interaction liquid chromatography
HPLC	High-performance liquid chromatography
IC	Ion chromatography
IPC	Ion pairing chromatography
IPR	Ion pair reagent
LC	Liquid chromatography
LSER	Linear solvation energy relationship
MS	Mass spectrometry
MS/MS	Tandem mass spectrometry
MSI	Multisegment injection
NMR	Nuclear magnetic resonance
NP	Normal phase
Q	Quadrupole mass analyzer
RPLC	Reversed-phase liquid chromatography
TBA	Tributylamine
TOF	Time-of-flight mass analyzer
UHPLC	Ultra-high-performance liquid chromatography
UPLC	Ultra-performance liquid chromatography

1 Introduction

The fast developing field of metabolomics has provided invaluable information to improve understanding of complex phenotypes in diverse biological systems, with practical applications in many areas, including medicine, nutrition, agriculture, biotechnology, environmental sciences, and forensics [1, 2]. As end products of gene expression and protein activity, which also reflect the microbiome and environmental exposure, metabolites have enormous potential as indicators of disease states, as well as individual responses to therapies, diet, and lifestyle, complementing other omics approaches to reveal underlying biochemical mechanisms [3, 4]. In contrast to genomics and proteomics, which analyze more limited sets of nucleic acids or proteins/peptides/amino acids, metabolomics has unique analytical challenges when trying to comprehensively analyze thousands of metabolites with widely differing physicochemical properties (e.g., molecular mass, polarity, solubility, stability, and volatility), which are present in concentrations that may differ by 7–9 orders of magnitude in biological samples [3]. Moreover, the complete extension of the metabolome remains largely uncharacterized, which makes even more challenging the design of analytical methods with optimal coverage [5]. In this context, many metabolomics studies have either used a combination of complementary techniques running in parallel or selected a single analytical protocol to cover the largest possible number of metabolites within predefined chemical classes or biochemical pathways [5, 6].

A growing interest has been directed to analytical platforms that assess polar and ionizable metabolites (e.g., amino acids and derivatives, short peptides, organic acids, biogenic amines, nucleotides, nucleosides, nucleobases, phosphorylated compounds, carbohydrates, etc.), which are involved in several fundamental metabolic pathways in animals, plants, and microorganisms [5, 7, 8]. Although nuclear magnetic resonance (NMR) spectroscopy has been used in many metabolomics studies, with continuous efforts to improve sensitivity, resolution, and data analysis [9], mass spectrometry (MS) is currently the analytical platform of choice in metabolomics, especially when hyphenated with a separation technique. The large majority of studies in this field has been performed by reversed-phase liquid chromatography (RPLC), adopting either the conventional instrumentation or (ultra-) high-performance LC systems (UPLC or UHPLC) [10]. Typical octadecyl-derived silica (C18) columns used for metabolite profiling in RPLC-MS methods provide extensive coverage of nonpolar and moderately polar compounds, which allows the determination of a fair number of metabolites in complex biological samples. However, polar metabolites tend to be poorly retained, which results in inefficient separation, with elution at or near the column void volume; in addition, a lack of isomeric resolution, as well as ion suppression when using electrospray ionization (ESI), is frequently reported. As a result, retention of polar compounds has been considered one of the most challenging aspects of LC-MS-based metabolomics [10, 11].

Considering the foremost importance of polar and ionizable compounds in the central carbon and energy metabolisms, alternative analytical approaches have been proposed to improve metabolic coverage for those metabolites, in order to allow a more complete metabolic characterization of biological systems. This includes the development of novel stationary phases and the use of alternative chromatographic modes, as well as complementary techniques, such as capillary electrophoresis (CE) and gas chromatography (GC) with metabolite derivatization. Therefore, the present chapter discusses the main MS-based analytical platforms for metabolomics analysis of polar and ionizable compounds, including GC, hydrophilic interaction liquid chromatography (HILIC), CE, ion pairing chromatography (IPC), and ion chromatography (IC). Basic principles for each platform are briefly presented, including a description of important analytical features, as well as relative strengths, limitations, and future trends. It is worth mentioning that this chapter is not intended to comprehensively compile all recent related publications. Instead, selected applications in clinical, plant, food, and/or forensic metabolomics are presented to exemplify the analytical approaches described here. Tables listing recent applications in metabolomics include only the less commonly used platforms, such as IPC-MS and IC-MS, whereas recent review articles are cited in the text for applications using GC-MS, HILIC-MS, and CE-MS. Although the scientific literature presents many analytical platforms in metabolomics, very few manuscripts have focused specifically on polar and ionizable metabolites [7, 8, 12].

2 Gas Chromatography

GC-MS is one of the most mature analytical platforms and its use in metabolomics has been a straightforward option [3, 13]. Even before the introduction of the term "metabolomics" about 20 years ago, GC-MS had been already used for targeted metabolite profiling as described in publications from the 1970s [13, 14]. High separation efficiency, sensitivity, and good repeatability of retention times are some strengths that contribute to make GC-MS a valuable platform in metabolomics. In addition, a large body of electron ionization (EI)-MS spectral libraries is available for metabolite identification (e.g., FiehnLib, NIST Mass Spectral Library, etc.), which contain complementary information about retention times and/or retention indices [15]. EI-MS provides highly reproducible fragmentation patterns that are transferable between instruments and laboratories, which greatly facilitates metabolite annotation [3, 6]. Another advantage of GC-MS is the lower cost associated with instrument acquisition, operation, and maintenance when compared to other chromatographic platforms, as low-resolution MS instruments can be used, due to the consistent fragmentation patterns and the availability of spectral libraries.

Polar and ionizable metabolites can be determined by GC-MS after sample extraction followed by chemical derivatization, which is necessary to enhance volatility and to secure thermal stability at temperatures that reach up to 350 °C in the instrument [6]. Although there are many derivatization reactions available (e.g., fluoroesterification, cloroformation, etc.), a two-step process that includes an alkyl oximation followed by a trimethylsilylation has been the most commonly used option in metabolomics [16]. This combination of reactions promotes the derivatization of several classes of metabolites, including amino acids, organic acids, sugars, amines, and phosphorylated compounds, in addition to more hydrophobic molecules (e.g., fatty acids and sterols), and allows their concurrent inspection in a single chromatographic run and ionization mode [17]. However, extended protocols for sample pretreatment and chemical derivatization contribute to limit the analytical throughput while introducing experiment-derived variations (e.g., loss of very volatile and heat-labile metabolites and poor derivatization recovery), which may overshadow biologically originated metabolic changes [16, 17]. For instance, sugar phosphates and nucleotides are easily lost during derivatization, whereas asparagine and glutamine have low derivatization efficiency, and arginine is known to be converted into the same trimethylsilylation derivative product as ornithine [15]. As a result, coverage of polar/ionic metabolites by GC-MS is largely dependent on the derivatization performance, and data obtained for problematic metabolites should be considered with particular attention to avoid erroneous interpretations. Furthermore, the presence of multiple derivative products for a single metabolite is a common problem that complicates statistical analysis and peak annotation [16, 17]. In this latter case, Mastrangelo et al. [15] recommend summing the abundances of all the multiple derivative products for a metabolite, although another option would be the selection of only one derivative product and elimination of the others from the mass list.

Extensive sample pretreatment procedures, derivatization steps that may take 2–16 h, and run times of 20–60 min make GC-MS a relatively low-throughput technique for metabolomics [6, 14, 18]. In addition, the manual offline derivatization of sample batches prior to chromatographic analysis can introduce undesired nonbiological variability for samples analyzed at different times, due to derivatization reactions that remain progressing for several hours (e.g., metabolites containing hydroxyl, amine, ketone or carboxyl groups), as well as degradation of more unstable products that may happen within the same time span [19, 20]. The use of automated systems for online derivatization right before injection has been proposed to overcome these problems by standardizing the time between derivatization and GC-MS analysis. Sample derivatization can be automated by a robotic system (Fig. 1) that may contain a vortex or a shaker, an incubator with temperature control, a sample and reagent tray, a wash station, and one or two syringe modules, depending on the volume used for each step of the procedure, which are used to move the samples between the different stations. All the steps of sample preparation can be programed, so that each derivatized sample can be analyzed shortly after the derivatization process, which could improve the repeatability of the sample preparation process. For instance, Zarate et al. [20] have reported better repeatability and higher peak intensity for many metabolites in wine and plasma samples when using an automated system for derivatization, with a protocol that allowed preparation and analysis of 23 samples in about 28 h (~72 min/sample). However, the authors observed that the manual approach still provided better repeatability for some classes of metabolites, including amino acids and organic acids. A similar comparison was performed by Abbiss et al. [19] using human urine samples, which revealed larger variability in peak areas, as well as worse peak resolution for the automated online protocol. This indicates that further developments are still needed to improve the robustness and efficiency of automated robotic systems.

Although single quadrupole systems (GC-Q-MS) are used in the large majority of GC-MS metabolomics studies, time-of-flight mass analyzers (GC-TOF-MS) have attracted increasing attention in the last decade. In addition to higher mass resolution and sensitivity, GC-TOF-MS provides faster acquisition rates that allow more accurate determination of peak shapes, and facilitates deconvolution of co-eluting peaks [6]. For this reason, GC-TOF-MS has been recommended as a more suitable option over GC-Q-MS in untargeted studies, especially for complex samples with low abundance metabolites, such as plasma or serum [6, 21]. Dunn et al. [6] reported a GC-TOF-MS protocol for metabolomics analysis in human plasma and urine samples, with detection of 100–200 molecular features, although the authors stated that the actual number of metabolites is likely lower due to the presence of multiple derivatized species. Another promising tool for metabolomics is the use of two-dimensional GC-MS (GC×GC-MS), known for considerably increasing peak capacity, resolution, and sensitivity [4]. GC×GC combines

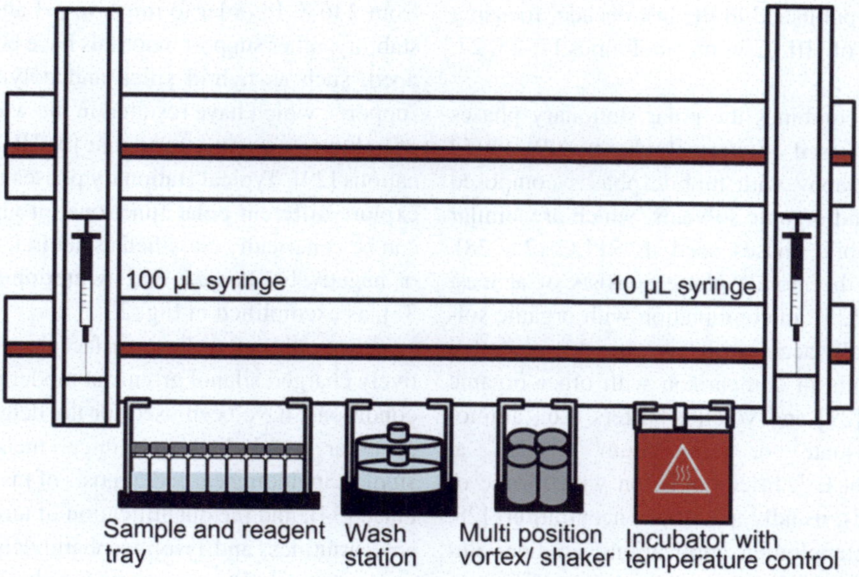

Fig. 1 Schematic representation of a robotic system for automated sample derivatization prior to GC-MS analysis

two columns with orthogonal separation properties, using a modulator that transfers the effluent from the first column to the second one in small concentrated portions [22]. In this case, a TOF-MS detector is commonly used to provide the necessary fast-acquisition rates. Most applications of GC×GC-MS in metabolomics combine a regular nonpolar capillary fused-silica column (typically coated with 5% diphenyl/95% dimethyl polysiloxane) with a mid-polarity column (e.g., 50% phenyl/50% dimethylpolysiloxane), although other combinations have been previously tested [22]. For instance, Miyazaki et al. [23] studied serum from neonatal calves after colostrum ingestion and detected ~180 metabolites with GC-MS and ~1400 metabolites with GC×GC-MS using a nonpolar column in the first dimension, followed by a mid-polarity column, which allowed the resolution of metabolite classes such as oligosaccharides that co-eluted with other compounds in the first dimension. Nevertheless, only a small number of metabolomics studies have been conducted so far using GC×GC-MS, which reflects the relatively expensive instrumentation, the need to evaluate its long-term robustness, and further improvements required in software for data deconvolution [22, 24].

3 Hydrophilic Interaction Liquid Chromatography

The need to improve the separation and detection of polar compounds in metabolomics has led to the development of novel separation approaches in chromatography, including the resurgence of hydrophilic interaction liquid chromatography (HILIC), which has demonstrated to provide excellent coverage for highly polar compounds [25]. The term HILIC was first used by Alpert [26] in a paper published in 1990, which demonstrated efficient separation of many compounds, including peptides, phosphorylated amino acids, and proteins. Thereafter, HILIC-MS began to be used complementarily to RPLC-MS in metabolomics [27], with the advantage that both modes can be performed using the same instrumentation, although with different column chemistries and operational conditions. Indeed, many reviews have been published in the last decade, focusing on the use of HILIC in metabolomics [7, 25, 27, 28].

HILIC combines the polar stationary phases commonly used in normal phase (NP) liquid chromatography with mobile phases composed of water and organic solvents, which are similar to the mobile phases used in RPLC [25, 28]. Typically, the mobile phase consists of at least 5% water [25], in combination with organic solvents (usually acetonitrile, as it has demonstrated better results in comparison with other organic solvents) [29] and volatile buffers (i.e., ammonium formate or ammonium acetate at 5–20 mmol L^{-1}, in combination with formic or acetic acids, usually at 0.1% concentration) [28, 30]. In metabolomics applications, gradient elution is preferred due to the complexity of the biological samples. Additionally, proper optimization of buffer concentration and pH can considerably improve chromatographic resolution and peak shapes [29].

Although the separation mechanism in HILIC is not yet fully understood, it is believed to be based on the partitioning of the analyte between a semi-immobilized aqueous phase on the stationary phase particles and the organic solvent-rich mobile phase. In addition, other specific interactions between the analyte and the stationary phase can contribute to retention, including ion-ion, hydrogen bonding, dipole-dipole, and weak hydrophobic interactions [25]. In the last years, linear solvation energy relationship (LSER) studies have been applied to reveal the nature of the solute-stationary phase interactions. LSER studies demonstrate that interactions are closely dependent on the type of analyte, as well as the composition of the stationary and mobile phases [31], which makes difficult the characterization of a single separation mechanism, as opposed to what is observed in RPLC, for example. The combination of multiple types of interactions can offer an explanation for the broad and/or tailing peaks often observed in HILIC separations [32].

Silica gel and chemically bonded silica-based columns are the most commonly used stationary phases in HILIC [29]. However, silica supports are usually stable when working in the pH range from 2 to 8. In order to improve pH and thermal stability, other support materials have been developed, such as hybrid silica and polymer-based supports, which have resulted in the wide variety of columns currently available for HILIC applications [29]. Typical stationary phases for HILIC explore different polar functional groups, which can be generically classified as neutral, positively or negatively charged, and zwitterionic [29, 30, 33], as exemplified in Fig. 2.

Silica columns, whose surfaces present negatively charged silanol groups at moderately basic conditions, have been used for the determination of polar metabolites in clinical metabolomics studies, including early diagnosis of insulin resistance [34], and the quantification of amino acids, acylcarnitines, and lysophosphatidylcholines, as well as metabolites associated with risk factors for cardiovascular diseases [35]. Meanwhile, positively charged stationary phases, such as aminopropyl, have been applied successfully in untargeted metabolomics to understand acute respiratory distress syndrome [36], and for kidney cancer screening using urine samples [37]. In a targeted analysis, an aminopropyl column was used for the quantification of sugar phosphates, acyl-CoA, nucleotides, amino acids, and carboxylic acids that are part of the bacterial central metabolism, as applied to cultures of *Methylobacterium extorquens* [38]. Recently, Li et al. [39] have developed a method for multi-targeted metabolomics using a positively charged polymer-based aminopropyl column to measure 610 metabolites, from 63 metabolic pathways, and 95 stable isotope standards in a single 45-min separation, when applying negative and positive ionization modes with rapid polarity switching. In this case, 468 metabolites were detected in quality control plasma samples, with good system stability for over 800–1000 injections.

Some metabolomics studies have compared different HILIC columns and separation conditions [32, 40–42]. Zwitterionic columns have shown outstanding performance [32]. For example, different zwitterionic stationary phases were evaluated by Zhang and Watson [43] in human urine for clinical studies and by Zhang et al. [44] in *Leishmania* parasites extracts. In both cases,

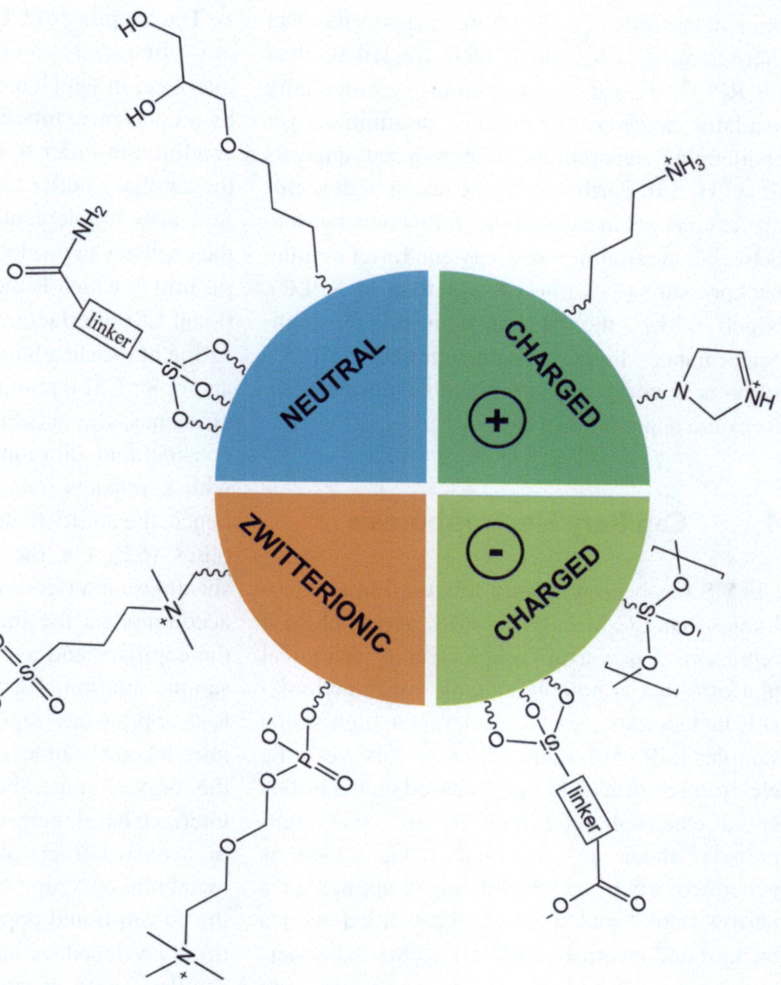

Fig. 2 Examples of stationary phases currently available for HILIC-MS metabolomics, including neutral (e.g., diol and amide), zwitterionic (e.g., sulfobetaine and phosphorylcholine), positively charged (e.g., amine and imidazole), and negatively charged (e.g., silica and alkyl carboxylic acid) functional groups

the authors intended to determine the largest number of polar metabolites when performing untargeted metabolomics studies. In another work, three different HILIC columns (i.e., diol, zwitterionic, and amide) were compared, in order to develop and validate a multi-targeted method for metabolomics applications in biological fluids. The developed method was capable of quantifying more than 100 metabolites in a single run with efficiency and accuracy, employing a neutral amide column [45].

Although HILIC-MS is known to be less reproducible than RPLC-MS, suitable results have been obtained for metabolomics analysis when consecutive injections of quality control samples were performed to stabilize the system before the batch analysis is run [32, 45]. Additionally, longer equilibration times between sample injections are required for achieving satisfactory repeatability [25, 46]. New developments in LC technologies have also helped to make HILIC-MS a more useful technique for the determination of polar metabolites, including the use of UPLC/UHPLC systems that lead to more efficient separations and shorter analysis times. HILIC phases that are currently not available in the sub-2 μm format have also been used in UPLC/UHPLC systems [47, 48], which still show benefits in terms of separation efficiency, due to the typical smaller extra-column volumes in comparison to conventional HPLC systems. Naturally, the availability of a greater variety of HILIC phases as sub-2 μm and superficially porous particles also constitutes important developments in HILIC. Therefore, despite the fact that retention is less predictable for HILIC than for RPLC, the variety of columns commercially available nowadays offers new possibilities for challenging separations with reduced analysis time [31, 46]. Furthermore, the use of high acetonitrile contents in the mobile phase contributes to better concentration sensitivity and lower column backpressure (2–3 times lower than in RPLC), which allows the use of conventional high-performance liquid chromatography (HPLC) systems in combination with sub-2 μm particles for some applications [46].

4 Capillary Electrophoresis

CE-MS has been relatively less used in metabolomics than LC-MS and GC-MS; nevertheless, it represents a powerful complementary analytical platform that is notably suitable for ionic/ionizable metabolites, as well as tolerant to high-saline samples [49, 50]. Separation in this versatile electromigration technique is based on the differential electrophoretic mobility of ionic compounds under an electrical field, which is generated when a high voltage is applied to a narrow-bore fused-silica capillary filled with a background electrolyte (BGE). Remarkably narrow peaks and higher separation efficiency are characteristic of CE separations in comparison to chromatographic techniques. Furthermore, ionizable metabolites can be determined with minimal sample pretreatment in volume-limited biological samples, while consuming much lower volumes of solvent than LC-MS [51, 52]. On the other hand, CE-MS is often considered a less robust technique, with appreciable shifts in migration times, lower concentration sensitivity, and challenges related to anion determination [49, 50]. Considerable efforts have been made to address these problems, both instrumental and condition related, which has contributed to the gradual growth and establishment of CE-MS among the metabolomics analytical platforms, as highlighted in regularly published reviews of recent CE-MS applications in metabolomics [49–53].

The coupling of CE with MS (Fig. 3) has been most frequently conducted by a sheath liquid interface, in which a coaxial liquid is introduced by a concentric tube at the end of the separation capillary, in order to close the electric circuit at the capillary outlet [54]. The sheath liquid interface also supplements the nanoscale flow from the capillary to produce a flow rate in the order of μL min^{-1}, which is more appropriate for conventional ESI interfacing [49, 54]. Although optimization of the sheath liquid composition, flow rate, and other ESI conditions can enhance ionization efficiency, the sheath liquid interface results in considerable dilution of the capillary effluent, which impacts concentration sensitivity and, hence, the ability to detect low abundance metabolites [52]. On the other hand, a number of sheathless interfaces have been proposed, which accommodate the inherent low flow rates from the capillary and avoid and/or greatly minimize sample dilution [52, 54–56]. Among the sheathless approaches reported to date, the recently introduced sheathless porous tip interface [57] is the only commercially available option. This interface has demonstrated to improve sensitivity by over two orders of magnitude and expand the metabolic coverage by tenfold in comparison to the sheath liquid approach [52, 58]. The porous tip is produced by etching the outlet end of the capillary with hydrofluoric acid, whereas the electrical contact at the capillary outlet is obtained by introducing the porous tip into the ESI needle, which is then filled with a conductive liquid without direct contact with the capillary effluent [52, 57], as depicted in Fig. 3. In this case, nano-ESI conditions are required to maintain a stable spray when working with nanoscale flow rates from the capillary. The sheathless interface is commercialized in the form of a nonstandard and expensive capillary assembly, composed of a porous tip capillary that is preinstalled in a protective cartridge and in the ESI sprayer, which is connected to a conductive liquid capillary and liquid cooling tubes. Although the capillary assembly was designed to maximize interface robustness by minimizing direct handling of a fragile porous tip capillary, the costs involved may not be amenable to many laboratories, especially when consider-

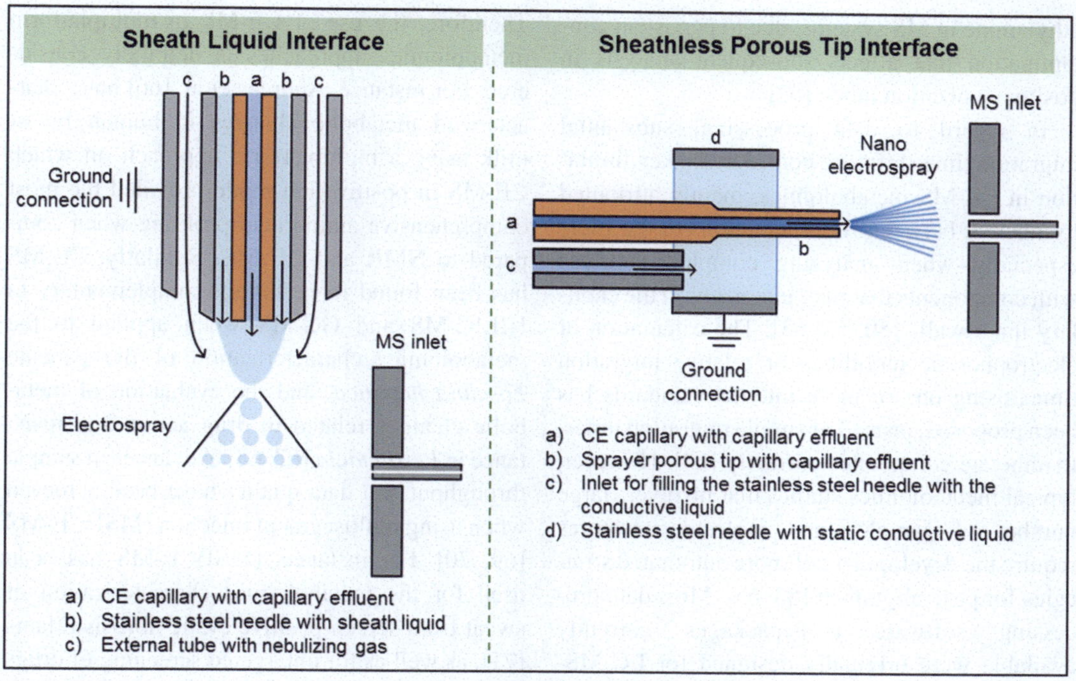

Fig. 3 Schematic representation of a sheath liquid and a porous tip sheathless interface in CE-MS

ing that the whole assembly has to be replaced after analyzing up to 200 samples [52]. Overall, the porous tip sheathless interface represents a very promising option to improve analytical sensitivity in CE-MS. However, it still requires further studies to investigate its long-term performance [52], as well as to improve the interface design, in order to make it economically more accessible. For this reason, the sheath liquid interface remains the coupling of choice in CE-MS to date, where a stable spray obtained by optimized ESI conditions is used in combination with a well-positioned capillary in the sprayer needle tip by a trained analyst [50]. An alternative approach to boost sensitivity when using a sheath liquid interface consists in the use of online sample preconcentration techniques, which have been applied to several metabolomics studies using CE-MS for cationic metabolites [49, 53, 59].

For anionic compounds, CE-MS has been more challenging, although several approaches have been reported to date [53]. A reversed polarity configuration (anode at the capillary outlet), combined with a high-pH BGE, leads to corrosion of the stainless steel ESI needle and blockage of the capillary outlet [53, 60]. As an alternative, Soga et al. [60] have proposed the replacement of the stainless steel with a platinum ESI needle, in combination with the use of cationically coated capillaries under reversed polarity, which directs the EOF toward the outlet. A more straightforward configuration that allows the use of the regular stainless steel ESI needle with a bare fused-silica capillary has been described by Yamamoto et al. [61], which combines normal polarity with a pressure-assisted separation for faster and robust metabolite profiling of anions in a moderately basic BGE (pH 8.5) composed of ammonium bicarbonate. In this study, exposure of the capillary to more strongly alkaline ammonia-based BGEs was demonstrated to cause aminolysis of the polyimide outer coating, which increases the incidence of capillary breakage and compromises method robustness [61]. Another option is the use of amine-based BGEs (e.g., triethylamine), which have been reported to improve detection sensitivity for anionic metabolites under normal polarity [49]. However, caution should be taken when using triethylamine in MS systems, due to persistent contamination that affects subsequent analysis in positive ionization mode [62].

In regard to data processing, substantial migration time shifts are considered a key limitation in CE-MS metabolomics, mainly attributed to inherent between-run fluctuations in the EOF, especially when analyzing complex matrices with components that may interact with the capillary inner walls [50, 53, 63]. The calculation of electrophoretic mobilities or relative migration times using one or more internal standards has been proposed, instead of simple migration times, to annotate compounds correctly [64]. However, typical metabolomics studies that involve a large number of samples and molecular features require the development of more automated strategies for peak alignment [53, 63]. Most data processing software tools/packages currently available were originally designed for LC-MS-derived data with lower retention time variabilities and therefore are not suitable to correct for the larger and irregular time shifts observed in CE-MS [53]. In order to overcome this limitation, some algorithms have already been proposed to improve migration time alignment in CE-MS, including MasterHands (Metabolome Analysis and Screening Tool for Easy and Rapid Handling of Sample data) from Keio University [65] and the open-source software msalign2 [63], which were specially developed for CE-MS accurate mass metabolomics data. Additionally, several metabolomics studies have used the Molecular Feature Extraction (MFE) tool of MassHunter software followed by Mass Profiler Professional (Agilent Technologies), which allows for migration time correction prior to alignment. Nevertheless, a considerable fraction of studies using CE-MS in metabolomics still give no details about the software tools used in data processing (28%, as reported by García et al. [50]), which limits replication of conditions by other researchers.

Overall, CE-MS has demonstrated great potential as a complementary analytical platform for ionizable metabolites, often covering a unique set of compounds that are poorly separated and/or detected by other analytical techniques. Therefore, the use of CE-MS in multiplatform metabolomics approaches is definitely considered. For instance, Andreas et al. [66] have characterized metabolic changes in human breast milk using a multiplatform approach, in which CE-MS in positive ion mode provided the most comprehensive amino acid profiling when compared to NMR and GC-MS. Similarly, CE-MS has been found to be highly complementary to HILIC-MS and GC-MS when applied to the metabolomics characterization of the parasite *Fasciola hepatica*, and the evaluation of metabolic changes related to drug action and resistance in *Leishmania* [67, 68]. Advances in sample throughput and data quality have been achieved when using multisegment injection (MSI)CE-MS [69, 70]. For instance, (MSI)CE-MS has been used for the metabolomics characterization of sweat from screen-positive cystic fibrosis infants [71], as well as for untargeted screening of drugs of abuse and their metabolites in human urine [72]. Efforts have been made to develop novel CE-MS interfacing approaches for improved sensitivity, robustness, and affordability, as well as to streamline data processing/peak alignment in metabolomics.

5 Ion Pairing Chromatography

As it was mentioned before, LC-MS has been considered the major analytical technique for metabolic profiling, due to its high sensitivity, relatively extensive metabolite coverage, and robustness in routine analysis [10, 73]. Among the many chromatographic modes, RPLC with C18 columns remains the most widely used mode in metabolomics, which promotes the separation of nonpolar and moderately polar metabolites that interact with hydrophobic stationary phases and hydrophilic mobile phases [10]. Polar and ionic compounds, however, are poorly retained, which results in inadequate separation for many important metabolites, including amino acids, organic acids, amines, sugars, nucleotides, and phosphorylated compounds. The addition of ion pair reagents (IPR) to the mobile phase has proven to be a suitable alternative to improve

Table 1 Summary of selected applications of ion pairing chromatography-mass spectrometry (IPC-MS) in metabolomics studies

Metabolomics type/ metabolites	LC-MS system	Matrix	Column type	Mobile phase	Other conditions	Sample preparation	Method performance	Ref.
Targeted metabolomics/138 in standards and >76 in samples: carboxylic acids, amino acids, sugar phosphates, nucleotides, aromatic anions	UPLC-ESI-MS/MS (Waters Acquity UPLC and Thermo TSQ Quantum Ultra triple quadrupole MS)	*Saccharomyces cerevisiae*, *Escherichia coli*, *Bacillus subtilis*, murine liver, soil, tomato	Waters Acquity T3 end-capped C18 (1.8 μm, 2.1 × 150 mm)	A: 10 mmol/L TBA, 15 mmol/L HOAc, 5% v/v MeOH. B: IPA (gradient: 0–50% B)	Flow rate = 0.4–0.15 mL/min, temperature = 40 °C, injection volume = 10 μL, total time = 36 min	Extraction with 60% EtOH in water with 10 mmol/L NH_4OAc, pH 7.2 at 78 °C, dried under vacuum and reconstituted	RT shifts <40 s for 3000 injections and between lots of columns, LOD = 1.7 to 16 nmol/L, requires dedicated instrumentation due to contamination with ion pair reagent	[73]
Targeted metabolomics/at least 68 nucleotides and nucleosides and 24 coenzyme A esters	LC-ESI-MS (unspecified LC system and Thermo Finnigan LTQ linear ion trap MS)	*Bacillus subtilis*, *Escherichia coli*	Chrompack Inertsil ODS-3 (5 μm, 3 × 100 mm), guard column Chrompack Inertsil ODS-3 (5 μm)	A: 5 mmol/L hexylamine, pH 6.3 adjusted with HOAc. B: 90% MeOH, 10% 10 mmol/L NH_4OAc, pH 8.5 adjusted with ammonia (gradient: 0–60% B)	Flow rate = 0.4 mL/min, temperature = 30 °C, injection volume = 10 μL, total time = 45 min, flow reduced to 100 μL/min prior to MS detection	Extracted with $CHCl_3$ at −45 °C, filtered (10 kDa molecular weight cutoff), lyophilized, reconstituted	LOD = 0.01 to 0.2 μg/mL, repeatability and reproducibility RSD ≤ 10%, inter-batch RSD ≤ 12%, $R^2 < 0.995$	[74]

(continued)

Table 1 (continued)

Metabolomics type/ metabolites	LC-MS system	Matrix	Column type	Mobile phase	Other conditions	Sample preparation	Method performance	Ref.
Targeted metabolomic/72 cationic (amines, amino acids, carnitines) and anionic metabolites (carboxylic acids, sugar phosphates, nucleotides, nucleosides, nucleobases, coenzyme A esters)	LC-ESI-MS/MS (Waters Acquity UPLC and Waters TQD triple quadrupole MS)	Human hepatic cells expressing the oncogene *EIF5A2*	Waters Acquity UPLC BEH C18 (1.7 μm, 2.1 × 100 mm)	Cations: A: MeOH. B: 3% MeOH in water with 0.02% HFBA and 0.1% HOAc (gradient: 0–95% A) Anions: C: MeOH. D: 3% MeOH in water with 10 mmol/L TBA and 15 mmol/L HOAc (gradient: 0–95% C)	Flow rate = 0.2 mL/min, temperature = 25 °C, total time = 20.5 min for cations and 31.5 min for anions	Extracted with cold 80% MeOH in water and cold water, dried under N_2, reconstituted	LOD = 0.1 to 12 nmol/L, LOQ = 0.3 to 36.2 nmol/L, repeatability RSD = 0.9 to 14.9%, $R^2 > 0.990$, dedicated columns required for cationic and anionic metabolites	[75]
Targeted metabolomics/59 carboxylic acids and phosphorylated compounds	UPLC-ESI-MS/MS (Thermo UltiMate 3000 UHPLC and Thermo TSQ Quantum Ultra AM triple quadrupole MS)	DB-1 human melanoma cells	Phenomenex Synergi Polar-RP phenyl ether linked (4 μm, 2 × 150 mm)	A: 5 mmol/L DIPEA and 200 mmol/L HFIP. B: MeOH and 5 mmol/L DIPEA and 200 mmol/L HFIP (gradient: 100–10% A)	Flow rate = 0.2 mL/min, temperature = 55 °C, injection volume = 5 μL, total time = 25 and 15 min for derivatized and non-derivatized samples, respectively	Extracted with ice-cold 4:1 MeOH/water, supernatant aliquoted to 2 tubes: (1) diluted with 50 mmol/L NH_4CO_3 for underivatized redox cycling metabolites and (2) derivatized with phenylhydrazine for other metabolites	Isomeric resolution, increased sensitivity, and less strongly interfering ions in the MS when switching from negative to positive ion mode	[76]

Metabolite profiling/54 metabolites in standards and 46 in samples: phosphorylated sugars, nucleotides, coenzyme A thioesters, organic acids	Nano-HPLC-nano-ESI-MS (Eksigent nano LC Ultra system and Thermo LQT-Orbitrap MS)	*Methylobacterium extorquens* AM1 cells	Thermo Hypersil gold C18 (5 μm, 0.1 × 150 mm)	A: 1.7 mmol/L TBA in 1.5 acetic acid, pH adjusted to 9.4 with 6 mol/L NH_4OH. B: MeOH (gradient: 3–90% B)	Flow rate = 0.5 μL/min, injection volume = 1 μL, total time = 45 min	Extraction with cold ACN with 0.1 mol/L FA at −20 °C, freeze-dried, and reconstituted	LOD = 0.1 to 10 fmol, good peak shapes, only 50 ng of cell dry weight needed, no isomeric resolution obtained	[77]
Metabolite profiling/nucleotide cofactors, sugar phosphates, and organic acids	LC-ESI-QTOF (Dionex UltiMate 3000 Rapid Separation LC and Bruker micrOTOF-Q II MS)	*Drosophila* species	Phenomenex Kinetex C18 (1.7 μm, 2.1 × 100 mm)	A: 10 mmol/L DAA, pH adjusted to 4.95 with HOAc. B: MeOH (gradient: 95–10% A)	Flow rate = 0.4 mL/min, temperature = 50 °C, injection volume = 10 μL, total time = 31.50 min	Extracted with MeOH/water/$CHCl_3$ and diluted in 10 mmol/L DAA, pH 4.95	LOD = 0.022 to 19.5 ppm, intra-day RSD mostly <15%, inter-day RSD <20%, $R^2 \geq 0.95$, ion suppression or enhancement for some metabolites, significant matrix effect for 3 of 5 tested metabolites	[78]
Targeted metabolomics/10 phosphorylated sugars and 6 neutral sugars (other compounds tested, but poor performance obtained)	HILIC-MS (Waters Acquity UPLC system and Waters Micromass QTOF MS)	Rice tissue samples	Waters Acquity UPLC bridge ethylene hybrid (BEH) amide (1.7 μm, 2.1 × 100 mm)	A: 5 mmol/L MePO. B: ACN with 3.5 mmol/L TEA and 1.3 mmol/L FA (gradient: 90–50% B)	Flow rate = 0.35 mL/min, temperature = 40 °C, injection volume = 10 μL, total time = 17 min	Extraction with $CHCl_3$, ACN, and water, dried under air stream, reconstituted in water, SPE, dried under air stream, reconstituted in mobile phase	MePO avoided tailing of phosphorylated compounds	[79]

(continued)

Table 1 (continued)

Metabolomics type/ metabolites	LC-MS system	Matrix	Column type	Mobile phase	Other conditions	Sample preparation	Method performance	Ref.
Targeted metabolomics/20 amino acids and 5 metabolically related compounds	LC-MS/MS (Shimadzu LC-20 AD HPLC system and Waters Quattro Micromass MS)	*Arabidopsis thaliana* seeds	Waters Symmetry C18 (3.5 μm, 2.1 × 100 mm)	A: 1 mmol/L perfluoroheptanoic acid. B: ACN (gradient: 98–60% A)	Flow rate = 0.3 mL/min, temperature = 30 °C, injection volume = 10 μL, total time = 6.0 min	Extraction in water at 90 °C and supernatant filtered through 0.45 μm low-binding hydrophilic filter	Intra- and inter-assay RSD < 5% (except for Gly, with RSD up to 18%), recoveries = 62 to 94%, suppression by matrix = 31 to 65%, accuracy = 77 to 133%	[80]
Targeted metabolomics/8 nucleoside triphosphates and other polar metabolites (carbohydrate derivatives, carboxylic acids, acyl coenzyme A derivatives)	LC-MS/MS (Shimadzu Nexera HPLC and AB SCIEX QTRAP 6500 MS)	Cell extracts	Waters Atlantis T3 (3.0 μm, 2.1 × 100 mm)	A: 100 mmol/L HFIP and 8.6 mmol/L TEA, pH 8.3. B: 10% ACN in mobile phase A	Flow rate = 0.5 mL/min, temperature = 45 °C, injection volume = 10 μL, total time = 10 min	Extraction in ice-cold 60:40 MeOH/water, supernatant filtered (3 kDa molecular weight cutoff), dried under vacuum, reconstituted in mobile phase A	Intra-day precision <10%, inter-day precision <13%, accuracy (spike recovery) = 87 to 115%	[81]

ACN acetonitrile, *BEH* bridged ethylene hybrid, *DAA* diamyl ammonium acetate, *DIPEA* diisopropylethylamine, *EtOH* ethanol, *FA* formic acid, *HFBA* heptafluorobutyric acid, *HFIP* hexafluoroisopropanol, *HOAc* acetic acid, IPA isopropanol, *LOD* limit of detection, *LOQ* limit of quantification, *MeOH* methanol, *MePO* methylphosphonic acid, NH_4OAc ammonium acetate, *PBS* phosphate-buffered saline, *RSD* relative standard deviation, *RT* retention time, *SPE* solid-phase extraction, *TBA* tributylamine, *TEA* triethylamine retention of polar metabolites in typical RPLC separations [74]. This approach, called ion pairing chromatography (IPC), has been applied in metabolic profiling studies of polar/ionizable metabolites in many types of samples (e.g., cells, tissues, plants, animals), as described in selected applications listed in Table 1. Despite being a relatively mature technique that adequately offers retention capability for a wide range of ionic metabolites, IPC-MS has not been used in metabolomics as extensively as some other chromatographic modes, possibly due to ion suppression and MS contamination with IPR species [73].

Chromatographic separation by IPC involves a mixed retention mechanism that remains a topic for discussion [76]. A combination of complex ion interactions is recognized to play a central role in the separation, including (1) the formation of non-charged ion pairs of analytes with IPR, which become relatively nonpolar and can partition into the stationary phase, and (2) the strong interaction of the hydrophobic portion of the IPR with the stationary phase, combined with the electrostatic interaction of analytes with the charged portion of the IPR [82]. In order to optimize separations by IPC, the IPR type and concentration, as well as the mobile phase pH, are considered relevant parameters [73, 74]. Traditional IPRs (e.g., tetraalkylammonium salts for anionic analytes and alkyl/aryl sulfonates for cations) are incompatible with MS detection, due to their low volatility and tendency to accumulate in the system [74]. Alternatively, more volatile compounds, such as alkyl amines and perfluorinated carboxylic acids, have been proposed as IPRs for the IPC-MS of anionic and cationic determinations, respectively [82]. Figure 4 depicts the main IPRs used in IPC-MS metabolomics studies to date. For instance, Tang et al. [75] have used heptafluorobutyric acid (HFBA) for cationic metabolites (amines, amino acids, and carnitines) and tributylamine (TBA) for anionic metabolites (carboxylic acids, sugar phosphates, nucleo-compounds, and coenzyme A esters) as volatile IPRs when doing targeted metabolomics of hepatic cell lines by UHPLC using C18 columns. Interestingly, Guo et al. [76] have observed that the concentration of the IPR diisopropylethylamine (DIPEA), optimized for standard mixtures containing carboxylic acids and phosphorylated compounds, was insufficient when applied to human melanoma cell extracts. This highlights the importance of optimizing the chromatographic conditions for each sample matrix.

Despite the use of volatile IPRs, ion suppression and MS contamination remain the most critical limitations of IPC-MS. Ion suppression derives from the formation of strong ion pairs that persist in the ion source, affecting the ESI signal [83]. Additionally, alkyl amines, used as IPRs for anionic analytes, tend to adsorb strongly on front parts of the MS system, leading to persistent carryover when the instrument is subsequently used in the positive ion mode [62]. For this reason, it is generally required to have a dedicated LC-MS instrument to perform IPC-MS for anionic metabolites, a largely inconvenient option [73]. The proper choice of IPRs may help to attenuate this problem, as reported by Guo et al. [76], who determined anionic metabolites using DIPEA, an IPR that prevented strong interference when switching from negative to positive ion mode. Establishing a routine cleaning protocol to remove IPRs from the MS instrument is another approach that has been used to minimize contamination [78]. Still, further development is necessary to make IPC-MS a more practical technique in metabolomics.

6 Ion Chromatography

In the last decades, IC has been widely used to determine inorganic and small organic ions in well-established applications in the environmental, food, clinical, pharmaceutical, and forensic fields [84–86]. The introduction of eluent suppression, formerly called "stripping" by Small et al. [87] in 1975, was a key step in the development of IC as a modern high-performance liquid chromatography technique. As a result, the use of conductometric ion detection without interference from high-abundance eluent electrolytes became possible. Eluent suppression was also fundamental for the advancement of

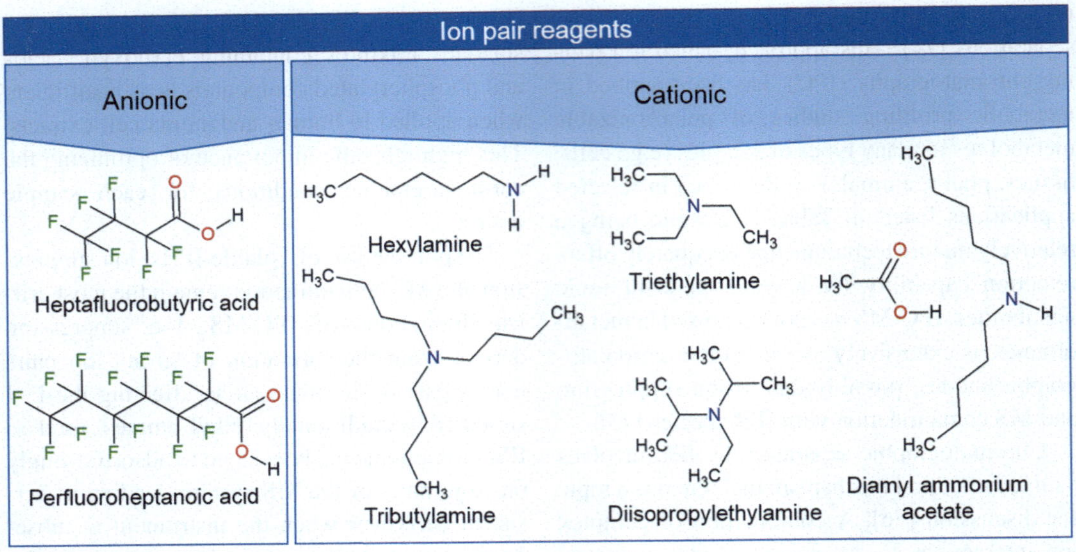

Fig. 4 The most commonly used IPRs in IPC-MS metabolomics studies for anionic and cationic metabolites

IC-MS. Although there are applications of non-suppressed IC-MS, which combine low capacity columns and weakly ionized eluents that are compatible with MS detection (e.g., formate and acetate), these conditions have limited performance for low-abundance analytes in complex matrices [84]. For this reason, suppressed IC-MS with non-volatile and strongly acidic or basic mobile phases (e.g., methanesulfonic acid, NaOH, KOH) has been more suitable for multiple analyte determination, which can be used with high capacity columns [84]. In metabolomics, IC-MS remains scarcely explored, despite its enormous potential to expand the metabolic coverage for ionic/ionizable metabolites, especially for anionic compounds that are even more challenging when using other analytical platforms [88].

IC methods, which are mainly based on ion exchange mechanism, use columns containing ionic groups (e.g., carboxylate, sulfonate, quaternary ammonium, alkyl amine) that are covalently bound to a solid support (usually polymeric resins, like polyvinyl or styrene-divinylbenzene) [84, 89]. Separation is based on ion exchange of oppositely charged analytes on the surface of the stationary phase. The analyte charge and hydrated radius largely influence their retention, and elution occurs by competitive displacement of the analyte by co-ions present in the mobile phase [84]. The selection of columns with high ion exchange capacity is important in untargeted metabolomics, when the objective is the separation of a large number of analytes in complex biological samples. Anion exchange columns containing quaternary ammonium groups on ethylvinylbenzene-divinylbenzene polymeric resins have been used in most metabolomics studies involving IC-MS so far, as listed in Table 2. To the best of our knowledge, cationic IC-MS has not been explored in metabolomics studies to date, although there are applications involving the determination of amines in environmental [90] and food samples [91], whose conditions can potentially be further optimized for metabolomics.

The coupling between IC and MS, schematically represented in Fig. 5, can be performed using a few different setups, depending on the column diameter. The IC-MS system typically contains a separation ion exchange column followed by an eluent suppressor unit, which is regenerated using a solution transported by a peristaltic pump. Standard bore columns (4 mm i.d.) are generally operated at flow rates around 1 mL min^{-1} and require the use of a T-junction (T1 in Fig. 5) that partially diverts the column effluent prior to introduction into the MS interface

Table 2 Summary of applications of ion chromatography-mass spectrometry (IC-MS) in metabolomics studies

Metabolomics type/metabolites	IC mode	IC-MS system	Matrix	Column type	Mobile phase	Other conditions	Sample preparation	Method performance	Ref.
Targeted metabolomics/28 organic acids	Anion exchange	Dionex ICS-5000 IC and Thermo Quantiva TSQ triple quadrupole MS	Mice quadriceps muscle	IonPac AS-11-HC (4 μm, 2 × 250 mm), guard column Ion Pac AG-11-HC (4 μm, 2 × 50 mm)	5–100 mmol/L KOH	Flow rate = 350 μL/min, temperature = 35 °C, total time = 20 min	Extraction of lyophilized samples with ACN/0.3% FA and derivatization for keto-acids	LOQ = 0.25 to 50 μmol/L, inter-day precision = 0.8 to 19.4%, accuracy = 82 to 109%	[92]
Targeted metabolomics/6 organic acids, 11 sugar monophosphates, 9 sugar diphosphates	Anion exchange	Dionex ICS-5000+ IC and Thermo Q Exactive HF hybrid quadrupole-Orbitrap MS	Head and neck cancer cells and cancer stem-like cells	IonPac AS-11-HC (4 μm, 2 × 250 mm)	10–85 mmol/L KOH	Flow rate = 380 μL/min, makeup liquid = 2 mmol/L HOAc in MeOH, total time = 23 min	Extraction with ice cold MeOH/water and $CHCl_3$ under liquid N_2 and centrifugation	Linear range = 0.1 to 10,000 pg/μL	[93]
Semi-targeted metabolomics/phosphorylated compounds	Anion exchange	Dionex ICS-5000 IC and Thermo LTQ Orbitrap Velos	*Trypanosoma brucei* cells	IonPac AS-19 (for standards) and IonPac AS-20 (for extracts)	8–100 mmol/L KOH	Flow rate = 10 μL/min, makeup liquid = MeOH, total time = 60 min	Extraction with $CHCl_3$, MeOH and water at 4 °C	Not mentioned	[94]
Untargeted metabolomics/phosphorylated compounds, sugars, organic acids, nucleo-compounds	Anion exchange	Dionex ICS-4000 IC and Thermo Q Exactive Orbitrap MS	Head and neck human cancer cells	IonPac AS-11-HC (4 μm, 0.4 × 250 mm)	2–70 mmol/L KOH	Flow rate = 25 μL/min, temperature = 35 °C, makeup liquid = 2 mmol/L HOAc in MeOH, total time = 45 min	Extraction with ice cold $MeOH/CHCl_3$ under liquid N_2	LOD = 0.04 to 0.5 nmol/L	[95]

(continued)

Table 2 (continued)

Metabolomics type/metabolites	IC mode	IC-MS system	Matrix	Column type	Mobile phase	Other conditions	Sample preparation	Method performance	Ref.
Targeted metabolomics/22 amino acids	Anion exchange	Dionex ICS 3000 IC with unspecified MS system	Urine from gastric cancer patients	AminoPac PA10Pac CS18 (2 × 250 mm), guard column IonPac CG18 (2 × 50 mm)	40–200 mmol/L NaOH and 400–700 mmol/L NaOAc	Flow rate = 250 μL/min, temperature = 30 °C, injection volume = 25 μL, total time = 40 min	Not mentioned	Not mentioned	[96]
Targeted and untargeted metabolomics/45 anions: organic acids, phosphorylated compounds, nucleotides, carbohydrates	Anion exchange	Dionex Integrion HPIC and Thermo Q Exactive HF Orbitrap MS	Cancer cells	IonPac AS11-HC (4 μm, 2 × 250 mm), guard column AG11-HC (4 μm, 2 × 50 mm)	10–100 mmol/L KOH	Flow rate = 380 μL/min, temperature = 30 °C, injection volume = 5 μL, makeup liquid = MeOH, total time = 25 min	Extraction with ice cold MeOH/water, dried under reduced pressure, and reconstituted	LOD = low nmol/L range, average RSD < 5% for 5 replicates within 40 h	[88]
Metabolic flux analysis/6 phosphorylated metabolites	Anion exchange	Dionex ICS 2000 IC and Applied Biosystems 4000 Q Trap MS	*Escherichia coli* cell extracts	IonPac AS11 (13 μm, 2 × 250 mm), guard column IonPac AG11 (13 μm, 2 × 50 mm)	0.5–40 mmol/L KOH and 0.1–1.6% MeOH	Flow rate = 350 μL/min, temperature = 22 °C, injection volume = 15 μL, total time = 70 min	Extraction with boiling water	Improved accuracy and sensitivity for MRM in comparison with MIM experiments	[97]
Semi-targeted metabolomics/44 anions: phosphorylated compounds and organic acids	Anion exchange	Dionex ICS-4000 IC and Waters Xevo QT-S triple quadrupole MS	Human kidney Hek293 cell line	IonSwift MAX-100 (1.6 μm, 0.25 × 250 mm), guard column IonSwift MAX-100G (0.25 × 50 mm)	6–140 mmol/L KOH	Flow rate = 12 μL/min, temperature = 40 °C, injection volume = 5 μL, makeup liquid = 90% ACN in water with 0.01% NH_4OH, total time = 45 min (separation) + 21 min (wash)	Extraction with ACN/water at 0 °C, freeze-dried, reconstituted, and filtered (3 kDa molecular weight cutoff)	LOD = 1 to 100 nmol/L, LOQ = 1 to 500 nmol/L, average RSD = 8.5%	[98]

Targeted metabolomics/ glycolytic intermediates: 8 phosphorylated metabolites	Anion exchange	Waters Alliance IC and Micromass Quattro-LC triple quadrupole MS	*Saccharomyces cerevisiae* cell extracts	IonPac AS11 (13 µm, 4 × 250 mm), guard column IonPac AG11 (13 µm, 4 × 50 mm)	3–150 mmol/L NaOH	Flow rate = 1000 µL/ min, room temperature, injection volume = 10 µL, flow split, total time = 42 min	Quenching with cold MeOH/water, extraction with boiling EtOH, evaporation, and reconstitution in water	LOD = 1.5 to 15.7 nmol/g d.w., LOQ = 5.0 to 52.3 nmol/g d.w., repeatability = 3.6 to 7.2%, average recoveries = 86 to 98%	[99]
Targeted metabolomics/ plant cell wall precursors: 16 sugar phosphates and nucleotide sugar phosphates	Anion exchange	Waters Acquity UPLC and Waters Quattro-Premier triple quadrupole MS	*Arabidopsis thaliana* T87 cell cultures and *Trigonella foenum-graecum* endosperm tissue	IonPac AS11 (13 µm, 2 × 250 mm), guard column IonPac AG11 (13 µm, 2 × 50 mm)	0.5–50 mmol/L NaOH	Flow rate = 350 µL/ min, room temperature, total time = 85.1 min	Extraction with boiling water, filtered with syringe filter, evaporated, reconstituted	LOD = 0.3 to 34 pmol, LOQ = 1 to 111 pmol, repeatability = 0.5 to 16.2%, recovery = 86 to 115%	[100]
Untargeted metabolomics/ organic acids, amino acids, nucleotides, phosphorylated compounds, peptides	Anion exchange	Dionex ICS-4000 IC and Thermo Q Exactive Orbitrap MS	Human saliva from gout patients	IonPac AS11-HC (4 µm, 0.4 × 250 mm)	2–95 mmol/L KOH	Flow rate = 25 µL/ min, temperature = 35 °C, total time = 33.1 min	Extraction with ice cold MeOH/$CHCl_3$ and centrifugation	Not mentioned	[101]
Semi-targeted metabolomics/24 organic acids, sugars, and nucleotides	Anion exchange	Thermo ICS-5000+ IC and Q Exactive HF Orbitrap MS	Basil (*Ocimum basilicum*) leaves	IonPac AS11-HC (4 µm, 2 × 250 mm), guard column IonPac AG-11-HC (4 µm, 2 × 50 mm)	0–100 mmol/L KOH	Flow rate = 250 µL/ min, temperature = 30 °C, injection volume = 10 µL, total time = 37 min	Reverse iontophoresis of leaves with phosphate saline buffer (pH 7)	Not mentioned	[102]

(continued)

Table 2 (continued)

Metabolomics type/metabolites	IC mode	IC-MS system	Matrix	Column type	Mobile phase	Other conditions	Sample preparation	Method performance	Ref.
Polar metabolites/ glycolytic, Krebs cycle, glutathione synthesis, and pentose phosphate pathway metabolites	Anion exchange	Dionex ICS-5000+ IC and Thermo Orbitrap Fusion Tribrid MS	Lung tissue from lung cancer patients	IonPac AS11-HC (4 μm, 2 × 250 mm), guard column IonPac AG-11-HC (4 μm, 2 × 50 mm)	1–100 mmol/L KOH	Flow rate = 380 μL/min, temperature = 35 °C, injection volume = 5 μL, makeup liquid = 60 μL/min MeOH, total time = 52.5 min	Extraction of frozen, pulverized tissue with ACN/water/$CHCl_3$, lyophilization of polar extract and reconstitution	Not mentioned	[103]

ACN acetonitrile, *d.w.* dry weight, *EtOH* ethanol, *FA* formic acid, *HOAc* acetic acid, *LOD* limit of detection, *LOQ* limit of quantification, *MeOH* methanol, *MIM* multiple ion monitoring, *MRM* multiple reaction monitoring, *MSA* methanesulfonic acid, *NaOAc* sodium acetate, *RSD* relative standard deviation

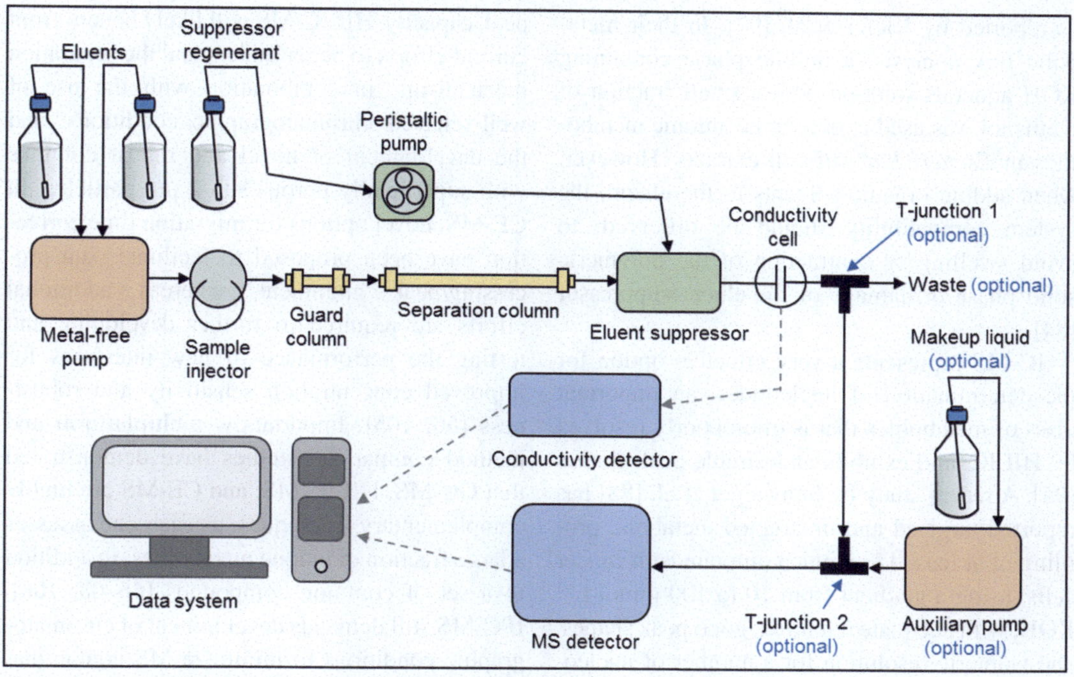

Fig. 5 Schematic representation of a generic IC-MS system with eluent suppression and simultaneous conductivity detection

to avoid overloading of the ion source with the aqueous solution. For instance, van Dam et al. [99] have used a standard bore column for targeted determination of phosphorylated intracellular metabolites in yeast cell extracts, with a post-column flow splitter to reduce the flow to ~100 μL min^{-1} before MS detection. In contrast, the use of a splitter is not necessary for microbore columns (2 mm i.d.), which are typically used with flow rates between 100 and 700 μL min^{-1} [84]. For example, Alonso et al. [100] have developed a tandem mass spectrometry IC-MS/MS method for targeted metabolomics analysis of intracellular cell wall precursors in plants (*Arabidopsis thaliana* and *Trigonella foenum-graecum*), using a microbore column with a flow rate of 350 μL min^{-1} without a splitter. In addition, capillary columns (≤1 mm i.d.) can be used with much lower flow rates (~10 μL min^{-1}), reported to improve sensitivity, while minimizing eluent consumption and waste generation. However, the high backpressure produced requires the use of IC systems that support backpressures of up to 5000 psi, while standard IC instruments are typically operated under 3000 psi. In this latter case, a flow splitter is equally not necessary, but a makeup liquid is frequently utilized to supplement the flow entering the ion source. A makeup liquid is also often used in IC-MS for standard bore and microbore columns to improve sensitivity by increasing the volatility of the aqueous column effluent. That is the case in the study reported by Petucci et al. [92], in which targeted metabolomics of 28 organic acids was performed in mice muscle samples, using a microbore anion exchange column at 350 μL min^{-1}, without flow splitter. In this case, a sheath liquid composed of methanol was introduced through a low dead volume T-junction mixer (T2 in Fig. 5), using a post-column auxiliary pump. Similarly, head and neck cancer cells have been analyzed for untargeted metabolomics by IC-MS using a capillary column, with a makeup liquid composed of 2 mmol L^{-1} acetic acid in methanol [95]. Another approach to increase the volatility of the column effluent consists in the introduction of organic solvent to the mobile phase, instead of using a makeup liquid, as reported by Kiefer et al. [97]. In their metabolic flux analysis, a mobile phase containing KOH aqueous solution with a small fraction of methanol was used to determine anionic metabolites in *Escherichia coli* cell extracts. However, when adding organic solvents to the eluent, the system compatibility should be observed, to avoid swelling or contraction of the polymeric solid phase or damage of the eluent suppressor [84].

IC-MS represents a very effective option for the determination of nucleotides, an important class of metabolites that is often poorly resolved by HILIC, and exhibits undesirable peak shapes [94]. A recent study by Schwaiger et al. [88] has reported targeted and untargeted metabolic profiling of at least 45 anionic compounds in cancer cells, using a gradient from 10 to 100 mmol L^{-1} KOH, with adequate retention, good peak shapes, and isomeric resolution for a number of nucleotides. Sugars and sugar phosphates are another class of metabolites that have been adequately resolved by IC-MS. Although carbohydrates are not usually considered as ionic/ionizable compounds, they are substantially ionized in strongly basic solutions, which makes them suitable for determination by IC-MS with eluent suppression [104]. Although IC-MS has been scarcely used in metabolomics, it represents an emerging technique in the field to complement currently used methods, with great potential to expand the metabolic coverage for ionic/ionizable metabolites.

7 Conclusion and Future Trends

The present chapter described the main MS-based techniques used in metabolomics for polar/ionizable compounds, including GC-MS, HILIC-MS, CE-MS, IPC-MS, and IC-MS. The main strengths and limitations of these techniques are summarized in Table 3, which also highlights some trends. Recent developments of GC-MS for metabolomics have focused on the use of automated derivatization systems, as well as the development of high-resolution methods using TOF-MS detectors and GC×GC-MS to improve peak capacity. HILIC-MS will likely benefit from current efforts to better understand the separation mechanisms, in combination with the use of well-selected chromatographic conditions, and the development of novel and robust columns with superficially porous sub-2 μm particles. In CE-MS, novel options for migration time correction have been proposed to facilitate data processing/peak alignment, whereas additional efforts are required to further developing and testing the performance of new interfaces for improved concentration sensitivity and robustness [50, 105]. Importantly, multiplatform and method comparison studies have demonstrated that GC-MS, HILIC-MS, and CE-MS are highly complementary techniques, as each one assesses a large fraction of unique metabolites, in addition to a set of common compounds [66–68, 106]. IPC-MS still demands development of chromatographic conditions to minimize MS contamination and ion suppression caused by IPRs [78, 107]. Finally, future studies on IC-MS will likely focus on optimizing analytical conditions to determine anionic and cationic metabolites, including the evaluation of long-term performance for studies involving larger sample sizes. At present, it is still a scarcely explored technique for metabolomics.

In addition to the platforms discussed here, other approaches have recently started to be used in metabolomics for polar/ionizable compounds, including aqueous normal phase chromatography with silica hydride columns, mixed-mode chromatography (e.g., combined mechanisms for HILIC-IC, RPLC-IC, RPLC-HILIC), porous graphitic columns, and supercritical fluid chromatography by modulating the polarity of carbon dioxide with organic solvents to be used as the mobile phase to allow the determination of more polar compounds [9, 11, 12, 108].

In conclusion, this chapter highlights the complementarity among analytical platforms used to scan polar/ionizable metabolites in metabolomics studies. A discussion and summary of their main strengths and limitations serve as a starting point to facilitate the selection of techniques to be used in metabolomics, when considering specific needs regarding sensitivity, throughput, financial

Table 3 Summary of strengths, limitations, and future trends for the main analytical platforms used in metabolomics for polar/ionizable metabolites

	GC-MS	HILIC-MS	CE-MS	IPC-MS	IC-MS
Strengths	Well-established High resolving power Simple metabolite annotation (reproducible MS spectra and large body of databases) Lower cost	Broad coverage Variety of column chemistries High sensitivity Uses common LC systems (RPLC-MS can be performed using the same instrument)	High resolving power Minimal volumes of consumed reagents Amenable to limited sample amounts	Performed by simple addition of IPR to common RPLC mobile phases Uses common LC systems and RPLC columns	Post-column eluent suppression expands the choice of mobile phases Great performance demonstrated for anionic metabolites
Limitations	Derivatization is required (limited throughput, multiple derivative products, introduction of nonbiological variation)	Less robust when compared to RPLC-MS Longer column equilibration times Poor peak shapes for some compounds	Less robust when compared to RPLC-MS and HILIC-MS Interface-related sensitivity	A dedicated instrument is often required Careful cleaning protocol to minimize MS contamination Ion suppression	Specific IC system required for compatibility with extreme pH and ionic strength (not amenable to conventional LC systems)
Future trends	Improvements in automated derivatization systems GC-TOF-MS GC×GC-MS	Better understanding of separation mechanisms Use of well-selected chromatographic conditions Development of novel column chemistries, UHPLC methods, and the availability of more HILIC phases as sub-2 µm and superficially porous particles	New algorithms and protocols for peak alignment Development and evaluation of long-term performance of novel CE-MS interfaces	Development of chromatographic conditions that are more compatible with ESI-MS	Optimization of method conditions for metabolomics Evaluation of method performance in larger cohorts

resources, or amount of available sample. Advances currently under development for each methodology indicate future trends that will likely help to improve robustness, sensitivity, and metabolic coverage of polar/ionizable compounds.

Acknowledgments The authors wish to acknowledge the Sao Paulo Research Foundation (FAPESP) and the National Council for Scientific and Technological Development (CNPq) of Brazil for financial support and doctoral and post-doctoral fellowships (FAPESP 2017/27059-6, FAPESP 2017/12149-0, CNPq 152064/2016-8, CNPq 306813/2018-1, and CNPq 141064/2018-8).

References

1. Kuehnbaum NL, Britz-Mckibbin P (2013) New advances in separation science for metabolomics: Resolving chemical diversity in a post-genomic era. Chem Rev 113:2437–2468. https://doi.org/10.1021/cr300484s
2. Castillo-Peinado LS, Luque de Castro MD (2016) Present and foreseeable future of metabolomics in forensic analysis. Anal Chim Acta 925:1–15. https://doi.org/10.1016/j.aca.2016.04.040
3. Dunn WB, Ellis DI (2005) Metabolomics: current analytical platforms and methodologies. Trends Anal Chem 24:285–294. https://doi.org/10.1016/j.trac.2004.11.021
4. Putri SP, Yamamoto S, Tsugawa H, Fukusaki E (2013) Current metabolomics: technological

advances. J Biosci Bioeng 116:9–16. https://doi.org/10.1016/j.jbiosc.2013.01.004
5. Patti GJ (2011) Separation strategies for untargeted metabolomics. J Sep Sci 34:3460–3469. https://doi.org/10.1002/jssc.201100532
6. Dunn WB, Broadhurst D, Begley P, Zelena E, Francis-McIntyre S, Anderson N, Brown M, Knowles JD, Halsall A, Haselden JN, Nicholls AW, Wilson ID, Kell DB, Goodacre R (2011) Procedures for large-scale metabolic profiling of serum and plasma using gas chromatography and liquid chromatography coupled to mass spectrometry. Nat Protoc 6:1060–1083. https://doi.org/10.1038/nprot.2011.335
7. Rojo D, Barbas C, Rupérez FJ (2012) LC-MS metabolomics of polar compounds. Bioanalysis 4:1235–1243. https://doi.org/10.4155/bio.12.100
8. Liu Z, Rochfort S (2014) Recent progress in polar metabolite quantification in plants using liquid chromatography-mass spectrometry. J Integr Plant Biol 56:816–825. https://doi.org/10.1111/jipb.12181
9. Kohler I, Giera M (2017) Recent advances in liquid-phase separations for clinical metabolomics. J Sep Sci 40:93–108. https://doi.org/10.1002/jssc.201600981
10. Gika HG, Theodoridis GA, Plumb RS, Wilson ID (2014) Current practice of liquid chromatography-mass spectrometry in metabolomics and metabonomics. J Pharm Biomed Anal 87:12–25. https://doi.org/10.1016/j.jpba.2013.06.032
11. Callahan DL, De Souza D, Bacic A, Roessner U (2009) Profiling of polar metabolites in biological extracts using diamond hydride-based aqueous normal phase chromatography. J Sep Sci 32:2273–2280. https://doi.org/10.1002/jssc.200900171
12. Siegel D, Permentier H, Reijngoud DJ, Bischoff R (2014) Chemical and technical challenges in the analysis of central carbon metabolites by liquid-chromatography mass spectrometry. J Chromatogr B 966:21–33. https://doi.org/10.1016/j.jchromb.2013.11.022
13. Fiehn O (2008) Extending the breadth of metabolite profiling by gas chromatography coupled to mass spectrometry. Trends Anal Chem 27:261–269. https://doi.org/10.1016/j.trac.2008.01.007
14. Pasikanti KK, Ho PC, Chan ECY (2008) Gas chromatography/mass spectrometry in metabolic profiling of biological fluids. J Chromatogr B 871:202–211. https://doi.org/10.1016/j.jchromb.2008.04.033
15. Mastrangelo A, Ferrarini A, Rey-Stolle F, García A, Barbas C (2015) From sample treatment to biomarker discovery: a tutorial for untargeted metabolomics based on GC-(EI)-Q-MS. Anal Chim Acta 900:21–35. https://doi.org/10.1016/j.aca.2015.10.001
16. Koek MM, Jellema RH, van der Greef J, Tas AC, Hankemeier T (2011) Quantitative metabolomics based on gas chromatography mass spectrometry: status and perspectives. Metabolomics 7:307–328. https://doi.org/10.1007/s11306-010-0254-3
17. Xu F, Zou L, Ong CN, Zou L (2010) Experiment-originated variations, and multi-peak and multi-origination phenomena in derivatization-based GC-MS metabolomics. Trends Anal Chem 29:269–280. https://doi.org/10.1016/j.trac.2009.12.007
18. Kind T, Wohlgemuth G, Lee DY, Lu Y, Palazoglu M, Shahbaz S, Fiehn O (2009) FiehnLib: mass spectral and retention index libraries for metabolomics based on quadrupole and time-of-flight gas chromatography/mass spectrometry. Anal Chem 81:10038–10048. https://doi.org/10.1021/ac9019522
19. Abbiss H, Rawlinson C, Maker GL, Trengove R (2015) Assessment of automated trimethylsilyl derivatization protocols for GC–MS-based untargeted metabolomic analysis of urine. Metabolomics 11:1908–1921. https://doi.org/10.1007/s11306-015-0839-y
20. Zarate E, Boyle V, Rupprecht U, Green S, Villas-Boas SG, Baker P, Pinu FR (2017) Fully automated trimethylsilyl (TMS) derivatisation protocol for metabolite profiling by GC-MS. Meta 7:1–13. https://doi.org/10.3390/metabo7010001
21. Begley P, Francis-McIntyre S, Dunn WB, Broadhurst DI, Halsall A, Tseng A, Knowles J, Goodacre R, Kell DB (2009) Development and performance of a gas chromatography-time-of-flight mass spectrometry analysis for large-scale nontargeted metabolomic studies of human serum. Anal Chem 81:7038–7046. https://doi.org/10.1021/ac9011599
22. Almstetter MF, Oefner PJ, Dettmer K (2012) Comprehensive two-dimensional gas chromatography in metabolomics. Anal Bioanal Chem 402:1993–2013. https://doi.org/10.1007/s00216-011-5630-y
23. Miyazaki T, Okada K, Yamashita T, Miyazaki M (2017) Two-dimensional gas chromatography time-of-flight mass spectrometry-based serum metabolic fingerprints of neonatal calves before and after first colostrum ingestion. J Dairy Sci 100:4354–4364. https://doi.org/10.3168/jds.2017-12557
24. van Stee LLP, Brinkman UAT (2016) Peak detection methods for GC×GC: an overview. Trends Anal Chem 83:1–13. https://doi.org/10.1016/j.trac.2016.07.009
25. Tang D-Q, Zhou L, Yin X-X, Ong CN (2016) HILIC-MS for metabolomics: an attractive and complementary approach to RPLC-MS. Mass Spectrom Rev 35:574–600. https://doi.org/10.1002/mas
26. Alpert AJ (1990) Hydrophilic-interaction chromatography for the separation of peptides, nucleic acids and other polar compounds. J Chromatogr A 499:177–196. https://doi.org/10.1016/S0021-9673(00)96972-3
27. Spagou K, Tsoukali H, Raikos N, Gika H, Wilson ID, Theodoridis G (2010) Hydrophilic interaction chromatography coupled to MS for metabonomic/metabolomic studies. J Sep Sci 33:716–727. https://doi.org/10.1002/jssc.200900803
28. Cubbon S, Antonio C, Wilson J, Thomas-Oates J (2010) Metabolomic applications of HILIC–LC–

MS. Mass Spectrom Rev 29:671–684. https://doi.org/10.1002/mas.20252
29. Jandera P, Janás P (2017) Recent advances in stationary phases and understanding of retention in hydrophilic interaction chromatography. a review. Anal Chim Acta 967:12–32. https://doi.org/10.1016/j.aca.2017.01.060
30. Jandera P (2008) Stationary phases for hydrophilic interaction chromatography, their characterization and implementation into multidimensional chromatography concepts. J Sep Sci 31:1421–1437. https://doi.org/10.1002/jssc.200800051
31. Guo Y (2015) Recent progress in fundamental understanding of hydrophilic interaction chromatography (HILIC). Analyst 140:6452–6466. https://doi.org/10.1039/C5AN00670H
32. Wernisch S, Pennathur S (2016) Evaluation of coverage, retention patterns, and selectivity of seven liquid chromatographic methods for metabolomics. Anal Bioanal Chem 408:6079–6091. https://doi.org/10.1007/s00216-016-9716-4
33. Jian W, Edom RW, Xu Y, Weng N (2010) Recent advances in application of hydrophilic interaction chromatography for quantitative bioanalysis. J Sep Sci 33:681–697. https://doi.org/10.1002/jssc.200900692
34. Pantophlet AJ, Roelofsen H, De Vries MP, Gerrits WJJ, Van Den Borne JJGC, Vonk RJ (2017) The use of metabolic profiling to identify insulin resistance in veal calves. PLoS One 12:1–12. https://doi.org/10.1371/journal.pone.0179612
35. Wang M, Yang R, Dong J, Zhang T, Wang SS, Zhou W, Li H, Zhao H, Zhang L, Wang SS, Zhang C, Chen W (2015) Simultaneous quantification of cardiovascular disease related metabolic risk factors using liquid chromatography tandem mass spectrometry in human serum. J Chromatogr B 1010:144–151. https://doi.org/10.1016/j.jchromb.2015.12.019
36. Evans CR, Karnovsky A, Kovach MA, Standiford TJ, Burant CF, Stringer KA (2014) Untargeted LC-MS metabolomics of bronchoalveolar lavage fluid differentiates acute respiratory distress syndrome from health. J Proteome Res 13:640–649. https://doi.org/10.1021/pr4007624
37. Kind T, Tolstikov V, Fiehn O, Weiss RH (2007) A comprehensive urinary metabolomic approach for identifying kidney cancer. Anal Biochem 363:185–195. https://doi.org/10.1016/j.ab.2007.01.028
38. Yang S, Sadilek M, Lidstrom ME (2010) Streamlined pentafluorophenylpropyl column liquid chromatography-tandem quadrupole mass spectrometry and global 13C-labeled internal standards improve performance for quantitative metabolomics in bacteria. J Chromatogr A 1217:7401–7410. https://doi.org/10.1016/j.chroma.2010.09.055
39. Li K, Naviaux JC, Bright AT, Wang L, Naviaux RK (2017) A robust, single-injection method for targeted, broad-spectrum plasma metabolomics. Metabolomics 13:122. https://doi.org/10.1007/s11306-017-1264-1
40. Contrepois K, Jiang L, Snyder M (2015) Optimized analytical procedures for the untargeted metabolomic profiling of human urine and plasma by combining hydrophilic interaction (HILIC) and reverse-phase liquid chromatography (RPLC)–mass spectrometry. Mol Cell Proteomics 14:1684–1695. https://doi.org/10.1074/mcp.M114.046508
41. Sampsonidis I, Witting M, Koch W, Virgiliou C, Gika HG, Schmitt-Kopplin P, Theodoridis GA (2015) Computational analysis and ratiometric comparison approaches aimed to assist column selection in hydrophilic interaction liquid chromatography-tandem mass spectrometry targeted metabolomics. J Chromatogr A 1406:145–155. https://doi.org/10.1016/j.chroma.2015.06.008
42. Tufi S, Lamoree M, de Boer J, Leonards P (2015) Simultaneous analysis of multiple neurotransmitters by hydrophilic interaction liquid chromatography coupled to tandem mass spectrometry. J Chromatogr A 1395:79–87. https://doi.org/10.1016/j.chroma.2015.03.056
43. Zhang T, Watson DG (2016) Evaluation of the technical variations and the suitability of a hydrophilic interaction liquid chromatography-high resolution mass spectrometry (ZIC-pHILIC-Exactive orbitrap) for clinical urinary metabolomics study. J Chromatogr B 1022:199–205. https://doi.org/10.1016/j.jchromb.2016.04.017
44. Zhang R, Watson DW, Wang L, Westrop GD, Coombs GH, Zhang T (2014) Evaluation of mobile phase characteristics on three zwitterionic columns in hydrophilic interaction liquid chromatography mode for liquid chromatography-high resolution mass spectrometry based targeted metabolite profiling of Leishmania parasites. J Chromatogr A 1362:168–179. https://doi.org/10.1016/j.chroma.2014.08.039
45. Virgiliou C, Sampsonidis I, Gika HG, Raikos N, Theodoridis GA (2015) Development and validation of a HILIC-MS/MS multitargeted method for metabolomics applications. Electrophoresis 36:2215–2225. https://doi.org/10.1002/elps.201500208
46. Nováková L, Havlíková L, Vlčková H (2014) Hydrophilic interaction chromatography of polar and ionizable compounds by UHPLC. Trends Anal Chem 63:55–64. https://doi.org/10.1016/j.trac.2014.08.004
47. Xu Y, Yang L, Yang F, Xiong Y, Wang Z, Hu Z (2012) Metabolic profiling of fifteen amino acids in serum of chemical-induced liver injured rats by hydrophilic interaction liquid chromatography coupled with tandem mass spectrometry. Metabolomics 8:475–483. https://doi.org/10.1007/s11306-011-0333-0
48. Ebshiana AA, Snowden SG, Thambisetty M, Parsons R, Hye A, Legido-Quigley C (2015) Metabolomic method: UPLC-q-ToF polar and non-polar metabolites in the healthy rat cerebellum using an in-vial dual extraction. PLoS One 10:e0122883. https://doi.org/10.1371/journal.pone.0122883
49. Ramautar R, Somsen GW, de Jong GJ (2017) CE-MS for metabolomics: developments and applications in

the period 2014–2016. Electrophoresis 38:190–202. https://doi.org/10.1002/elps.201400388
50. García A, Godzien J, López-Gonzálvez Á, Barbas C (2017) Capillary electrophoresis mass spectrometry as a tool for untargeted metabolomics. Bioanalysis 9:99–130. https://doi.org/10.4155/bio-2016-0216
51. Begou O, Gika HG, Wilson ID, Theodoridis G (2017) Hyphenated MS-based targeted approaches in metabolomics. Analyst 142:3079–3100. https://doi.org/10.1039/C7AN00812K
52. Zhang W, Hankemeier T, Ramautar R (2017) Next-generation capillary electrophoresis–mass spectrometry approaches in metabolomics. Curr Opin Biotechnol 43:1–7. https://doi.org/10.1016/j.copbio.2016.07.002
53. Hirayama A, Wakayama M, Soga T (2014) Metabolome analysis based on capillary electrophoresis-mass spectrometry. Trends Anal Chem 61:215–222. https://doi.org/10.1016/j.trac.2014.05.005
54. Bonvin G, Schappler J, Rudaz S (2012) Capillary electrophoresis-electrospray ionization-mass spectrometry interfaces: fundamental concepts and technical developments. J Chromatogr A 1267:17–31. https://doi.org/10.1016/j.chroma.2012.07.019
55. González-Ruiz V, Codesido S, Rudaz S, Schappler J (2018) Evolution in the design of a low sheath-flow interface for CE-MS and application to biological samples. Electrophoresis 39:853–861. https://doi.org/10.1002/elps.201700328
56. Peuchen EH, Zhu G, Sun L, Dovichi NJ (2017) Evaluation of a commercial electro-kinetically pumped sheath-flow nanospray interface coupled to an automated capillary zone electrophoresis system. Anal Bioanal Chem 409:1789–1795. https://doi.org/10.1007/s00216-016-0122-8
57. Moini M (2007) Simplifying CE-MS operation. 2. Interfacing low-flow separation techniques to mass spectrometry using a porous tip. Anal Chem 79:4241–4246. https://doi.org/10.1021/ac0704560
58. Gulersonmez MC, Lock S, Hankemeier T, Ramautar R (2016) Sheathless capillary electrophoresis-mass spectrometry for anionic metabolic profiling. Electrophoresis 37:1007–1014. https://doi.org/10.1002/elps.201500435
59. Grochocki W, Markuszewski MJ, Quirino JP (2016) Different detection and stacking techniques in capillary electrophoresis for metabolomics. Anal Methods 8:1216–1221. https://doi.org/10.1039/C5AY02493E
60. Soga T, Igarashi K, Ito C, Mizobuchi K, Zimmermann HP, Tomita M (2009) Metabolomic profiling of anionic metabolites by capillary electrophoresis mass spectrometry. Anal Chem 81:6165–6174. https://doi.org/10.1021/ac900675k
61. Yamamoto M, Ly R, Gill B, Zhu Y, Moran-Mirabal J, Britz-McKibbin P (2016) Robust and high-throughput method for anionic metabolite profiling: preventing polyimide aminolysis and capillary breakages under alkaline conditions in capillary electrophoresis-mass spectrometry. Anal Chem 88:10710–10719. https://doi.org/10.1021/acs.analchem.6b03269
62. Rutters H, Mohring T, Rullkotter J, Griep-Raming J, Metzger JO (2000) The persistent memory effect of triethylamine in the analysis of phospholipids by liquid chromatography/mass spectrometry. Rapid Commun Mass Spectrom 14:122–123. https://doi.org/10.1002/rcm.1393
63. Nevedomskaya E, Derks R, Deelder AM, Mayboroda OA, Palmblad M (2009) Alignment of capillary electrophoresis-mass spectrometry datasets using accurate mass information. Anal Bioanal Chem 395:2527–2533. https://doi.org/10.1007/s00216-009-3166-1
64. Nowak PM, Woźniakiewicz M, Gładysz M, Janus M, Kościelniak P (2017) Improving repeatability of capillary electrophoresis – a critical comparison of ten different capillary inner surfaces and three criteria of peak identification. Anal Bioanal Chem 409:4383–4393. https://doi.org/10.1007/s00216-017-0382-y
65. Sugimoto M, Wong DT, Hirayama A, Soga T, Tomita M (2010) Capillary electrophoresis mass spectrometry-based saliva metabolomics identified oral, breast and pancreatic cancer-specific profiles. Metabolomics 6:78–95. https://doi.org/10.1007/s11306-009-0178-y
66. Andreas NJ, Hyde MJ, Gomez-Romero M, Lopez-Gonzalvez MA, Villaseñor A, Wijeyesekera A, Barbas C, Modi N, Holmes E, Garcia-Perez I (2015) Multiplatform characterization of dynamic changes in breast milk during lactation. Electrophoresis 36:2269–2285. https://doi.org/10.1002/elps.201500011
67. Saric J, Want EJ, Duthaler U, Lewis M, Keiser J, Shockcor JP, Ross GA, Nicholson JK, Holmes E, Tavares MFM (2012) Systematic evaluation of extraction methods for multiplatform-based metabotyping: application to the Fasciola hepatica metabolome. Anal Chem 84:6963–6972
68. Canuto GAB, Castilho-Martins EA, Tavares MFM, Rivas L, Barbas C, López-Gonzálvez Á (2014) Multi-analytical platform metabolomic approach to study miltefosine mechanism of action and resistance in Leishmania. Anal Bioanal Chem 406:3459–3476. https://doi.org/10.1007/s00216-014-7772-1
69. Kuehnbaum NL, Kormendi A, Britz-Mckibbin P (2013) Multisegment injection-capillary electrophoresis-mass spectrometry: A high-throughput platform for metabolomics with high data fidelity. Anal Chem 85:10664–10669. https://doi.org/10.1021/ac403171u
70. DiBattista A, McIntosh N, Lamoureux M, Al-Dirbashi OY, Chakraborty P, Britz-McKibbin P (2017) Temporal signal pattern recognition in mass spectrometry: a method for rapid identification and accurate quantification for inborn errors of metabolism with quality assurance. Anal Chem. ASAP article. https://doi.org/10.1021/acs.analchem.7b01727
71. Macedo AN, Mathiaparanam S, Brick L, Keenan K, Gonska T, Pedder L, Hill S, Britz-McKibbin P (2017)

The sweat metabolome of screen-positive cystic fibrosis infants: Revealing mechanisms beyond impaired chloride transport. ACS Cent Sci 3:904–913. https://doi.org/10.1021/acscentsci.7b00299
72. Dibattista A, Rampersaud D, Lee H, Kim M, Britz-mckibbin P (2017) High throughput screening method for systematic surveillance of drugs of abuse by multisegment injection-capillary electrophoresis-mass spectrometry. Anal Chem 89:11853–11861. https://doi.org/10.1021/acs.analchem.7b03590
73. Buescher JM, Moco S, Sauer U, Zamboni N (2010) Ultrahigh performance liquid chromatography-tandem mass spectrometry method for fast and robust quantification of anionic and aromatic metabolites. Anal Chem 82:4403–4412. https://doi.org/10.1021/ac100101d
74. Coulier L, Bas R, Jespersen S, Verheij E, van der Werf J, Hankemeier T (2006) Simultaneous quantitative analysis of metabolites using ion-pair liquid chromatography-electrospray ionization mass spectrometry. Anal Chem 78:6573–6582. https://doi.org/10.1021/ac0607616
75. Tang Z, Cao T, Lin S, Fu L, Li S, Guan X-Y, Cai Z (2016) Characterization of oncogene-induced metabolic alterations in hepatic cells by using ultrahigh performance liquid chromatography-tandem mass spectrometry. Talanta 152:119–126. https://doi.org/10.1016/j.talanta.2016.01.056
76. Guo L, Worth AJ, Mesaros C, Snyder NW, Glickson JD, Blair IA (2016) Diisopropylethylamine/hexafluoroisopropanol-mediated ion-pairing ultra-high-performance liquid chromatography/mass spectrometry for phosphate and carboxylate metabolite analysis: utility for studying cellular metabolism. Rapid Commun Mass Spectrom 30:1835–1845. https://doi.org/10.1002/rcm.7667
77. Kiefer P, Delmotte N, Vorholt JA (2011) Nanoscale ion-pair reversed-phase HPLC−MS for sensitive metabolome analysis. Anal Chem 83:850–855. https://doi.org/10.1021/ac102445r
78. Knee JM, Rzezniczak TZ, Barsch A, Guo KZ, Merritt TJS (2013) A novel ion pairing LC/MS metabolomics protocol for study of a variety of biologically relevant polar metabolites. J Chromatogr B 936:63–73. https://doi.org/10.1016/j.jchromb.2013.07.027
79. Mathon C, Barding GA, Larive CK (2017) Separation of ten phosphorylated mono-and disaccharides using HILIC and ion-pairing interactions. Anal Chim Acta 972:102–110. https://doi.org/10.1016/j.aca.2017.03.029
80. Gu L, Jones AD, Last RL (2007) LC-MS/MS assay for protein amino acids and metabolically related compounds for large-scale screening of metabolic phenotypes. Anal Chem 79:8067–8075. https://doi.org/10.1007/978-1-61779-445-2
81. Wu J, Zhang Y, Wiegand R, Wang J, Bepler G, Li J (2015) Quantitative analysis of intracellular nucleoside triphosphates and other polar metabolites using ion pair reversed-phase liquid chromatography coupled with tandem mass spectrometry. J Chromatogr B 1006:167–178. https://doi.org/10.1016/j.jchromb.2015.10.030
82. Cecchi T (2008) Ion pairing chromatography. Crit Rev Anal Chem 38:161–213. https://doi.org/10.1080/10408340802038882
83. Gao S, Bhoopathy S, Zhang ZP, Wright DS, Jenkins R, Karnes HT (2006) Evaluation of volatile ion-pair reagents for the liquid chromatography-mass spectrometry analysis of polar compounds and its application to the determination of methadone in human plasma. J Pharm Biomed Anal 40:679–688. https://doi.org/10.1016/j.jpba.2005.05.022
84. Barron L, Gilchrist E (2014) Ion chromatography-mass spectrometry: a review of recent technologies and applications in forensic and environmental explosives analysis. Anal Chim Acta 806:27–54. https://doi.org/10.1016/j.aca.2013.10.047
85. Michalski R (2009) Applications of ion chromatography for the determination of inorganic cations. Crit Rev Anal Chem 39:230–250. https://doi.org/10.1080/10408340903032453
86. López-Ruiz B (2000) Advances in the determination of inorganic anions by ion chromatography. J Chromatogr A 881:607–627. https://doi.org/10.1016/S0021-9673(00)00244-2
87. Small H, Stevens TS, Bauman WC (1975) Novel ion exchange chromatographic method using conductimetric detection. Anal Chem 47:1801–1809. https://doi.org/10.1021/ac60361a017
88. Schwaiger M, Rampler E, Hermann G, Miklos W, Berger W, Koellensperger G (2017) Anion-exchange chromatography coupled to high-resolution mass spectrometry: a powerful tool for merging targeted and non-targeted metabolomics. Anal Chem 89:7667–7674. https://doi.org/10.1021/acs.analchem.7b01624
89. Liu J-M, Liu C-C, Fang G-Z, Wang S (2015) Advanced analytical methods and sample preparation for ion chromatography techniques. RSC Adv 5:58713–58726. https://doi.org/10.1039/C5RA10348G
90. Hermans C, Jonkers ACA, De Bokx PK (2010) Determination of amines in the presence of excess ammonia by ion chromatography – mass spectrometry. J Chromatogr Sci 48:544–548
91. Saccani G, Tanzi E, Pastore P, Cavalli S, Rey M (2005) Determination of biogenic amines in fresh and processed meat by suppressed ion chromatography-mass spectrometry using a cation-exchange column. J Chromatogr A 1082:43–50. https://doi.org/10.1016/j.chroma.2005.05.030
92. Petucci C, Zelenin A, Culver JA, Gabriel M, Kirkbride K, Christison TT, Gardell SJ (2016) Use of ion chromatography/mass spectrometry for targeted metabolite profiling of polar organic acids. Anal Chem 88:11799–11803. https://doi.org/10.1021/acs.analchem.6b03435
93. Hu S, Wang J, Ji EH, Christison T, Lopez L, Huang Y (2015) Targeted metabolomic analysis of head and neck cancer cells using high performance ion

chromatography coupled with a Q exactive HF mass spectrometer. Anal Chem 87:6371–6379. https://doi.org/10.1021/acs.analchem.5b01350

94. Burgess K, Creek D, Dewsbury P, Cook K, Barrett MP (2011) Semi-targeted analysis of metabolites using capillary-flow ion chromatography coupled to high-resolution mass spectrometry. Rapid Commun Mass Spectrom 25:3447–3452. https://doi.org/10.1002/rcm.5247
95. Wang J, Christison TT, Misuno K, Lopez L, Huhmer AF, Huang Y, Hu S (2014) Metabolomic profiling of anionic metabolites in head and neck cancer cells by capillary ion chromatography with orbitrap mass spectrometry. Anal Chem 86:5116–5124. https://doi.org/10.1021/ac500951v
96. Fan J, Hong J, Hu JD, Chen JL (2012) Ion chromatography based urine amino acid profiling applied for diagnosis of gastric cancer. Gastroenterol Res Pract 2012:474907. https://doi.org/10.1155/2012/474907
97. Kiefer P, Nicolas C, Letisse F, Portais JC (2007) Determination of carbon labeling distribution of intracellular metabolites from single fragment ions by ion chromatography tandem mass spectrometry. Anal Biochem 360:182–188. https://doi.org/10.1016/j.ab.2006.06.032
98. Kvitvang HFN, Kristiansen KA, Bruheim P (2014) Assessment of capillary anion exchange ion chromatography tandem mass spectrometry for the quantitative profiling of the phosphometabolome and organic acids in biological extracts. J Chromatogr A 1370:70–79. https://doi.org/10.1016/j.chroma.2014.10.029
99. van Dam JC, Eman MR, Frank J, Lange HC, Van Dedem GWK, Heijnen SJ (2002) Analysis of glycolytic intermediates in Saccharomyces cerevisiae using anion exchange chromatography and electrospray ionization with tandem mass spectrometric detection. Anal Chim Acta 460:209–218. https://doi.org/10.1016/S0003-2670(02)00240-4
100. Alonso AP, Piasecki RJ, Wang Y, LaClair RW, Shachar-Hill Y (2010) Quantifying the labeling and the levels of plant cell wall precursors using ion chromatography tandem mass spectrometry. Plant Physiol 153:915–924. https://doi.org/10.1104/pp.110.155713
101. Cui L, Liu J, Yan X, Hu S (2017) Identification of metabolite biomarkers for gout using capillary ion chromatography with mass spectrometry. Anal Chem 89:11737–11743. https://doi.org/10.1021/acs.analchem.7b03232
102. Sánchez MIG, Mccullagh J, Guy RH, Compton RG (2017) Reverse iontophoretic extraction of metabolites from living plants and their identification by ion-chromatography coupled to high resolution mass spectrometry. Phytochem Anal 28:195–201. https://doi.org/10.1002/pca.2660
103. Fan TW, Warmoes MO, Sun Q, Song H, Turchan-cholewo J, Martin JT, Mahan A, Higashi RM, Lane AN (2016) Distinctly perturbed metabolic networks underlie differential tumor tissue damages induced by immune modulator β -glucan in a two-case ex vivo non-small-cell lung cancer study. Cold Spring Harb Mol Case Stud 2:a000893. https://doi.org/10.1101/mcs.a000893
104. Bruggink C, Wuhrer M, Koeleman CAM, Barreto V, Liu Y, Pohl C, Ingendoh A, Hokke CH, Deelder AM (2005) Oligosaccharide analysis by capillary-scale high-pH anion-exchange chromatography with on-line ion-trap mass spectrometry. J Chromatogr B 829:136–143. https://doi.org/10.1016/j.jchromb.2005.10.006
105. Ramautar R (2016) CE-MS in metabolomics: status quo and the way forward. Bioanalysis 8:371–374. https://doi.org/10.4155/bio-2016-0001
106. Kok MGM, Somsen GW, De Jong GJ (2015) Comparison of capillary electrophoresis-mass spectrometry and hydrophilic interaction chromatography-mass spectrometry for anionic metabolic profiling of urine. Talanta 132:1–7. https://doi.org/10.1016/j.talanta.2014.08.047
107. Lu W, Bennett BD, Rabinowitz JD (2008) Analytical strategies for LC-MS-based targeted metabolomics. J Chromatogr B 871:236–242. https://doi.org/10.1016/j.jchromb.2008.04.031
108. Lemasson BE, Bertin S, Hennig P, Lesellier E, West C (2017) Mixed-mode chromatography – a review. LCGC 30:22–33

Metabolomics Data Treatment: Basic Directions of the Full Process

Hans Rolando Zamora Obando, Gustavo Henrique Bueno Duarte, and Ana Valéria Colnaghi Simionato

Abstract

The present chapter describes basic aspects of the main steps for data processing on mass spectrometry-based metabolomics platforms, focusing on the main objectives and important considerations of each step. Initially, an overview of metabolomics and the pivotal techniques applied in the field are presented. Important features of data acquisition and preprocessing such as data compression, noise filtering, and baseline correction are revised focusing on practical aspects. Peak detection, deconvolution, and alignment as well as missing values are also discussed. Special attention is given to chemical and mathematical normalization approaches and the role of the quality control (QC) samples. Methods for uni- and multivariate statistical analysis and data pretreatment that could impact them are reviewed, emphasizing the most widely used multivariate methods, i.e., principal components analysis (PCA), partial least squares-discriminant analysis (PLS-DA), orthogonal partial least square-discriminant analysis (OPLS-DA), and hierarchical cluster analysis (HCA). Criteria for model validation and softwares used in data processing were also approached. The chapter ends with some concerns about the minimal requirements to report metadata in metabolomics.

Keywords

Untargeted metabolomics · Mass spectrometry · Chromatography · Data treatment · Data processing · Data analysis · Statistical analysis · Software tools

Abbreviations

ANN	Artificial Neural Network
CE-MS	Capillary Electrophoresis-Mass Spectrometry
COW	Correlation-Optimized Warping
CV	Coefficient of Variation
DI FT-ICR MS	Direct-Infusion Fourier-Transform Ion-Cyclotron-Resonance Mass Spectrometry
DTW	Dynamic Time Warping

H. R. Zamora Obando · G. H. B. Duarte
Department of Analytical Chemistry, Institute of Chemistry, University of Campinas, Campinas, SP, Brazil

A. V. C. Simionato (✉)
Analytical Chemistry Department, University of Campinas, Institute of Chemistry, Campinas, São Paulo, Brazil
e-mail: avsimionato@unicamp.br

A. V. Colnaghi Simionato (ed.), *Separation Techniques Applied to Omics Sciences*, Advances in Experimental Medicine and Biology 1336, https://doi.org/10.1007/978-3-030-77252-9_12

GA	Genetic Algorithm
GC-MS	Gas Chromatography-Mass Spectrometry
HCA	Hierarchical Cluster Analysis
HILIC	Hydrophilic Interaction Chromatography
IC	Intensity Count
kNN	k-Nearest Neighbors
LC-MS	Liquid Chromatography-Mass Spectrometry
LDA	Linear Discriminant Analysis
LOESS	Lowest Point of Smoothed Spectrum
MS	Mass Spectrometry
NMR	Nuclear Magnetic Resonance
NOMIS	Normalization Using the Optimal Selection of Multiple Internal Standards
OPLS-DA	Orthogonal Partial Least-Square Discriminant Analysis
PARAFAC	Parallel Factor Analysis
PC	Principal Component
PCA	Principal Components Analysis
PLS-DA	Partial Least Squares-Discriminant Analysis
PQN	Probabilistic Quotient Normalization
PTW	Parametric Time Warping
QCs	Quality Control Samples
RAFFT	Rapid Fast Fourier Transform
RF	Random Forest
ROC	Receiver Operating Characteristic Curve
ROI	Region of Interest
S/N	Signal-Noise Ratio
SIMCA	Soft Independent Modeling of Class Analogy
SOM	Self-Organization Map
SVM	Support Vector Machine
TOF-MS	Time of Flight-Mass Spectrometry
XIC	Extracted Ion Chromatogram

1 Introduction

Metabolomics is one of the "omics" sciences, which systematically studies a wide and heterogeneous group of compounds of relatively low molecular weight metabolites (<1500 Da) [1] in biological systems – cells, tissues, organs, and individuals in their totality – in the face of an environmental and/or genetic disturbance. Therefore, metabolomics may explain related and mutually affected processes in both directions (genomics ↔ environmental processes) that result in phenotypic changes in biological systems.

In general, metabolomics is applied to human and environmental related samples. More specifically, it has been used for discovery and improvement of therapies, optimization of biotechnological processes, improvement in food quality [2], forensic medicine, precision medicine [3], phytochemistry [4, 5], microbiology [6, 7], ecology, and agronomy [8].

Nuclear magnetic resonance (NMR) spectroscopy techniques [9, 10] and separation techniques coupled with mass spectrometry (MS) such as gas chromatography (GC-MS), liquid chromatography (LC-MS), and capillary electrophoresis (CE-MS) have been used for metabolomics analyses [11, 12]. The results provided by these analytical platforms are not exclusive, but complementary, requiring a multi-platform approach for a thorough study [13, 14]. In fact, since 2001, research combining NMR and MS-based methods has increased exponentially [15].

MS-based techniques are highly sensitive and selective. GC-MS allows access to the volatile portion of the metabolome by means of the headspace technique. However, other less volatile compounds, such as lipids, organic acids, and carbohydrates, require derivatization steps prior to analysis, introducing a source of errors by loss of volatile components. Accordingly, LC-MS is characterized by high robustness, sensitivity, and selectivity. In addition, it has the flexibility to analyze a wide range of metabolites due to the availability of numerous stationary and mobile phases [16], allowing the use of hydrophilic interaction chromatography (HILIC) [17], which results in a more comprehensive analysis of metabolites by reaching ionic and polar compound classes. CE-MS has been implemented more recently and has been slowly expanding along with technological, instrumental, and computer development improvement. For example,

CE-MS coupling is still challenging, and it has a critical impact on performance and experimental reproducibility. In addition, free software for data processing are still unavailable, although crucial for electrophoretic signal alignment (due to migration time variability), restricting statistical analysis [12].

Thorough metabolomics studies are performed by several steps, which are described in Fig. 1. The first step corresponds to obtaining and preparing samples, where the metabolites of interest (targeted metabolomics) or the largest possible number of metabolites (untargeted metabolomics) will be extracted. Afterwards, samples are analyzed qualitatively or quantitatively [16]. The following steps include theoretical work, such as raw data cleanup – for subsequent statistical analysis, identification of possible metabolites (annotation), and respective biological interpretation, in order to describe the present state of the system under investigation [18].

The steps to follow during data preprocessing depend on the instrumental platform on which the analysis is performed. Figure 2 shows these differences, which basically correspond to the type of information acquired by the instruments, i.e., retention times (t_r) and mass-charge ratio (*m/z*), for mass spectrometry-based analysis, and chemical shift (δ), in the case of NMR ones.

As mentioned above, data processing is a key step in statistical analysis, since relevant information and conclusions of the metabolomics research depend on the appropriate procedure [19]. Therefore, the present chapter will address, in a general way, the main steps of data processing on mass spectrometry-based platforms, describing the objectives and important considerations of each step. For more information about data processing in NMR metabolomics, the reader is encouraged to look for some literature reviews [20]. Finally, some chemometric tools for data analysis will be described.

2 Data Processing

2.1 Data Preprocessing

Data preprocessing has two objectives: (a) reducing data and noise amount and making comparable spectra [21], and (b) converting the acquired

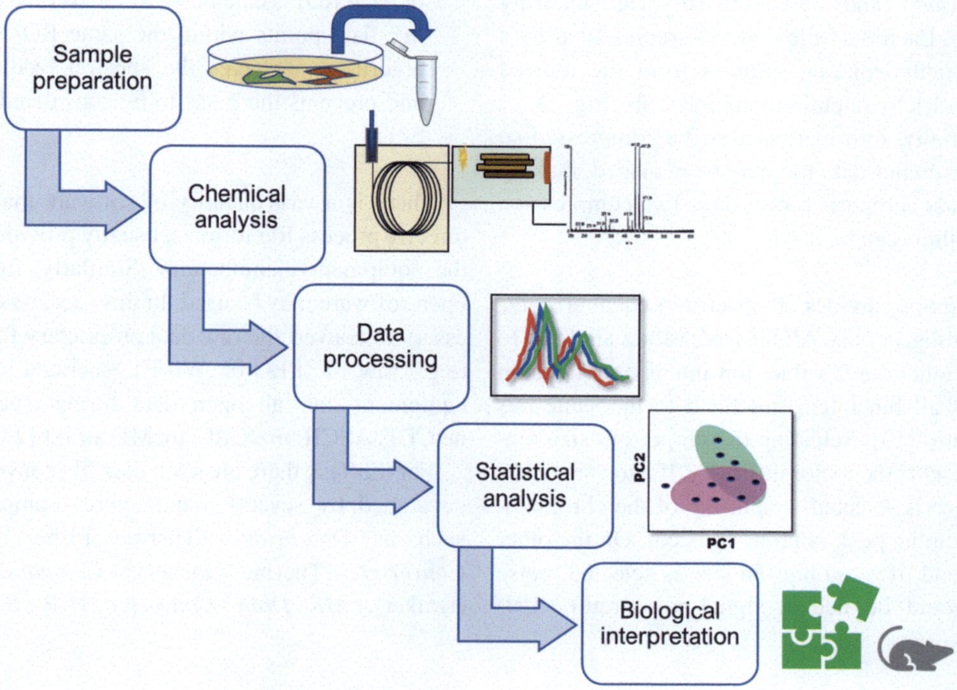

Fig. 1 Overview of workflow in metabolomics studies

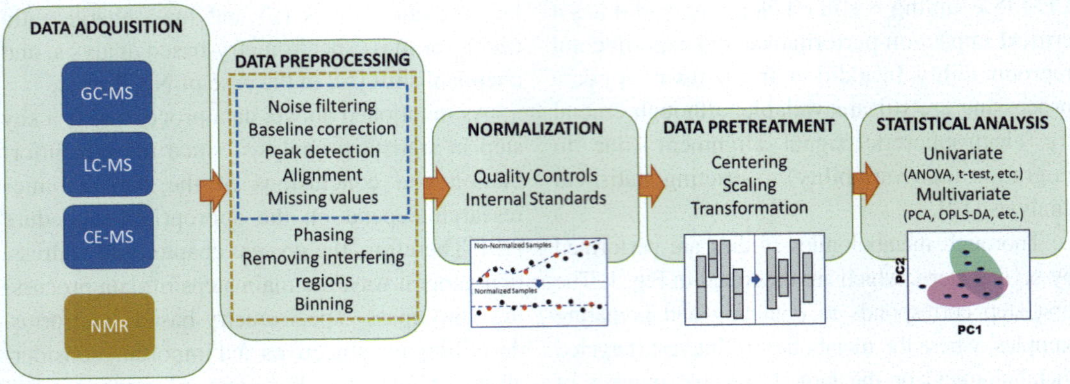

Fig. 2 General process of data processing in metabolomics, from data acquisition to statistical analysis

raw data by the instrument into a table where samples are disposed in rows and *features* (pairs $m/z - t_r$), in columns, for statistical analysis purpose. The following steps describe the usual algorithms to achieve this conversion.

2.1.1 Data Acquisition and Data Compression

A huge quantity of three-dimensional data information from separation techniques such as LC, GC, or CE coupled with a mass spectrometer are generated, i.e., retention time (t_r), mass-charge ratio (m/z), and signal intensity (IC, intensity count). Each data file could be represented by a histogram depicting impacts from the ionized molecules by retention time intervals (Fig. 3).

Initially, information must be compressed so that a smaller data file may be managed, making an easier computer processing. Two compression algorithms can be used:

(a) *Binning* divides all spectra in equal size m/z ratios, or *bins*. All bins m/z values are fixed to a unique m/z value. Ion intensity is the sum of all bin intensities fixed to the same m/z ratio [19]. Selecting the proper bin size represents the major difficulty. If a too small bin size is selected, a splitting of the chromatographic peak is prone to occur. On the other hand, if a too big bin size is selected, noise would be added, signal-noise ratio (S/N) would be decreased, and low-intensity peaks would not be detected, resulting in loss of spectral resolution [22].

(b) *Region of interest* (ROI) exploits the centroid mode MS data to identify regions with high probability of containing chromatographic peaks. These ROIs are characterized by the density and consistency of m/z data points surrounded by regions of "data void." Then, ROIs are vectorized and converted into a matrix (t_r, m/z). Each ROI is grouped by similarity at specific retention time, and the m/z of each ROI is calculated as an average of all m/z data points within the same ROI. This algorithm preserves the spectral resolution and prevents the noise to be considered [22, 23].

There is a vast quantity of software that can directly process file formats, usually provided by the equipment manufacturer. Similarly, free or open software may be used. In this case, it is necessary to convert the raw data proprietary format (e.g., .raw, .d, .PEG or .WIFF) generated by the equipment into an open data format such as netCDF, ASCII, mzXML, mzML, or txt [23].

For this task there are open data file converters developed by several manufacturer companies such as *Databridge* (Thermo Fisher), *File Converter* (Thermo Fisher), *CompassXport* (Bruker), *MS Data Converter* (AB Sciex),

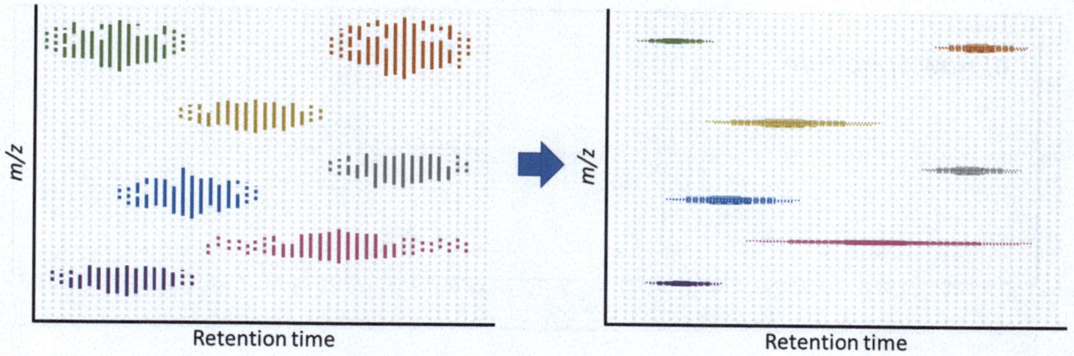

Fig. 3 Representation of binning process. (**a**) Position of data points in a small region of *m/z* and t_r of the raw data profile. (**b**) 2D image generated by binning the values of *m/z* at the resolution of a bitmap

ChemStation (Agilent), MassHunter Qualitative Analysis (Agilent), MassLynx (Waters), or research groups (e.g., *OpenMS* (https://www.openms.de/), Proteowizard MSConvert (http://proteowizard.sourceforge.net/download.html)) [24]. Conversion of these file formats is a key step to use open data analysis software, for instance, *XCMS*, *MZmine*, *MetAlign*, or *MetaboAnalyst* [25, 26]. A more extended list of preprocessing and statistical analysis software is presented later in this chapter.

2.1.2 Noise Filtering and Baseline Correction

Noise filtering aims to remove noise caused by effects generated by the sample matrix, instrumental interference, etc., from the real signals. Noise filtering in LC-MS and CE-MS is more complicated than in GC-MS due to the presence of chemical noise – apart from instrumental noise – caused by buffer composition, formation of pipelines, and column bleeding, among others, which causes baseline oscillation [19, 25, 26].

Algorithms for noise filtering are classified into four methods: wavelet transform, median filtering, polyfit, and rubber band [27]. Some examples of algorithms are mean filter, moving mean filter, Savitzky-Golay method, wavelet transform, and lowest point of smoothed spectrum (LOESS) [19, 28]. The results of applying two of these algorithms are presented in Fig. 4, which shows that a specific algorithm selection will affect the peak shape and peak intensity/area, resulting in differences in the subsequent data treatment steps.

Baseline correction seeks to correct the baseline drift, which originates from analysis chemical noise [25]. For this purpose baseline shape must be determined and then subtracted from the spectrum [19]. It can be performed either manually, indicating the points where peaks begin and end – a time-consuming and subjective method – or through algorithms, such as those used for noise filtering.

Whatever noise filtering and baseline correction method is selected, special attention must be directed to an excess of adjustment (leading to a loss of information and peak distortion) or an insufficient adjustment (harming statistical analysis due to possible false positives) [28].

2.1.3 Peak Detection and Deconvolution

Peak detection aims at detecting signals that potentially correspond to a compound, avoiding false positives – as well as their quantification [7, 19, 25]. This is an important step in the alignment, normalization, and identification of metabolites and may be carried out using some algorithms, such as those described below [19]:

(a) *Vectorized peak detection.* Peaks are exploited and detected in each dimension (*m/z* and t_r). Discrimination is carried out by looking for maximums in each dimension with a function of wavelet additive decompo-

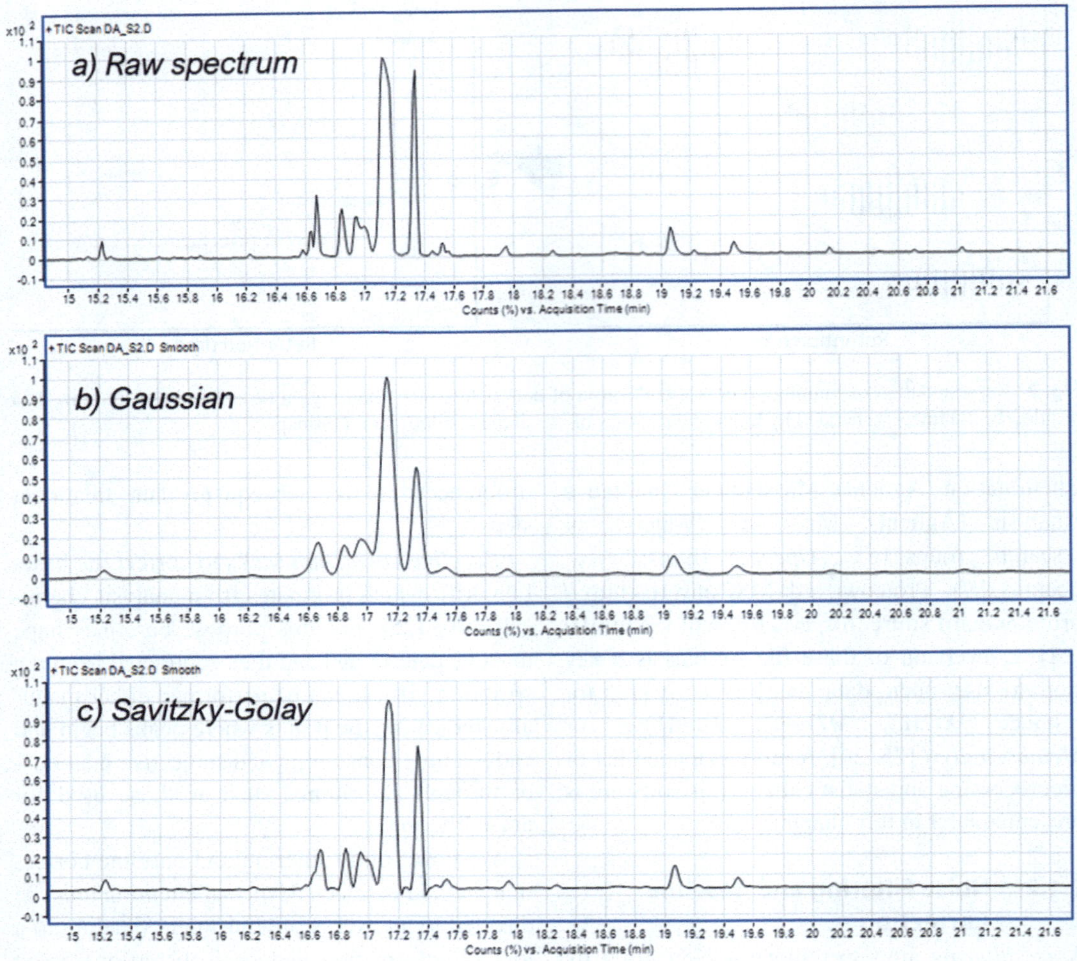

Fig. 4 Examples of smoothing raw data in a chromatogram. (**a**) Raw data, (**b**) Gaussian, and (**c**) Savitzky-Golay filter determined at 5 points

sition, smoothing over the dimension of time and peak identification corroborated over multiple scans, and finally, assembling by isotopic patterns that appear, maximize, and disappear concomitantly.

(b) *Extracted ion chromatogram* (XIC) *slices*. The *m/z* dimension is divided into small *m/z* intervals, and each XIC is individually processed by a second-order Gaussian time filter to find peak inflections in the data or finding areas above a threshold area determined by an average or median chromatogram.

(c) *Model fitting against the original raw signal*. In this strategy, implemented by Hermansson et al. [29], the most intense peak is determined by adjusting a three-dimensional signal model – based on a generic mass-dependent isotopic pattern – which is subtracted from the entire chromatogram. This procedure is repeated until the highest remaining peak is twice as intense as the noise intensity.

After peak detection, the following step corresponds to *deconvolution*, which resolves peak superposition (Fig. 5). Detected ions may correspond to the same molecule, so the deconvolution methods attribute the different ions to the corresponding compound. Deconvolution algorithms are based on the principle that different ion frag-

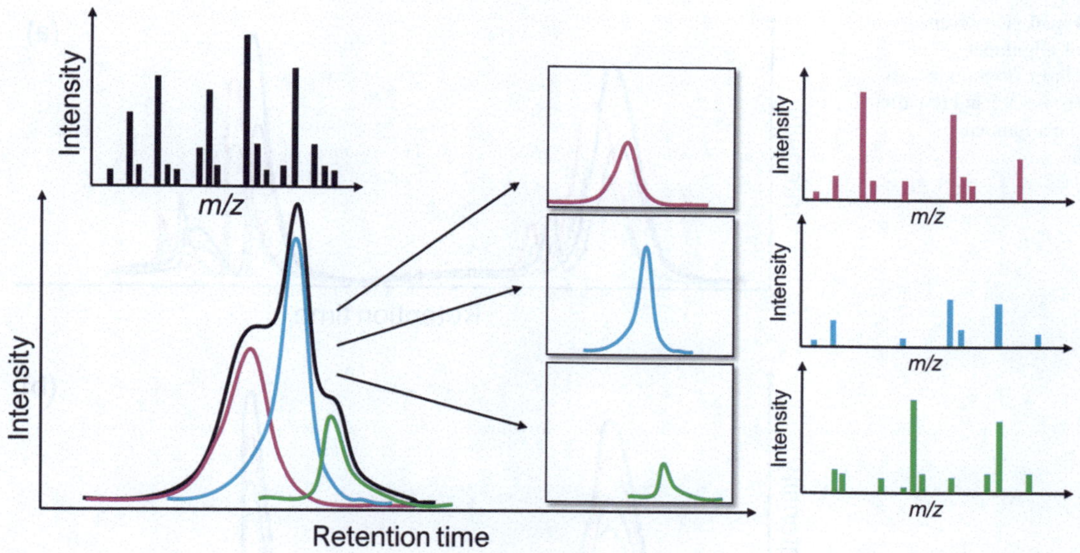

Fig. 5 Deconvolution diagram. To the left are the original signals as the superimposition of the mass spectrum and the chromatogram; to the right are the separate chromatographic peaks with the respective mass spectra

ments present the same retention times, as well as that the profiles of multiple samples are correlated, and experience the same biological variation and systematic error. Nevertheless, metabolites may co-elute, generating an overlapped isotopic pattern [19].

AMDIS (developed by NIST), DRS (Agilent), AnalyazerPro (SpectralWorks) MS-DIAL, and CromaTOF (LECO) are the main software that perform automated peak detection and alignment. A relatively new free software tool is MS-DIAL [30], which may be used to treat MS/MS spectra.

Recently, methods based on chemometric tools have been developed for data analysis in LC-MS, such as multivariate curve resolution with alternating least squares (MCR-ALS) [23, 31]. This method isolates, resolves, and quantifies the sources of data variation, resulting in a set of components with elution and spectral profiles associated with different *m/z*. In addition, it aligns the spectrum in the *m/z* dimension, instead of time dimension, which may vary between sample analyses.

2.1.4 Peak Alignment

In spite of technological advances in LC-MS technology, changes in temperature, pH, flow fluctuations, and column clogging, among others, cause retention times to fluctuate as well [1, 28], requiring an alignment procedure to circumvent it. This procedure aims to correct the lags in retention times that correspond to the same peak [7], and combine the data from different samples [19] (Fig. 6).

Gorrochategui et al. [23] point out three aspects that must be considered in the alignment processes: (a) differences in retention times may be nonlinear, (b) one compound in a sample may have multiple corresponding features (in *m/z* and t_r) in other samples, and (c) some peaks may not appear in some samples.

Spectra pairwise alignment against a reference spectrum is an usual methodology [28]. Commonly, two approaches may be selected [32]: (a) adding standard compounds during sample preparation or (b) applying mathematical methods after data acquisition. In the former approach, the signals align with the pattern peaks by means of a linear adjustment by parts. On the other hand, the latter may be divided into (i) linear methods, which can use reference peaks – common to all samples, with relatively stable retention times and intensity – and then align the spectra to each other, and (ii) nonlinear methods. However, some authors classify nonlinear meth-

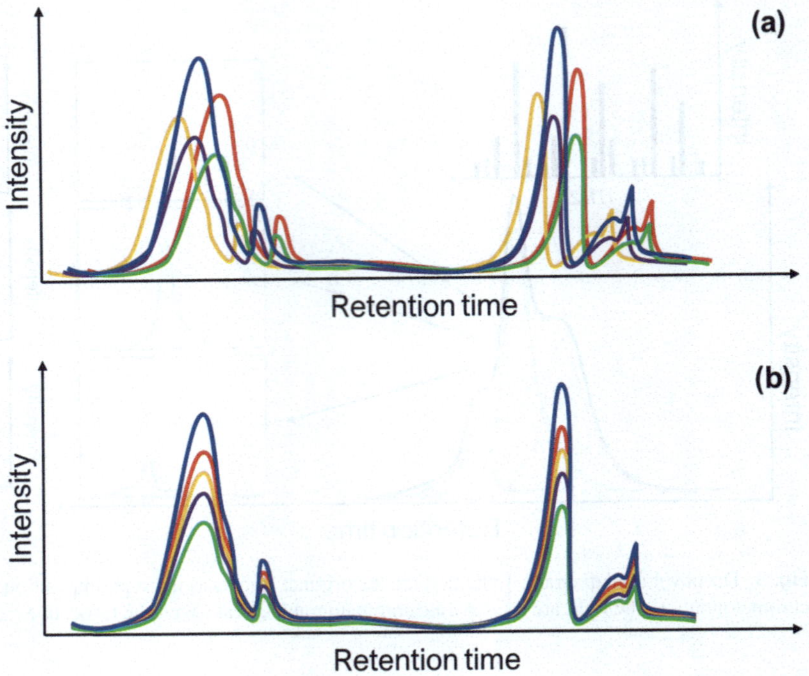

Fig. 6 Representation of alignment chromatogram process. (**a**) before and (**b**) after the alignment

ods into (a) those which use the raw data and construct a map that combines the retention time of each chromatographic run, generating only a common t_r, and (b) methods that group the detected features and generate a matrix (ions × peak area/intensity) for each sample (Fig. 7) [19].

One of the nonlinear methods for peak alignment is the correlation-optimized warping (COW), which is an algorithm that divides the signals to be aligned and the reference signals into N sections, followed by deformation (elongation or compression) in order to match the reference more accurately [33]. This algorithm is also useful for aligning GC × GC-TOF-MS data [34]. One of the disadvantages of the COW algorithm is that it requires long run times and high memory consumption of the computer system [33].

There are other algorithms such as: parametric time warping (PTW), similar to the previous one, except for using a polynomial correlation; dynamic time warping (DTW); and rapid alignment by rapid fast Fourier transform (RAFFT). Each of them presents some disadvantages [25], for example, DTW generates artifacts when mono-channel detectors are used and is sensitive to peak intensities [35], while in RAFFT, peak distortion, artifact appearance, and peak point removal occur.

2.1.5 Missing Values

Inevitably, some data is lost after data preprocessing due to peak detection and poor peak intensity or shape, among others [18, 36]. To circumvent data loss, samples should be analyzed in replicates, at the expense of more complicated data preprocessing and higher possibility of generating duplicate metabolites in the data table, due to detection of metabolites with the same *m/z* but different retention times.

Reinhold et al. [36] suggest a flowchart for dealing with missing data (Fig. 8) based on the variability of abundance. If the coefficient of variation (CV) is equal to or higher than 0.5 among the replicates for the same analyte, it would be labeled as a missing value. Instead, the mean or median value will be assigned (*imputation*) according to the number of replicates. However, the method to perform peak assignment will depend on the nature of missing values [37]. If a specific metabolite presents a high

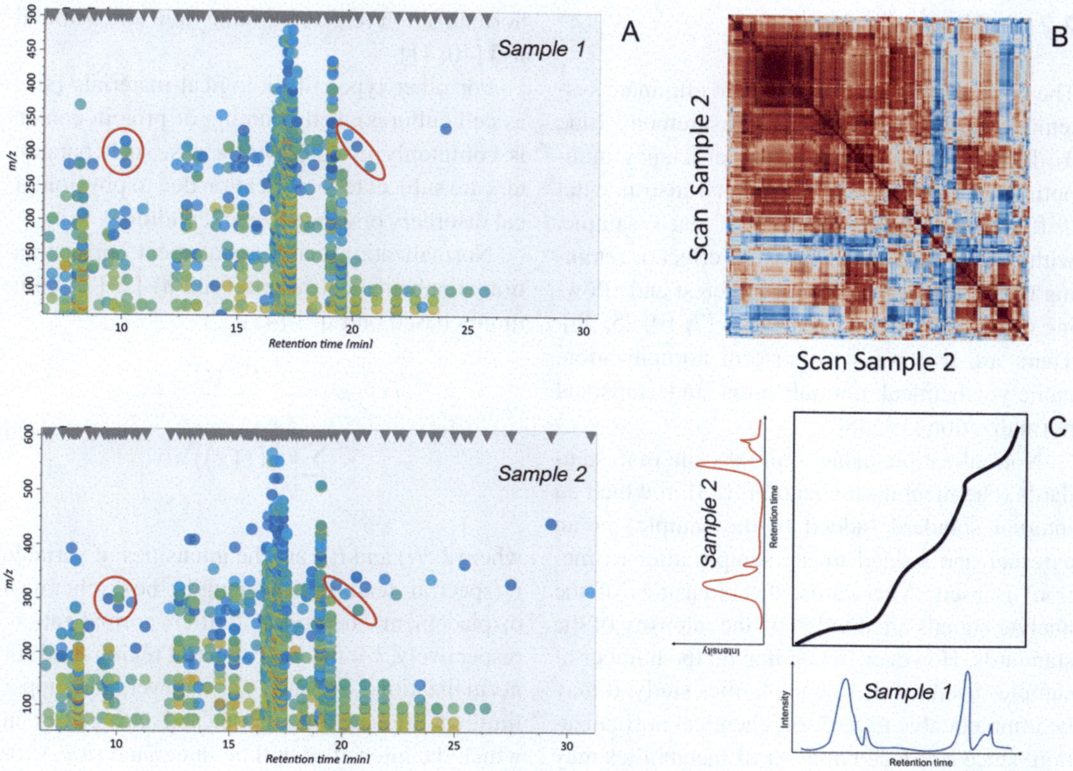

Fig. 7 Representation of alignment strategies. (**a**) Grouping the detected features, (**b**) using a correlation matrix from the raw data, or (**c**) alignment using the chromatographic profile

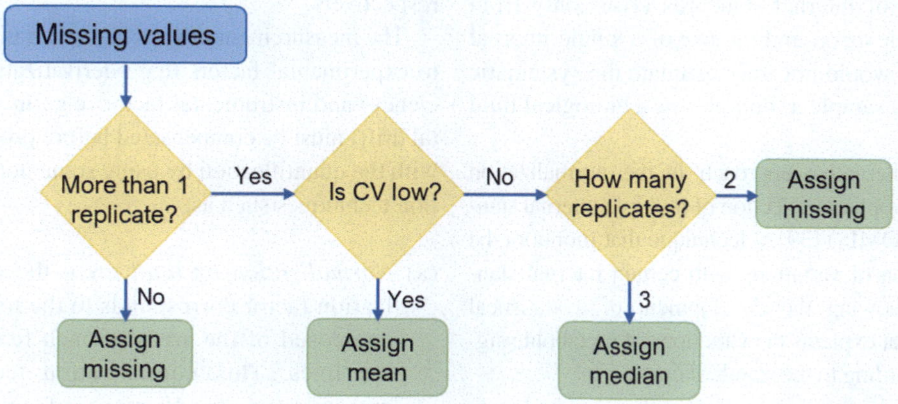

Fig. 8 Input/output diagram for processing missing data. (Adapted from [36])

percentage of missing values (>20%), it should be removed (filtered), although other cutoff percentages may be employed.

The impact of missing values on univariate and multivariate statistics results and comparison of methods for handling them were evaluated by Hrydziuszko and Viant [38], which obtained metabolomics data by direct-infusion Fourier-transform ion-cyclotron-resonance mass spectrometry (DI FT-ICR MS). For this study, k-nearest neighbors (KNN) method showed the best performance in estimating missing values.

2.2 Normalization

The normalization process aims to eliminate systematic variation among the measurements (due to differences in derivatization efficiency, non-optimized instrumental conditions, instrumental drift, individual composition of each sample, within-batch and between-batch effects), retaining the biological variation of interest and allowing comparison between samples [7, 19, 25, 36]. There are two ways to perform normalization, namely: chemical normalization and statistical normalization [19, 28].

Normalization using multiple internal standards (chemical normalization) [23] in which an internal standard (added to the sample) or an external one (added to the sample after extraction) is used. Afterwards, the intensities of the analyte signals are divided by the intensity of the standards. However, depending on the number of samples used for the metabolomics study, it may be impracticable to perform chemical normalization, since a reduced number of metabolites may be used as standards. Another disadvantage is that the representativeness and comprehensiveness of this type of normalization assume that the variance of internal standards arises only from systematic error, and the use of a single internal standard would not truly estimate the systematic error of a sample as complex as a biological fluid [19].

An alternative approach is the normalization using the optimal selection of multiple internal standards (NOMIS) [39], a technique that monitors the correlation of variations with certain internal standards, allowing the development of a statistical model that explains the variations of metabolite signals according to the standard ones.

There is an intrinsic physically regulated variation in sample metabolite concentrations used in metabolomics studies (such as blood serum/plasma, saliva, and cerebrospinal fluid, among others) [18]. However, urine must be particularly considered due to different dilutions within the collected samples, composition, and concentration variation. In order to overcome this problem, it is recommended to use different methods of normalization (e.g., creatinine ratio and osmolality) [40, 41].

For other types of biological materials (such as cell cultures), cell counting or protein content is commonly used. Nevertheless, such parameters are subject to modification due to physiological disorders or experimental conditions [1].

Normalization using a statistical approximation (mathematical normalization) [23] is commonly based on Eq. 1 [42]:

$$I(i) = \frac{I^{\text{old}}(i)}{\sum_k \left(\int_{j_k^l}^{j_k^u} \left(I(x) \right)^n dx \right)^{\frac{1}{n}}} \quad (1)$$

where $I^{\text{old}}(i)$ and $I(i)$ are the intensities of variable i (spectral feature, wavelength, box, chemical displacement) before and after normalization, respectively, k = index of spectral region used for normalization, j_k^l and j_k^u = lower and upper limit of the spectral region, and n = power in which the intensities will be integrated ($n = 1$ for the normalization integral). The numerator and denominator correspond to the individual signal intensity and the normalization factor, respectively.

The measurement of systematic variations due to experimental factors (e.g., derivatization efficiency) and instrumental factor (e.g., instrumental drift) must be compensated before proceeding with the quantification by using some normalization techniques such as:

(a) *Normalization by total area*: the normalization factor corresponds to the total area composed of the areas of each feature on the lines. This normalization technique presents the disadvantage of total area dependency on the concentrations of metabolites. However, the higher the intensity/peak area, the more sensitive to changes in the concentrations of metabolites will the total area be, i.e., variations in signal intensities depend only on the dilution of the sample and not on other factors. In fact, it has been demonstrated that it negatively affects the multivariate analysis [42].

(b) *Quantile normalization*: forces all samples to have the same intensity distribution or peak area. The method assigns a range (i, ii, iii, iv, etc.) to each feature in the rows of the data table, then sorts all ranges in ascending order, and calculates the median or average of the values with the same range, and the median or average value replaces the original values for each range. Finally, each feature returns to its original position in the table, generating a table with normalized values for each sample. The weakness of this method is the susceptibility to the extreme values that may appear among samples [18, 43].

The inconvenience of these normalization techniques is the observed failure when changes in metabolite concentration occur, because of systematic errors introduced in the experimental part or from biological differences in large-scale experiments [23].

(c) *Probabilistic Quotient Normalization* (PQN): unlike the method by area normalization, this method assumes that changes in concentrations of individual metabolites affect only some regions of the total spectrum. However, the sample dilution affects the spectrum in an integral manner. The method is based on the calculation of a dilution quotient between the signals of the sample spectrum and the signals of a reference spectrum calculated from the median of the signals of all samples [18]. This method is highly recommended in cases where the factor of sample size is important [23]. PQN has demonstrated better normalization performance with integral normalization and vector length normalization [42].

A commonly used method is using a quality control samples (QCs). QCs allow to visualize the global variation of the data according to that observed on the samples under investigation, in order to evaluate reproducibility, performance, and instrumental stability [7]. In addition, this step is necessary for exploratory data analysis. QCs are prepared by the mixture of small aliquots of every single sample, forming a "pool of samples". The QCs contain and reflect all the biological information, which is their main advantage when compared to the normalization using internal standards [2].

After data preprocessing, the quality of each feature assigned to each peak and sample is evaluated, following the steps below:

(a) Corroborating feature presence in QCs: features that are not present in a significant number of QCs are deleted from the feature table.
(b) Correcting the intensity drift: signal intensities throughout analysis experience a deviation due to instrumental factors. Such deviations are specific for each feature, making correction impossible by simple normalization. To correct this effect, algorithms, such as LOESS, apply a correction factor for each individual feature. Nevertheless, it is mandatory that no outlier is observed within QCs to perform such correction.
(c) Evaluating the repeatability: the relative standard deviation of each feature in relation to the QCs must be less than 20% for LC-MS and 30% for GC-MS. If it exceeds these threshold percentages, the feature should be deleted from the table. Those threshold values will depend on the number of samples.
(d) Evaluating the linearity: if a series of dilutions of QCs samples are used, the linearity has to be evaluated. The recommended range of coefficient of determination values (R^2) is 0.5–0.7. Thus, the features with low R^2 and negative beta coefficients (intercepts) should be deleted from the table. It is recommended that the R^2 values be established and inspected based on their distribution and sample size.

2.3 Data Pretreatment

One of the objectives in a metabolomics study is to find differences between metabolomes of a target and a control group. For this purpose univariate (t-test, ANOVA, etc.) and multivariate statistical techniques (PCA, PLS, etc.) are employed, requiring that the characteristics of each analyte be comparable, due to different magnitude orders of abundance values and variance heterogeneity [2]. Some examples of statistical techniques applied for metabolomics data pretreatment are data centering, data scaling, and data transformation, so that experimental errors are minimized and the analysis is focused on the relevant biological information [7].

(a) *Data centering* aims to center the variation of each variable on the same axis (zero) in order to focus on the same biological variation. The process is achieved by subtracting the average of the intensity/area of the peaks from each of the metabolites of the same sample [18].
(b) *Data scaling* aims to make proportional variances of each metabolite so that multivariate methods do not dismiss metabolites [18] because of their low concentration, with the possibility that it may be a metabolite of interest. Scaling methods divide each variable by a scaling factor, subdividing it into different methods (Table 1).

The auto-scaling method (also called unitary variance scaling) gives each variable (with high or low abundance) equal weights to be analyzed by multivariate methods. The disadvantage of using auto-scaling is to be affected by noise, since it will have a weight of equal importance as other metabolite signals; therefore, its use for noisy data pretreatment is not recommended [44]. The Pareto scaling reduces the importance of metabolites with higher abundance when compared to those centered on the average, but still retain a significant weight; this method is suggested when high intensity signal data are not appreciably affected by noise. The range scaling has the disadvantage of being susceptible to atypical values [18]. Finally, variable stability (VAST) scaling focuses on metabolites with high standard deviations, while level scaling is adequate to determine abundant biomarkers [44].

(c) *Data transformation*: statistical analysis methods assume that the variability of the noise during data acquisition is constant, i.e., it presents a homoscedastic behavior. However, this is not the case, but instead an increase in noise variability occurs with the increase of the signal. In addition, the distribution of features is usually asymmetric [18]. Therefore, a transformation of the data to compensate for this pair of effects must be carried out. Some frequently used transformation methods are logarithmic, Glog, and exponential transformation [45]. Logarithmic transformation allows the correction of heteroscedasticity as long as the standard deviation is constant and can convert multiplicative relationships into additives, but it is not useful in values of zero, unlike exponential transformation [25, 28].

As described above, choosing the appropriate method for data pretreatment would significantly affect the statistical analysis and, consequently, the research results [45]. Therefore, this step should be treated carefully.

Table 1 Scaling methods [44]

(a) Variance scaling	(b) Auto-scaling	(c) Pareto scaling
$x_{ij(CV)} = \frac{x_{ij}}{s_j^2}$	$x_{ij(a)} = \frac{x_{ij} - \overline{x}_j}{s_j}$	$x_{ij(P)} = \frac{x_{ij} - \overline{x}_j}{\sqrt{s_j}}$
(d) VAST scaling.	(e) Level scaling.	(f) Range scaling.
$x_{ij(\text{VAST})} = \frac{x_{ij} - \overline{x}_j}{\frac{s_j^2}{\overline{x}_j}}$	$x_{ij(\text{level})} = \frac{x_{ij} - \overline{x}_j}{\overline{x}_j}$	$x_{ij(ran)} = \frac{x_{ij} - \overline{x}_j}{x_{j(\max)} - x_{j(\min)}}$

3 Statistical Data Analysis

There is a wide variety of statistical methods to extract the information from data, to identify and characterize potential biological biomarkers, and more importantly, to identify trends and clusters. The following paragraph describes some methods used to achieve these objectives.

3.1 Univariate Statistical Analysis

Univariate statistics refers to statistical methods that relate the behavior of a Y-dependent variable to changes of an X-independent variable. Depending on the distribution of the data, there are two categories of univariate statistical analysis.

The first category is *parametric* statistics, which assumes a normal or Gaussian distribution and a homoscedastic distribution of variance between measurements. This category includes methods to compare experimental averages such as the t-test, the F test, ANOVA, χ^2-chi-square, and correlation and regression methods (such as Pearson's correlation coefficient, linear regression and curvilinear methods, detection of atypical values with Dixon's Q, Grubs' test, significance tests, etc.).

The second category corresponds to *non-parametric* statistics, which are not based on the assumption of normal data distribution, thus responding to statistical problems that could not be solved by parametric statistics. Examples of non-parametric methods include median, sign test, Wald-Wolfowitz test, Wilcoxon's ranged test, Tukey's rapid test, Kruskal-Wallis test, Friedman's test, ranged correlation, Spearman's correlation coefficient (r_s), bootstrapping, etc.

Despite the fact that this type of statistics is not adequate to observe tendencies and natural clustering in metabolomics data, owing to the high complexity, they are useful for comparing potential discriminant markers after group identification [46].

3.2 Multivariate Statistical Analysis

Metabolomics data treatment complexity resides not only on the large data amount but also on the interrelationship between the generated data and the huge quantity of variables that make up the totality of data. Therefore, statistical methods are required to extract information that allows grouping the data and verifying the complex interactions between them. Multivariate statistics offers a variety of methods to treat data of this nature. The selection of the appropriate method will depend on the hypothesis, and the complexity of the data set, as well as the possible relation within the investigated groups [28]. Multivariate statistics can generally be classified as:

(a) *Supervised models:* these classification methods build models of class prediction based on discrete categories known a priori by the researcher, for example, all those studies aimed at finding metabolic differences between a disease-treated group and a control group. The problem to be solved by the supervised methods is to establish which associations classify a sample in any of the groups, indicated from the measurements made in the training set. The predictive power of each model should be verified from a "test/validation set," which in principle should be made up of new samples. However, if it is not possible to obtain new samples, an internal cross validation must be used. K nearest neighbors (kNN), soft independent modeling of class analogy (SIMCA), linear discriminant analysis (LDA), and partial least squares-discriminant analysis (PLS-DA) are some examples of supervised methods [44].

(b) *Unsupervised methods*: unlike supervised methods, unsupervised methods perform the classification of samples without the restrictions of predetermined classes, which ensures obtaining nonbiased clusters and trends.

Principal components analysis (PCA), hierarchical cluster analysis (HCA), self-organizing map (SOM), parallel factor analysis 1 (PARAFAC1), and parallel factor analysis 2 (PARAFAC2) methods belong to this category of methods [25].

The following section describes succinctly the multivariate methods commonly used in metabolomics data analysis. A more comprehensive review of these chemometric methods can be found in the literature [47–49].

3.2.1 Principal Components Analysis (PCA)

PCA is the most widely used method in chemometric analysis. The principle of PCA is to create orthogonal axes, named principal components (PCs), from the projections of the original variables. The first principal component (PC1) is oriented in the direction of higher variability; the second principal component (PC2) is oriented to the second higher variability and so on. As a result, a decrease of the dimensionality of the data and a variability-based separation among the groups are obtained, allowing an easier interpretation of the results, that is, finding hidden relations between the variables as clusters, trends, atypical values, and/or outlier samples [28, 50].

The results of PCA are represented with two-dimensional spaces, i.e., the sample space (PCA score plot) and the feature space (PCA loading plot). The sample space distinguishes possible differences between the groups. Figure 9 shows the PCA analysis results of endogenous metabolites in non-obese diabetic/severe combined immunodeficiency (NOD/SCID) mice plasma metabolome. The features responsible for such separation are shown in the PCA loading plot. The identification of the extreme features is achieved by algorithms. Samples that are out of confidence ellipses would be identified as atypical samples, probably arising from technical or processing factors, such as the batch effect [50].

The limitation of PCA lies in the fact that it depends on the variability between the groups and within each group. If the variability between the groups is similar to the variability within the groups, it would not be possible to observe any separation; therefore, it would not be possible to identify features that ruled this separation [51].

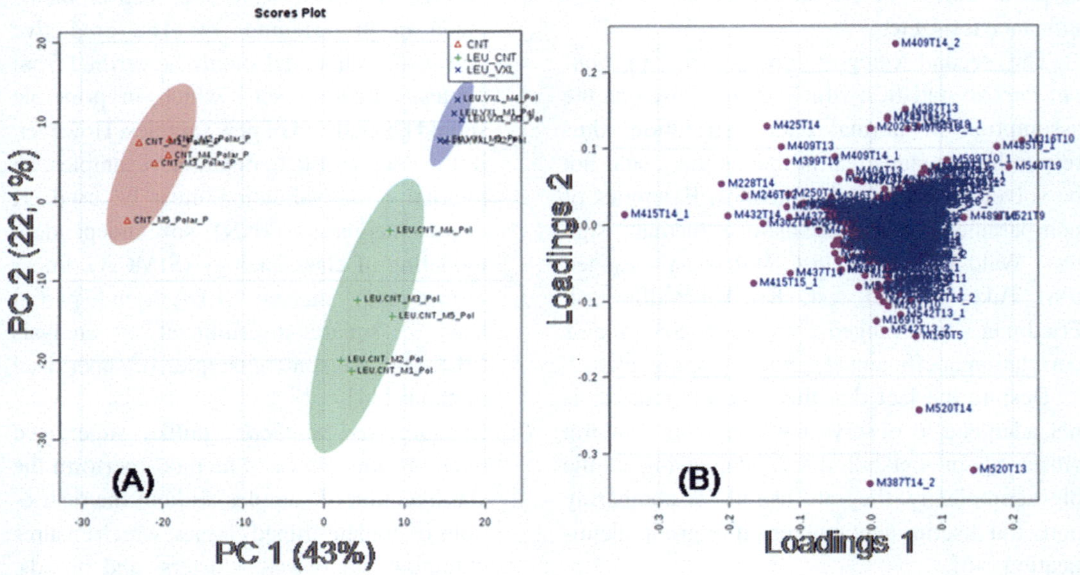

Fig. 9 PCA score plot (**a**) and loading plot (**b**) for the data obtained from the analyses of NOD/SCIID mice aqueous phase plasma samples by LC-MS with positive ionization mode. Healthy mice (pink), mice with leukemia (green), and mice with leukemia receiving treatment (blue) [77]

3.2.2 Partial Least Squares-Discriminant Analysis, PLS-DA

Partial least square (PLS) is a supervised method that performs a linear regression between a descriptive matrix **X** (the data) and a matrix **Y** (response), maximizing the covariance of latent variables, thus allowing to understand which metabolites (from matrix **X**) correlate better with the classes defined by the response (matrix **Y**) [50, 52]. It has the advantage of dealing with noisy data and missing values, besides offering a simple interpretation [27]. It is one of the most used methods in metabolomics since it reflects the largest difference between classification groups. When used to discriminate between two classes, it is called PLS-DA, and interpretation is performed similarly to PCA plots (Fig. 10) [50, 53].

3.2.3 Orthogonal Partial Least-Square Discriminant Analysis, OPLS-DA

The OPLS method is a variation of the PLS method, which improves interpretation by obtaining better information from the system. The method decomposes the data into relevant (predictive) and nonrelevant (orthogonal) information. Predictive information is related to characteristics, such as concentrations and groups, directly associated to the response matrix **Y**. In its turn, the orthogonal information contains variations due to biological and instrumental factors, but may also contain useful information related to other classifications, such as gender, age, batch effect, etc. [52, 54, 55]. The difference between OPLS and OPLS-DA is the same as in observed for PLS-DA. Therefore, in the former, classification is made on several classes, while in the latter, discrimination is performed into two different groups, although there is also the possibility of more groups [52]. Comparing with PLS-DA method, OPLS-DA achieves a better separation between groups, for the same data set.

3.2.4 Hierarchical Cluster Analysis (HCA)

HCA is an unsupervised method that clusters data pairs, according to their similarity. The similarity is a measurement based on the distance between data, which can be calculated as the Euclidean, Manhattan, or Mahalanobis distance, among others. The most common way to observe the groups is through a dendrogram tree (Fig. 11)

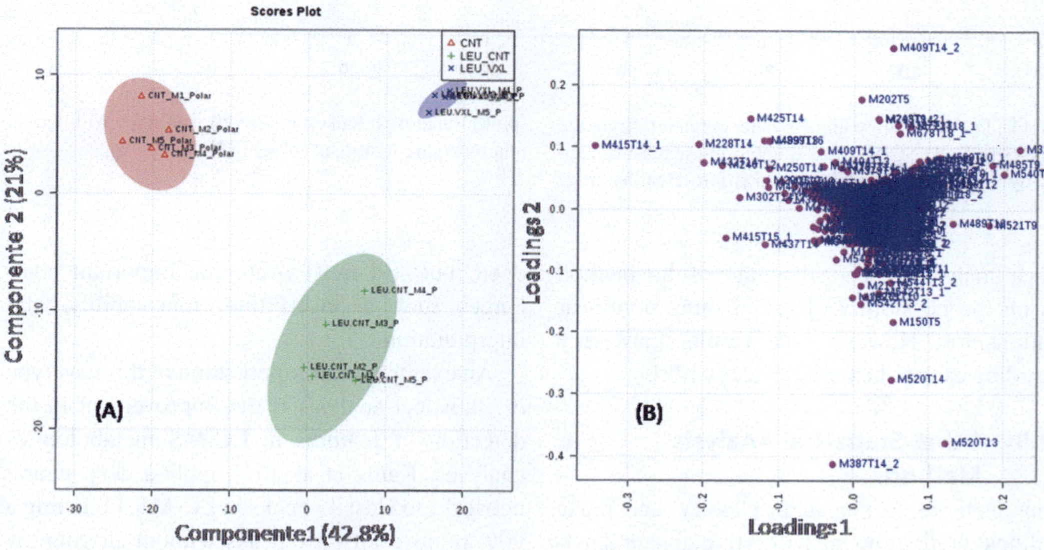

Fig. 10 Score plot (**a**) and loading plot (**b**) of PLS-DA for the data obtained from the analyses of NOD/SCIID mice aqueous phase plasma samples by LC-MS with positive ionization. Healthy mice (pink), mice with leukemia (green), and mice with leukemia receiving treatment (blue) [77]

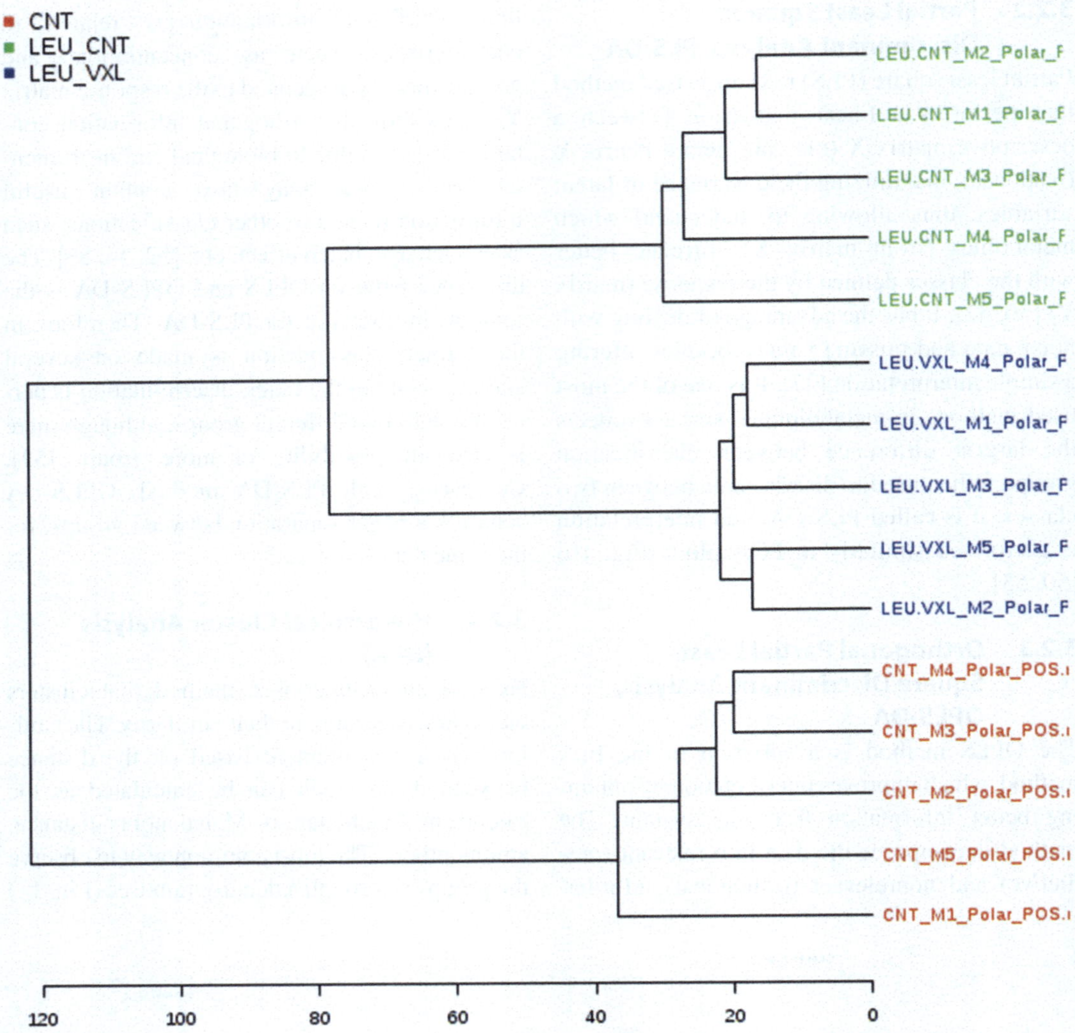

Fig. 11 Dendrogram with the data obtained from the analyses of NOD/SCIID mice aqueous phase plasma samples by LC-MS with positive ionization. Healthy mice (pink), mice with leukemia (green), and mice with leukemia receiving treatment (blue) [77]

or heat maps [56]. The advantage of this method lies on the possibility of establishing nonlinear relationships. However, the results may vary according to the clustering strategy [28].

3.2.5 Other Statistical Analysis Methods

Other methods to visualize, classify, and make nonlinear predictions such as artificial neural network (ANN), support vector machine (SVM), self-organization map (SOM), random forest (RF), or genetic algorithm (GA) have been developed, but still must overcome important challenges such as overfitting, repeatability, and interpretation.

An example of the application of this new type of statistical analysis is the improvement in the detection of features in LC-MS metabolomics analyses. Kantz et al. [57] applied deep neural networks to classify peaks in LC-MS, obtaining a 90% removal of false peaks without altering the number of positive signs in a global metabolomics study.

3.2.6 Model Validation

Model validation can be defined as the "process on deciding whether the results quantify hypothesized relationships between variables and responses and provide accurate estimation of the model prediction ability" [25]. Some criteria have been established to assess this predictive capability, including sensitivity, specificity, accuracy, receiver operating characteristic curve (ROC) [58], and the determination of cross-validation coefficient, Q^2 [25].

Multivariate models handle numerous and correlated complex data, so they must be validated by goodness of fit and predictive power [59]. There are two general types of validation: external validation and internal validation [60]. External validation is more recommended [60], but requires test samples, which cannot be the samples that were used to establish the model; therefore, it is subject to sample availability. If available, the samples are separated into a training group and a test group in a size ratio of 2:1. Internal validation may be performed by two methods, namely, cross-validation and permutation test [28].

Cross-validation methods are internal modeling methods used to validate the predictive capacity of models. Among the variety of methods, the K-fold method is the most widely used. This method divides the set of samples into one set for validation and another one for testing. The method is iterative, i.e., it varies the sample units that make up the training and testing set, and the total error of the model is calculated as the mean value of the errors [28].

On the other hand, the permutation test is used to validate the discriminatory capacity of the model. In this method, the class labels are randomly exchanged and assigned to each sample; then the model is recalculated. The test results generate a distribution of method performance that is evaluated by statistical significance, Q^2, and ROC [28]. For a validated model, the difference between correct and wrong models must be significant [25].

In conclusion, to perform data modeling, the subsequent methods should be followed: *i)* unsupervised methods with PCA, for exploratory studies; *ii)* supervised methods, such as PLS-DA and OPLS-DA, for sample classification and biomarker discovery; and iii) if the previous methods fail, nonlinear methods such as SVM and RF should be used to explore nonlinearity within the data. The parameters of each model must be adjusted and then validated [25].

4 Software

There is a wide range of computer programs available to the user, both in commercial and free format, to perform metabolomics data processing steps and metabolite identification [61]. The choice of a specific software will depend on the preferences of each researcher, whether for reasons of usability and versatility or because of the intrinsic characteristics of each software.

Tables 2 and 3 summarize some of the software used in data processing with updated links to access them.

5 Minimum Standards for Reporting on Data Processing in Metabolomics Studies

As observed, there is a great variety of options (statistical and computational) to perform metabolomics data treatment. Upon critical perspective, this condition generates two situations that should be highlighted. The wide diversity of methods and tools provides the analyst with multiple options to perform data treatment to different extents. However, experiment reproducibility must be ensured for their corresponding scientific verification.

This topic was addressed by the Chemical Analysis Working Group (CAWG, 2005) [74], which is part of the Metabolomics Standards Initiative (MSI) (http://metabolomicssociety.org/), with the objective of standardizing the information reported in metabolomics studies, the metadata. The importance of metadata is

Table 2 Commercial software for data processing in metabolomics

Name	Website	Platform
ChromaTOF (LECO)	https://www.leco.com/product/chromatof-software	LC-MS
Compound Discoverer (Thermo fisher Scientific)	https://www.thermofisher.com/order/catalog/product/OPTON-30929?SID=srch-srp-OPTON-30929	LC-MS
Mass Profiler Professional (Agilent Technologies, USA)	https://www.agilent.com/en/products/software-informatics/masshunter-suite/masshunter-for-life-science-research/mass-profiler-professional-software	LC-MS, GC-MS, and CE-MS
Metabolic Profiler (Bruker, USA)	https://www.bruker.com/es/products/mr/nmr/hyphenation/metabolic-profiler.html	MS and NMR
MetAlign (Plan Research International B. V.)	https://www.wur.nl/en/show/MetAlign-1.htm	LC-MS and GC-MS
MS Resolver (Pattern Recognition Systems, Norway)	http://www.prs.no/MS%20Resolver/MS%20Resolver.html	LC-MS and GC-MS
Progenesis QI (Waters, USA)	https://www.waters.com/waters/en_US/Progenesis-QI-Software/nav.htm?cid=134790655&locale=en_US	LC-MS
SIMCA (Sartorius Stedim Biotech)	https://umetrics.com/products/simca	LC-MS and GC-MS

Table 3 Open software for data processing in metabolomics

Name	Website	Platform
AMDIS (NIST)	https://chemdata.nist.gov/dokuwiki/doku.php?id=chemdata:amdis	LC-MS and GC-MS
MathDAMP [62]	https://mathdamp.iab.keio.ac.jp/	LC-MS, GC-MS, and CE-MS
MAVEN [63]	http://genomics-pubs.princeton.edu/mzroll/index.php	LC-MS
MetaboAnalyst [64]	https://www.metaboanalyst.ca/	LC-MS and GC-MS
Metandem [65]	http://metandem.com/web/	MS, nanoLC-MS, CE-MS/MS, and MALDI-MS
MET-COFEA [66]	http://bioinfo.noble.org/manuscript-support/met-cofea/	LC-MS and GC-MS
MET-IDEA [67]	http://bioinfo.noble.org/download/	LC-MS, GC-MS, and CE-MS
MS-DIAL [29, 30]	http://prime.psc.riken.jp/Metabolomics_Software/MS-DIAL/	GC-MS, GC-MS/MS, LC-MS, and LC-MS/MS
MSFACTs	Upon request	GC-MS and LC-MS
MZmine [68]	http://mzmine.github.io/	LC-MS and GC-MS
ROMANCE [69]	https://ispso.unige.ch/labs/fanal/romance	CE-MS
SMART [70]	http://www.stat.sinica.edu.tw/hsinchou/metabolomics/SMART.htm	LC-MS
TagFinder [71]	https://www.mpimp-golm.mpg.de/10871/Supplementary_Materials. For more information mail to: erban@mpimp-golm.mpg.de	LC-MS and GC-MS
XCMS [72, 73]	https://xcmsonline.scripps.edu/landing_page.php?pgcontent=mainPage	LC-MS and GC-MS

summarized in "Reporting of standard metadata provides a biological and empirical context for the data, facilitates experimental replication, and enables the re-interrogation and comparison of data by other" [74].

The CAWG proposals cover all the steps involved in a metabolomics study, i.e., sample preparation, experimental analysis, quality control, metabolite identification, and preprocessing of data. Regarding preprocessing of data, the metadata should indicate the general details describing the following methods: *i)* MS-based techniques (noise reduction, curve resolution for the chromatogram temporal alignment, peak picking, peak threshold, spectral deconvolution, and/or peak identifications) and *ii)* NMR-based techniques (phase correction method, frequency domain time conversion method, degree zero filling, apodization parameters, window functions, baseline corrections, first point multipliers, and any displacement in the free induction decay (FID)).

Compliance with these guidelines and minimum standards was assessed by Spice et al. [75, 76]. The authors reviewed the publications of the four largest metabolomics databases (MetaboLights, Metabolomics Workbench, MetaPhen, and MeRy-B), constituting a total of 339 data sets. According to the authors, even after 10 years of the guideline publication, compliance is not complete. Moreover, it highlights the existence of inconsistency and heterogeneity between the norms and requirements demanded by these databases.

References

1. Boccard J, Rudaz S (2019) Analysis of metabolomics data—a chemometrics perspective. In: Reference module in chemistry, molecular sciences and chemical engineering. Elsevier, Amsterdam, Netherlands, pp 1–23
2. Hendriks MMWB, van Eeuwijk FA, Jellema RH et al (2011) Data-processing strategies for metabolomics studies. TrAC – Trends Anal Chem 30:1685–1698. https://doi.org/10.1016/j.trac.2011.04.019
3. Madsen R, Lundstedt T, Trygg J (2010) Chemometrics in metabolomics – a review in human disease diagnosis. Anal Chim Acta 659:23–33. https://doi.org/10.1016/j.aca.2009.11.042
4. Shree M, Lingwan M, Masakapalli SK (2019) Metabolite profiling and metabolomics of plant systems using 1H NMR and GC-MS. In: Banerjee R, Kumar GV, Kumar SPJ (eds) OMICS-based approaches in plant biotechnology. John Wiley & Sons, Inc., Hoboken, pp 129–144
5. Duan L-X, Qi X (2015) Metabolite qualitative methods and the introduction of metabolomics database. In: Qi X, Chen X, Wang Y (eds) Plant metabolomics. Springer Netherlands, Dordrecht, pp 171–193
6. Tang J (2011) Microbial Metabolomics. Curr Genomics 12:391–403. https://doi.org/10.2174/138920211797248619
7. Belinato J, Bazioli J, Sussulini A et al (2019) Metabolômica microbiana: inovações e aplicações. Quim Nova 42:546–559. https://doi.org/10.21577/0100-4042.20170324
8. Dixon RA, Gang DR, Charlton AJ et al (2006) Applications of metabolomics in agriculture. J Agric Food Chem 54:8984–8994. https://doi.org/10.1021/jf061218t
9. Emwas A-H, Roy R, McKay RT et al (2019) NMR spectroscopy for metabolomics research. Meta 9:123. https://doi.org/10.3390/metabo9070123
10. Wishart DS (2019) NMR metabolomics: a look ahead. J Magn Reson 306:155–161. https://doi.org/10.1016/j.jmr.2019.07.013
11. Theodoridis GA, Gika HG, Want EJ, Wilson ID (2012) Liquid chromatography-mass spectrometry based global metabolite profiling: a review. Anal Chim Acta 711:7–16. https://doi.org/10.1016/j.aca.2011.09.042
12. Ramautar R, Somsen GW, de Jong GJ (2019) CE-MS for metabolomics: developments and applications in the period 2016–2018. Electrophoresis 40:165–179. https://doi.org/10.1002/elps.201800323
13. Buzatto AZ, de Sousa AC, Guedes SF et al (2014) Metabolomic investigation of human diseases biomarkers by CE and LC coupled to MS. Electrophoresis 35:1285–1307. https://doi.org/10.1002/elps.201300470
14. Tang H-Y, Chiu DT, Lin J-F et al (2017) Disturbance of plasma lipid metabolic profile in Guillain-Barre syndrome. Sci Rep 7:8140. https://doi.org/10.1038/s41598-017-08338-7
15. Marshall DD, Powers R (2017) Beyond the paradigm: combining mass spectrometry and nuclear magnetic resonance for metabolomics. Prog Nucl Magn Reson Spectrosc 100:1–16. https://doi.org/10.1016/j.pnmrs.2017.01.001
16. Canuto G, Costa JL, Cruz P et al (2017) Metabolômica: definições, estado-da-arte e aplicações representativas. Quim Nova 41:75–91. https://doi.org/10.21577/0100-4042.20170134
17. Tang D-Q, Zou L, Yin X-X, Ong CN (2016) HILIC-MS for metabolomics: an attractive and complementary approach to RPLC-MS. Mass Spectrom Rev 35:574–600. https://doi.org/10.1002/mas.21445
18. Karaman I (2017) Preprocessing and pretreatment of metabolomics data for statistical analysis. In: Sussulini A (ed) Metabolomics: from fundamentals to clinical applications. Springer International Publishing, Cham, pp 145–161
19. Katajamaa M, Orešič M (2007) Data processing for mass spectrometry-based metabolomics. J Chromatogr A 1158:318–328. https://doi.org/10.1016/j.chroma.2007.04.021

20. Euceda LR, Giskeodegård GF, Bathen TF (2015) Preprocessing of NMR metabolomics data. Scand J Clin Lab Invest 75:193–203. https://doi.org/10.3109/00365513.2014.1003593
21. Veltri P (2008) Algorithms and tools for analysis and management of mass spectrometry data. Brief Bioinform 9:144–155. https://doi.org/10.1093/bib/bbn007
22. Stolt R, Torgrip RJO, Lindberg J et al (2006) Second-order peak detection for multicomponent high-resolution LC/MS data. Anal Chem 78:975–983. https://doi.org/10.1021/ac050980b
23. Gorrochategui E, Jaumot J, Lacorte S, Tauler R (2016) Data analysis strategies for targeted and untargeted LC-MS metabolomic studies: overview and workflow. TrAC – Trends Anal Chem 82:425–442. https://doi.org/10.1016/j.trac.2016.07.004
24. Chambers MC, Maclean B, Burke R et al (2012) A cross-platform toolkit for mass spectrometry and proteomics. Nat Biotechnol 30:918–920
25. Yi L, Dong N, Yun Y et al (2016) Chemometric methods in data processing of mass spectrometry-based metabolomics: a review. Anal Chim Acta 914:17–34. https://doi.org/10.1016/j.aca.2016.02.001
26. Tian H, Li B, Shui G (2017) Untargeted LC–MS data preprocessing in metabolomics. J Anal Test 1:187–192. https://doi.org/10.1007/s41664-017-0030-8
27. Rowlands C, Elliott S (2011) Automated algorithm for baseline subtraction in spectra. J Raman Spectrosc 42:363–369. https://doi.org/10.1002/jrs.2691
28. Eliasson M, Rannar S, Trygg J (2011) From data processing to multivariate validation - essential steps in extracting interpretable information from metabolomics data. Curr Pharm Biotechnol 12:996–1004. https://doi.org/10.2174/138920111795909041
29. Hermansson M, Uphoff A, Käkelä R, Somerharju P (2005) Automated quantitative analysis of complex lipidomes by liquid chromatography/mass spectrometry. Anal Chem 77:2166–2175. https://doi.org/10.1021/ac048489s
30. Tsugawa H, Cajka T, Kind T et al (2015) MS-DIAL: data-independent MS/MS deconvolution for comprehensive metabolome analysis. Nat Methods 12:523–526. https://doi.org/10.1038/nmeth.3393
31. Março PH, Valderrama P, Alexandrino GL et al (2014) Multivariate curve resolution with alternating least squares: description, operation and aplications. Quim Nova 37:1525–1532. https://doi.org/10.5935/0100-4042.20140205
32. Chen T, Dai R (2015) Metabolomic data processing based on mass spectrometry platforms. In: Qi X, Chen X, Wang Y (eds) Plant metabolomics. Springer Netherlands, Dordrecht, pp 123–169
33. van Nederkassel AM, Daszykowski M, Eilers PHC, Vander HY (2006) A comparison of three algorithms for chromatograms alignment. J Chromatogr A 1118:199–210. https://doi.org/10.1016/j.chroma.2006.03.114
34. Mogollón NGS, de Lima PF, Gama MR et al (2014) State of the art two-dimensional liquid chromatography: fundamental concepts, instrumentation, and applications. Quim Nova 37:1680–1691. https://doi.org/10.5935/0100-4042.20140261
35. Zhang D, Huang X, Regnier FE, Zhang M (2008) Two-dimensional correlation optimized warping algorithm for aligning GCxGC-MS data. Anal Chem 80:2664–2671. https://doi.org/10.1021/ac7024317
36. Reinhold D, Pielke-Lombardo H, Jacobson S et al (2019) Pre-analytic considerations for mass spectrometry-based untargeted metabolomics data. In: D'Alessandro A (ed) High-throughput metabolomics: methods and protocols. Humana Press, New York, pp 323–340
37. Wei R, Wang J, Su M et al (2018) Missing value imputation approach for mass spectrometry-based metabolomics data. Sci Rep 8:1–10. https://doi.org/10.1038/s41598-017-19120-0
38. Hrydziuszko O, Viant MR (2012) Missing values in mass spectrometry based metabolomics: an undervalued step in the data processing pipeline. Metabolomics 8:161–174. https://doi.org/10.1007/s11306-011-0366-4
39. Sysi-Aho M, Katajamaa M, Yetukuri L, Orešič M (2007) Normalization method for metabolomics data using optimal selection of multiple internal standards. BMC Bioinformatics 8:93. https://doi.org/10.1186/1471-2105-8-93
40. Cook T, Ma Y, Gamagedara S (2020) Evaluation of statistical techniques to normalize mass spectrometry-based urinary metabolomics data. J Pharm Biomed Anal 177:112854. https://doi.org/10.1016/j.jpba.2019.112854
41. Wu Y, Li L (2016) Sample normalization methods in quantitative metabolomics. J Chromatogr A 1430:80–95. https://doi.org/10.1016/j.chroma.2015.12.007
42. Dieterle F, Ross A, Schlotterbeck G, Senn H (2006) Probabilistic quotient normalization as robust method to account for dilution of complex biological mixtures. Application in 1 H NMR Metabonomics. Anal Chem 78:4281–4290. https://doi.org/10.1021/ac051632c
43. Lee J, Park J, Lim MS et al (2012) Quantile normalization approach for liquid chromatography- mass spectrometry-based metabolomic data from healthy human volunteers. Anal Sci 28:801–805. https://doi.org/10.2116/analsci.28.801
44. Ferreira MMC (2015) Quimiometria: conceitos, métodos e aplicações. Editora da Unicamp, Campinas
45. van den Berg RA, Hoefsloot HCJ, Westerhuis JA et al (2006) Centering, scaling, and transformations: improving the biological information content of metabolomics data. BMC Genomics 7:1–15. https://doi.org/10.1186/1471-2164-7-142
46. Gougeon L, da Costa G, Guyon F, Richard T (2019) 1H NMR metabolomics applied to Bordeaux red wines. Food Chem 301:125257. https://doi.org/10.1016/j.foodchem.2019.125257

47. Kumar N, Bansal A, Sarma GS, Rawal RK (2014) Chemometrics tools used in analytical chemistry: an overview. Talanta 123:186–199. https://doi.org/10.1016/j.talanta.2014.02.003
48. Boccard J, Rudaz S (2014) Harnessing the complexity of metabolomic data with chemometrics. J Chemom 28:1–9. https://doi.org/10.1002/cem.2567
49. Brereton RG (2013) Chemometrics and statistics: multivariate classification techniques. Elsevier Inc., Oxford, UK
50. Bylesjö M (2015) Extracting meaningful information from Metabonomic data using multivariate statistics. In: Bjerrum JT (ed) Metabonomics: methods and protocols. Humana Press, New York, pp 137–146
51. Worley B, Powers R (2012) Multivariate analysis in metabolomics. Curr Metabol 1:92–107. https://doi.org/10.2174/2213235x130108
52. Pinto RC (2017) Chemometrics methods and strategies in metabolomics. In: Sussulini A (ed) Metabolomics: from fundamentals to clinical applications. Springer International Publishing, Cham, pp 163–190
53. Liu R, Zhang G, Sun M et al (2019) Integrating a generalized data analysis workflow with the single-probe mass spectrometry experiment for single cell metabolomics. Anal Chim Acta 1064:71–79. https://doi.org/10.1016/j.aca.2019.03.006
54. Ebbels TMD, Karaman I, Graça G (2019) Processing and analysis of untargeted multicohort NMR data. In: Gowda GAN, Raftery D (eds) NMR-based metabolomics: methods and protocols. Humana Press, New York, pp 453–470
55. Surowiec I, Johansson E, Stenlund H et al (2018) Quantification of run order effect on chromatography - mass spectrometry profiling data. J Chromatogr A 1568:229–234. https://doi.org/10.1016/j.chroma.2018.07.019
56. Peña-Bautista C, Roca M, Hervás D et al (2019) Plasma metabolomics in early Alzheimer's disease patients diagnosed with amyloid biomarker. J Proteome 200:144–152. https://doi.org/10.1016/j.jprot.2019.04.008
57. Kantz ED, Tiwari S, Watrous JD et al (2019) Deep neural networks for classification of LC-MS spectral peaks. Anal Chem 91:12407–12413. https://doi.org/10.1021/acs.analchem.9b02983
58. Fawcett T (2006) An introduction to ROC analysis. Pattern Recogn Lett 27:861–874. https://doi.org/10.1016/j.patrec.2005.10.010
59. Want E, Masson P (2011) Processing and analysis of GC/LC-MS-based metabolomics data. In: Metz TO (ed) Metabolic profiling. Humana Press, New York, pp 277–298
60. Truntzer C, Ducoroy P (2017) Statistical approach for biomarker discovery using label-free LC-MS data: an overview. In: Datta S, Mertens BJA (eds) Statistical analysis of proteomics, metabolomics, and Lipidomics data using mass spectrometry. Springer International Publishing, Cham, pp 177–201
61. Blaženović I, Kind T, Ji J, Fiehn O (2018) Software tools and approaches for compound identification of LC-MS/MS data in metabolomics. Meta 8:31. https://doi.org/10.3390/metabo8020031
62. Baran R, Kochi H, Saito N et al (2006) MathDAMP: a package for differential analysis of metabolite profiles. BMC Bioinformatics 7:1–9. https://doi.org/10.1186/1471-2105-7-530
63. Agrawal S, Kumar S, Sehgal R et al (2019) El-MAVEN: a fast, robust, and user-friendly mass spectrometry data processing engine for metabolomics. In: D'Alessandro A (ed) High-throughput metabolomics: methods and protocols. Humana Press, New York, pp 301–321
64. Xia J, Psychogios N, Young N, Wishart DS (2009) MetaboAnalyst: a web server for metabolomic data analysis and interpretation. Nucleic Acids Res 37:W652–W660. https://doi.org/10.1093/nar/gkp356
65. Hao L, Zhu Y, Wei P et al (2019) Metandem: an online software tool for mass spectrometry-based isobaric labeling metabolomics. Anal Chim Acta 1088:99–106. https://doi.org/10.1016/j.aca.2019.08.046
66. Zhang W, Chang J, Lei Z et al (2014) MET-COFEA: a liquid chromatography/mass spectrometry data processing platform for metabolite compound feature extraction and annotation. Anal Chem 86:6245–6253. https://doi.org/10.1021/ac501162k
67. Broeckling CD, Reddy IR, Duran AL et al (2006) MET-IDEA: data extraction tool for mass spectrometry-based metabolomics. Anal Chem 78:4334–4341. https://doi.org/10.1021/ac0521596
68. Pluskal T, Castillo S, Villar-Briones A, Orešič M (2010) MZmine 2: modular framework for processing, visualizing, and analyzing mass spectrometry-based molecular profile data. BMC Bioinformatics 11. https://doi.org/10.1186/1471-2105-11-395
69. González-Ruiz V, Gagnebin Y, Drouin N et al (2018) ROMANCE: a new software tool to improve data robustness and feature identification in CE-MS metabolomics. Electrophoresis 39:1222–1232. https://doi.org/10.1002/elps.201700427
70. Liang YJ, Lin YT, Chen CW et al (2016) SMART: statistical metabolomics analysis – an R tool. Anal Chem 88:6334–6341. https://doi.org/10.1021/acs.analchem.6b00603
71. Luedemann A, von Malotky L, Erban A, Kopka J (2011) TagFinder: preprocessing software for the fingerprinting and the profiling of gas chromatography–mass spectrometry based metabolome analyses. In: Hardy NW, Hall RD (eds) Plant metabolomics: methods and protocols. Humana Press, New York, pp 255–286
72. Smith CA, Want EJ, O'Maille G et al (2006) XCMS: processing mass spectrometry data for metabolite profiling using nonlinear peak alignment, matching, and identification. Anal Chem 78:779–787. https://doi.org/10.1021/ac051437y
73. Mahieu NG, Genenbacher JL, Patti GJ (2016) A roadmap for the XCMS family of software solutions in metabolomics. Curr Opin Chem Biol 30:87–93. https://doi.org/10.1016/j.cbpa.2015.11.009

74. Sumner LW, Amberg A, Barrett D et al (2007) Proposed minimum reporting standards for chemical analysis: Chemical Analysis Working Group (CAWG) Metabolomics Standards Initiative (MSI). Metabolomics 3:211–221. https://doi.org/10.1007/s11306-007-0082-2
75. Spicer RA, Salek R, Steinbeck C (2017) Comment: a decade after the metabolomics standards initiative it's time for a revision. Sci Data 4:2–4. https://doi.org/10.1038/sdata.2017.138
76. Spicer RA, Salek R, Steinbeck C (2017) Compliance with minimum information guidelines in public metabolomics repositories. Sci Data 4:1–8. https://doi.org/10.1038/sdata.2017.137
77. Duarte GHB (2016) Metabolomics by LC-ESI-QTOF-MS in NOD/SCID mice under chemoterapy treatment: potential biomarkers of leukemia. Master's thesis. Universidade Estadual de Campinas

Index

A. V. Colnaghi Simionato (ed.), *Separation Techniques Applied to Omics Sciences*, Advances in Experimental Medicine and Biology 1336, https://doi.org/10.1007/978-3-030-77252-9

D

H

V

W

The manufacturer's authorised representative in the EU is Springer
Nature Customer Service Centre GmbH, Europaplatz 3, 69115 Heidelberg,
Germany. If you have any concerns regarding our products, please
contact ProductSafety@springernature.com

Printed and bound by CPI Group (UK) Ltd, Croydon, CR0 4YY
05/10/2026
02228339-0011